# ASTRAL SCIENCES IN MESOPOTAMIA

# HANDBUCH DER ORIENTALISTIK
## HANDBOOK OF ORIENTAL STUDIES

ERSTE ABTEILUNG

## DER NAHE UND MITTLERE OSTEN
## THE NEAR AND MIDDLE EAST

HERAUSGEGEBEN VON

H. ALTENMÜLLER · B. HROUDA · B.A. LEVINE · R.S. O'FAHEY
K.R. VEENHOF · C.H.M. VERSTEEGH

---

VIERUNDVIERZIGSTER BAND

ASTRAL SCIENCES IN MESOPOTAMIA

TUTA SUB AEGIDE PALLAS · 1683 ·

# ASTRAL SCIENCES IN MESOPOTAMIA

BY

HERMANN HUNGER

AND

DAVID PINGREE

BRILL
LEIDEN · BOSTON · KÖLN
1999

This book is printed on acid-free paper.

**Library of Congress Cataloging-in-Publication Data**

Hunger, Hermann, 1942–
Astral sciences in Mesopotamia / by Hermann Hunger and David Pingree.
p. cm. — (Handbuch der Orientalistik. Erste Abteilung, Der Nahe und Mittlere Osten, ISSN 0169-9423 ; 44. Bd. = Handbook of Oriental studies. The Near and Middle East)
Includes bibliographical references and index.
ISBN 9004101276 (cloth : alk. paper)
1. Astronomy, Assyro—Babylonian. I. Pingree, David Edwin, 1933–
II. Title. III. Series : Handbuch der Orientalistik. Erste Abteilung, Nahe und der Mittlere Osten ; 44. Bd.
QB19.H86 1999
520'.935—dc21 99-37095
CIP

**Die Deutsche Bibliothek – CIP-Einheitsaufnahme**

**Handbuch der Orientalistik.** – Leiden ; Boston ; Köln : Brill.
Teilw. hrsg. von H. Altenmüller. Teilw. hrsg. von B. Spuler. – Literaturangaben
Teilw. mit Parallelt.: Handbook of oriental studies
Abt. 1, Der Nahe und Mittlere Osten = The Near and Middle East / hrsg. von H. Altenmüller ...
Teilw. hrsg. von B. Spuler
Bd. 44. Hunger, Hermann: Astral sciences in Mesopotamia. – 1999
**Hunger, Hermann:**
Astral sciences in Mesopotamia / by Hermann Hunger and David Pingree. – Leiden ; Boston ; Köln : Brill, 1999
(Handbuch der Orientalistik : Abt. 1, Der Nahe und Mittlere Osten ; Bd. 44)
ISBN 90-04-10127-6

ISSN 0169-9423
ISBN 90 04 10127 6



PRINTED IN THE NETHERLANDS

DEDICATION

This volume is dedicated to those scholars of the last hundred and twenty years who, by the clarity of their thoughts and the brilliance of their imaginations, made its composition possible—in particular Fathers Epping, Kugler, and Schaumberger, and Otto Neugebauer, Abraham Sachs, and Asger Aaboe. We are especially indebted to Neugebauer who, in his journey from Graz to Providence, either personally produced or inspired others to write the most penetrating analyses of Mesopotamian mathematical astronomy. Jayati gurur gajendraḥ!

CONTENTS

# PRELIMINARIES

## CHRONOLOGY[1]

### *Babylonia*

*First Dynasty of Babylon (based on Ammi-ṣaduqa 1 = -1701/0)*[2]

| | | | |
|---|---|---|---|
| Sumu-abum | -1949/8 | to | -1936/5 |
| Sumulael | -1935/4 | to | -1900/1899 |
| Sabium | -1899/8 | to | -1886/5 |
| Apil-Sin | -1885/4 | to | -1868/7 |
| Sin-muballiṭ | -1867/6 | to | -1848/7 |
| Ḫammu-rapi | -1847/6 | to | -1805/4 |
| Samsu-iluna | -1804/3 | to | -1767/6 |
| Abi-ešuḫ | -1766/5 | to | -1739/8 |
| Ammi-ditana | -1738/7 | to | -1702/1 |
| Ammi-ṣaduqa | -1701/0 | to | -1681/0 |
| Samsu-ditana | -1680/79 | to | -1650/49 |

*Second Dynasty of Isin*[3]

| | | | |
|---|---|---|---|
| Marduk-kabit-aḫḫešu | -1156/5 | to | -1139/8 |
| Itti-Marduk-balaṭu | -1138/7 | to | -1131/0 |
| Ninurta-nadin-šumi | -1130/29 | to | -1125/4 |
| Nebuchadnezzar I | -1124/3 | to | -1103/2 |
| Enlil-nadin-apli | -1102/1 | to | -1099/8 |
| Marduk-nadin-aḫḫe | -1098/7 | to | -1081/0 |
| Marduk-šapik-zeri | -1080/79 | to | -1068/7 |
| Adad-apla-iddina | -1067/6 | to | -1046/5 |
| Marduk-aḫḫe-eriba | -1045/4 | | |
| Marduk-zer-x | -1044/3 | to | -1033/2 |
| Nabu-šumu-libur | -1032/1 | to | -1025/4 |

---

[1] The year numbers refer to retrojected Julian years, counted astronomically: year 0 = 1 B.C., year -1 = 2 B.C., etc. Since the Babylonian year began in spring, each Babylonian year contains parts of two consecutive Julian years. For this reason, two year numbers appear in these tables. Sometimes it is of course possible to determine the boundary between reigns more precisely, but we have not deemed it necessary to investigate this in each case.

[2] According to Huber [1982].

[3] This and the remaining dynasties according to J. A. Brinkman in Oppenheim [1977] pp. 338-342, except that for the Parthians we have used K. Schippmann in Encyclopaedia Iranica II, 1986, pp. 525-536. Some of the entries could be modified according to later research, but for an overview this did not seem necessary to us.

*Mixed Dynasties*

| | | | |
|---|---|---|---|
| Nabu-naṣir | -746/5 | to | -733/2 |
| Nabu-nadin-zeri | -732/1 | to | -731/0 |
| Nabu-šuma-ukin II | -731/0 | | |
| Nabu-mukin-zeri | -730/29 | to | -728/7 |
| Tiglath-pileser (Assyrian) | -727/6 | to | -726/5 |
| Shalmaneser (Assyrian) | -725/4 | to | -721/0 |
| Merodach-Baladan II | -720/19 | to | -709/8 |
| Sargon (Assyrian) | -708/7 | to | -704/3 |
| Sennacherib (Assyrian) | -703/2 | to | -702/1 |
| Marduk-zakir-šumi II | -702 | | |
| Merodach-Baladan II | -702/1 | | |
| Bel-ibni | -701/0 | to | -699/8 |
| Aššur-nadin-šumi | -698/7 | to | -693/2 |
| Nergal-ušezib | -692/1 | | |
| Mušezib-Marduk | -691/0 | to | -688/7 |
| Sennacherib (Assyrian) | -687/6 | to | -680/79 |
| Esarhaddon (Assyrian) | -679/8 | to | -668/7 |
| Assurbanipal (Assyrian) | -667/6 | | |
| Šamaš-šum-ukin | -666/5 | to | -647/6 |
| Kandalanu | -646/5 | to | -626/5 |
| (interregnum) | -625/4 | | |

*Chaldaean Dynasty*

| | | | |
|---|---|---|---|
| Nabopolassar | -624/3 | to | -604/3 |
| Nebuchadnezzar II | -603/2 | to | -561/0 |
| Evil-Merodach | -560/59 | to | -559/8 |
| Neriglissar | -558/7 | to | -555/4 |
| Labaši-Marduk | -555/4 | | |
| Nabonidus | -554/3 | to | -538 |

*Assyria*

| | | | |
|---|---|---|---|
| Adad-nirari I | -1304/3 | to | -1273/2 |
| Shalmaneser I | -1272/1 | to | -1243/2 |
| Tukulti-Ninurta I | -1242/1 | to | -1206/5 |
| Aššur-nadin-apli | -1205/4 | to | -1202/1 |
| Aššur-nirari III | -1201/0 | to | -1196/5 |
| Enlil-kudurri-uṣur | -1195/4 | to | -1191/0 |
| Ninurta-apil-Ekur | -1190/89 | to | -1178/7 |
| Aššur-dan I | -1177/6 | to | -1132/1 |
| Ninurta-tukulti-Aššur | -1132/1 | | |
| Mutakkil-Nusku | -1132/1 | | |
| Aššur-reš-iši I | -1131/0 | to | -1114/3 |
| Tiglath-pileser I | -1113/2 | to | -1075/4 |
| Ašarid-apil-Ekur | -1074/3 | to | -1073/2 |
| Aššur-bel-kala | -1072/1 | to | -1055/4 |
| Eriba-Adad II | -1054/3 | to | -1053/2 |
| Šamši-Adad IV | -1052/1 | to | -1049/8 |
| Aššurnaṣirpal I | -1048/7 | to | -1030/29 |
| Shalmaneser II | -1029/8 | to | -1018/7 |
| Aššur-nirari IV | -1017/6 | to | -1012/1 |
| Aššur-rabi II | -1011/0 | to | -971/0 |
| Aššur-reš-iši II | -970/69 | to | -966/5 |
| Tiglath-pileser II | -965/4 | to | -934/3 |
| Aššur-dan II | -933/2 | to | -911/0 |
| Adad-nirari II | -910/9 | to | -890/89 |
| Tukulti-Ninurta II | -889/8 | to | -883/2 |
| Aššurnaṣirpal II | -882/1 | to | -858/7 |
| Shalmaneser III | -857/6 | to | -823/2 |
| Šamši-Adad V | -822/1 | to | -810/9 |
| Adad-nirari III | -809/8 | to | -782/1 |
| Shalmaneser IV | -781/0 | to | -772/1 |
| Aššur-dan III | -771/0 | to | -754/3 |
| Aššur-nirari V | -753/2 | to | -744/3 |
| Tiglath-pileser III | -743/2 | to | -726/5 |
| Shalmaneser V | -725/4 | to | -721/0 |
| Sargon II | -720/19 | to | -704/3 |
| Sennacherib | -703/2 | to | -680/79 |
| Esarhaddon | -679/8 | to | -668/7 |
| Assurbanipal | -667/6 | to | -626/5 |
| Aššur-etel-ilani | | | |
| Sin-šumu-lišir | | | |
| Sin-šarru-iškun | | to | -611 |

*Achaemenid Dynasty (in Babylonia)*

| | | | |
|---|---|---|---|
| Cyrus II | -537/6 | to | -529/8 |
| Cambyses II | -528/7 | to | -521/0 |
| Darius I | -520/19 | to | -485/4 |
| Xerxes | -484/3 | to | -464/3 |
| Artaxerxes I | -463/2 | to | -423/2 |
| Darius II | -422/1 | to | -404/3 |
| Artaxerxes II | -403/2 | to | -358/7 |
| Artaxerxes III | -357/6 | to | -337/6 |
| Arses | -336/5 | to | -335/4 |
| Darius III | -334/3 | to | -330/29 |

*Macedonian Dynasty*

| | | | |
|---|---|---|---|
| Alexander III | -329/8 | to | -322/1 |
| Philip Arrhidaeus | -321/0 | to | -315/4 |
| (Antigonus as general | -316/5 | to | -311/0) |
| Alexander IV | -315/4 | to | -311/0 |

*Seleucid Dynasty*

| | | | |
|---|---|---|---|
| Seleucus I Nicator | -310/9 | to | -280/79 |
| Seleucus I and Antiochus I Soter | -291/0 | to | -280/79 |
| Antiochus I and Seleucus | -279/8 | to | -266/5 |
| Antiochus I and Antiochus II Theos | -265/4 | to | -260/59 |
| Antiochus II Theos | -259/8 | to | -245/4 |
| Seleucus II Callinicus and Seleucus III Soter | -244/3 | to | -224/3 |
| Seleucus III Soter | -223/2 | to | -222/1 |
| Antiochus III | -221/0 | to | -186/5 |
| Antiochus III and Antiochus | -209/8 | to | -192/1 |
| Antiochus III and Seleucus IV Philopator | -188/7 | to | -186/5 |
| Seleucus IV Philopator | -186/5 | to | -174/3 |
| Antiochus IV Epiphanes | -174/3 | to | -163/2 |
| Antiochus IV and Antiochus | -174/3 | to | -169/8 |
| Antiochus V Eupator | -163/2 | to | -161/0 |
| Demetrius I Soter | -160/59 | to | -150/49 |
| Alexander I Balas | -149/8 | to | -145/4 |
| Demetrius II Nicator | -144/3 to -138/7 and -128/7 to -124/3 | | |
| Antiochus VI Epiphanes | -144/3 | to | -141/0 |
| Antiochus VII Sidetes | -138/7 | to | -128/7 |

*Arsacid Dynasty*

| | | | |
|---|---|---|---|
| Mithradates I | ca. -170 | to | -138/7 |
| Phraates II | -138/67 | to | ca. -127 |
| Artabanus II | ca. -126 | to | -123/2 |
| Mithradates II | ca. -123/2 | to | -87/6 |
| Gotarzes I | -90/89 | to | ca. -79 |
| Orodes I | ca. -79 | to | -77/6 |
| Sinatruces | -77/6 | to | -70/69 |
| Phraates III | -70/69 | to | -57/6 |
| Mithradates III | -57/6 | | |
| Orodes II | -57/6 | to | ca. -38 |
| Pacorus I | | to | -38/7 |
| Phraates IV | ca. -39 | to | -2/1 |
| Tiridates II | ca. -31 | to | ca. -24 |
| Phraates V (Phraataces) | -2/1 | to | 4/5 |
| Orodes III | 4/5 | to | 6/7 |
| Vonones I | 8/9 | to | 9/10 |
| Artabanus III | 10/11 | to | ca. 38 |
| Tiridates III | ca. 36 | | |
| Cinnamus | ca. 37 | | |
| Vardanes | ca. 39 | to | 45/6 |
| Gotarzes II | ca. 43 | to | 51/2 |
| Vonones II | ca. 51 | | |
| Vologases I | ca. 51 | to | ca. 76-80 |

*Eras: Beginnings of Year 1*

| | |
|---|---|
| Seleucid (Babylon) | 3 April -310 |
| Arsacid | 15 April -246 |

*Month Names*

| | Babylonian | Logograms |
|---|---|---|
| I | Nisannu | BAR |
| II | Aiaru | $GU_4$ |
| III | Simānu | SIG |
| IV | Du'ūzu | ŠU |
| V | Abu | IZI |
| VI | Ulūlu | KIN |
| VII | Tešrītu | $DU_6$ |
| VIII | Araḫsamnu | APIN |
| IX | Kislīmu | GAN |
| X | Ṭebētu | AB |
| XI | Šabāṭu | ZÍZ |
| XII | Addaru | ŠE |

*Metrology of angular distances/time*

| | | |
|---|---|---|
| 1 še (barleycorn) | = 0;0,50° | = 0;3,20 min. |
| 1 NINDA | = 0;1° | = 0;4 min. |
| 1 šu-si (finger) | = 0;5° | = 0;20 min. |
| 1 UŠ | = 1° | = 4 min. |
| 1 kùš (cubit) | = 2° (= 24 fingers) | = 8 min. |
| | = 2;30° (= 30 fingers) | = 10 min. |
| 1 danna = *bēru* | = 30° | = 2,0 min. = 2 h. |
| 1 day | = 6,0° | = 24,0 min. = 24 h. |
| 1 tithi | | = 0;2 mo. |

*Greek-letter Phenomena*

Superior Planets

Γ first visibility (in East)
Φ first station
Θ acronychal rising
Ψ second station
Ω first invisibility (in West)

Inferior Planets

Γ first visibility in East
Φ station in East (second station)
Σ first invisibility in East
Ξ first visibility in West
Ψ station in West (first station)
Ω first invisibility in West

## *Mathematical Formulae*

### *Linear Zigzag Function (System B)*

M = maximum
m = minimum
$\mu = \frac{M+m}{2}$
$\Delta$ = amplitude = M – m
d = difference
P = period of "tabulation function" = $\frac{2\Delta}{d}$
p = period of "true function" = $\frac{P}{P+1} = \frac{2\Delta}{2\Delta+d}$ $(P = \frac{p}{1-p})$

### *Step Function (System A)*

$\alpha_i$ = arc of ecliptic $(\alpha_1 + \alpha_2 + \ldots + \alpha_n = 360°)$
$w_i$ = synodic arc associated with $\alpha_i$
$P = \frac{\alpha_1}{w_1} + \frac{\alpha_2}{w_2} + \ldots + \frac{\alpha_n}{w_n}$

### *In ACT Planetary Theory*

$\overline{\Delta\lambda}$ = mean synodic arc = $\frac{6,0}{P}$; $P = \frac{6,0}{\overline{\Delta\lambda}}$
$\Pi$ = number of $\overline{\Delta\lambda}$'s in an ACT period of Y years = "number period"
Z = number of P's in Y years = "wave number"
$P = \frac{\Pi}{Z}$
$\Pi = PZ = \frac{6,0Z}{\overline{\Delta\lambda}}$
$Z = \frac{\Pi}{P} = \frac{\Pi\overline{\Delta\lambda}}{6,0}$

# ABBREVIATIONS

| | |
|---|---|
| ACh | Ch. Virolleaud, L'astrologie chaldéenne |
| ACT | O. Neugebauer, Astronomical Cuneiform Texts |
| AfO | Archiv für Orientforschung |
| AHES | Archive for the History of the Exact Sciences |
| AIHS | Archives internationaux de l'histoire des sciences |
| AJSL | American Journal of Semitic Languages and Literatures |
| AOAT | Alter Orient und Altes Testament |
| BiOr | Bibliotheca Orientalis |
| BPO | E. Reiner and D. Pingree, Babylonian Planetary Omens |
| BSOAS | Bulletin of the School of Oriental and African Studies |
| HAMA | O. Neugebauer, A History of Ancient Mathematical Astronomy |
| JAOS | Journal of the American Oriental Society |
| JCS | Journal of Cuneiform Studies |
| JEOL | Jaarbericht ... "Ex Oriente Lux" |
| JHA | Journal for the History of Astronomy |
| JNES | Journal of Near Eastern Studies |
| JRAS | Journal of the Royal Asiatic Society |
| KDVSMM | Det Kongelige Danske Videnskabernes Selskab, Matematisk-fysiske Meddelelser |
| KUB | Keilschrifturkunden aus Boğazköy |
| MAPS | Memoirs of the American Philosophical Society |
| MDOG | Mitteilungen der Deutschen Orient-Gesellschaft |
| NABU | Nouvelles assyriologiques brèves et utilitaires |
| OLZ | Orientalistische Literaturzeitung |
| PAPS | Proceedings of the American Philosophical Society |
| PBS | Publications of the Babylonian Section, The University Museum, Philadelphia |
| PSBA | Proceedings of the Society of Biblical Archaeology |
| QS | Quellen und Studien zur Geschichte der Mathematik und Astronomie |
| RA | Revue d'Assyriologie |
| SHAW | Sitzungsberichte der Heidelberger Akademie der Wissenschaften |
| SÖAW | Sitzungsberichte der Österreichischen Akademie der Wissenschaften |
| TAPS | Transactions of the American Philosophical Society |
| TCL | Textes cunéiformes du Louvre |

| | |
|---|---|
| WO | Die Welt des Orients |
| WZKM | Wiener Zeitschrift für die Kunde des Morgenlandes |
| ZA | Zeitschrift für Assyriologie |
| ZDMG | Zeitschrift der Deutschen Morgenländischen Gesellschaft |

# INTRODUCTION

This is a book about the origins and the development of the astral sciences in Mesopotamia, sciences which exercised an enormous and continuous influence on the contiguous cultures (Egypt, Greece, the Hittites, Syria, Palestine, Iran, and India) in antiquity, and on the numerous successors of these major civilizations in later periods. While we refer occasionally to these influences as illustrations of how others understood and used Mesopotamian astronomy, we have not attempted completeness in such references.

Nor have we attempted to deal as exhaustively with the astral omens as we have with astronomy. The project that Otto Neugebauer initiated in 1937 to produce in collaboration with others editions of cuneiform texts dealing with mathematical and observational astronomy and with the celestial omens has been substantially completed for the first two categories, and much has been learned about these texts and what they mean; this is summarized in the second section of this book. But, while some progress has been made in publishing reliable omen texts relating to the Moon, to the Sun, and to the constellations and planets, vast numbers of such tablets still await scholarly investigation. We have, therefore, simply indicated the general contents of this enormous literature, and refer the reader for the present to the commendable preliminary exposition of astral omens published recently by Koch-Westenholz [1995]. Future versions of this Handbuch should treat this important material thoroughly once it and its history has been more completely understood. Similarly, it should be possible in this successor volume to reflect a deep and comprehensive understanding of Babylonian astral magic which has been magisterially surveyed by Reiner [1995].

What this volume does attempt to cover is the astronomical material found in the tablets of both omen texts and purely astronomical texts from the earliest times—the Old Babylonian period of, probably, the first half of the second millennium B.C.—down to the latest—the period of the Parthian control of Mesopotamia in the late first century A.D. This massive amount of material is arranged both chronologically and topically within the chronological framework.

The texts of the first section, the tablets of *Enūma Anu Enlil*, are the primary sources of the Mesopotamian tradition of astral omens, but contain already the recognition of the periodicity of some celestial phenomena and a mathematical device, the linear zigzag function, for predicting them. The "Astrolabes" or "Three Stars Each" texts organize information concerning constellations; more and more sophisticated star lore along with intercalation schemes, crude planetary schemes, and time-keeping devices are described in MUL.APIN and related texts. Developing from one section of MUL.APIN are the *Ziqpu* Star Texts used for telling time in the first millennium B.C. and the related GU Text. A more complicated description of the interrelations of stars is provided by the as yet not understood DAL.BA.AN.NA Text. A pair of Time-keeping Texts conclude the section on early astronomy.

The next section, on observations and predictions from the last millennium B.C., begins with the material regarding solar, lunar, and planetary phenomena included in the Letters and Reports (regarding mostly celestial omens) written in the Sargonid period. These are followed by other observations also of the seventh century B.C. The Diaries are a major source of observational and computed data interspersed with some meteorological, economic, and historical information and a scattering of omens. They are preserved in a fragmentary sequence from Babylon from -651 to -60, but probably began to be compiled in -746 and continued to about 75 A.D. Following the Diaries, in the Seleucid and Parthian periods, are Normal Star Almanacs, Almanacs, and Goal-Year Texts. From the seventh century B.C. on we also have tablets concerning observations of the occurrences of characteristic planetary phenomena, the so-called Greek-letter phenomena, and of lunar and sometimes solar eclipses from -746 on on tablets that were probably compiled in the fifth and fourth centuries B.C.

The last section deals with the theoretical texts of the last five hundred years B.C.: the "Saros" Cycle, column Φ, predictions of eclipses, the origins of Φ, the use of time degrees, and lunar latitude. After this are discussed the nineteen-year cycle and the schemes for determining the dates of solstices, equinoxes, and the risings of Sirius. Then comes the development of more precise period-relations for the planets, the unique tablet dealing with planetary latitudes, and the subdivisions of the synodic arcs of the planets which represent the prelude to the Ephemerides of the Seleucid and Parthian periods. This section and the last culminate in a description of the fully developed mathematical astronomy of the ACT type.

In discussing each of these topics we have striven to reveal a number of points about them:

1. The provenance and the date of each tablet, whenever possible.
2. The history of its publication and of its interpretation.
3. The relationships to other cuneiform tablets.
4. What we believe to be the interpretation best fitting the level of astronomical knowledge and scientific procedure revealed by other cuneiform texts and, where discoverable, by texts composed in contemporary neighboring cultures.

The result, we hope, is a useful guide to the literature known to us in the Spring of 1999 and a summary of what is known about Mesopotamian astral sciences at this time. Many new tablets will be discovered that will cause our interpretations to be modified or rejected, and many scholars will present new interpretations that will challenge or at times refute our conclusions. We look forward to the advance of this important field, and only hope that we have facilitated that advance with this volume.

## CHAPTER ONE

# OMENS

People in Ancient Mesopotamia believed that the gods would indicate future events to mankind.[1] These indications were called "signs", in Sumerian (g)iskim, in Akkadian *ittu*. Such signs could be of very different kinds. They were to be found in the entrails of sacrificial animals, in the shapes of oil spreading after being dropped into water, in phenomena observed in the sky, in strange occurrences in everyday life. We can classify omens into two types: those that can be produced when they are wanted (e.g., to answer a question) and those that happen without human action provoking them. An example of the first type are omens from the inspection of the entrails of a sheep; to the second type belong all omens observed in the sky. Omens can also be classified according to their predictions: some omens concern the king, the country, or the city; others refer to private individuals and their fortunes. One thing is to be kept in mind: the gods send the signs; but what these signs announce is not unavoidable fate. A sign in a Babylonian text is not an absolute cause of a coming event, but a warning. By appropriate actions one can prevent the predicted event from happening. The idea of determinism is not inherent in this concept of sign. The knowledge about the signs is however based on experience: once it was observed that a certain sign had been followed by a specific event, it is considered known that this sign, whenever it is observed again, will indicate the same future event. So while there is an empirical basis for assuming a connection between sign and following event, this does not imply a notion of causality.

Almost all omens are formulated as conditional clauses, "If x happens, then y will happen."[2] The first part is called the protasis, the second part the apodosis.

The omen compendia which were composed from the beginning of the 2nd millennium onwards often contain impossible omens. When dealing with astronomical phenomena, it is usually easy to decide whether the protasis of an omen is possible or not; of course, the protasis has to be understandable. One reason for impossible protases could be the striving for

---

[1] For a discussion of omen literature, see Oppenheim [1977] pp. 206-227; Jeyes [1980] and [1991/1992]; Larsen [1987]; Koch-Westenholz [1995] 13-19.

[2] Exceptions are some liver omens which do not announce a future event but refer to some event in the past. Similarly, some of the Old Babylonian liver models are meant to show the appearance of the liver which had indicated some historical event. See J.-W. Meyer, Untersuchungen zu den Tonlebermodellen aus dem Alten Orient (Kevelaer/Neukirchen-Vluyn 1987).

completeness. If one wants to arrange omens systematically, gaps will become visible in the "documentation." It is likely that such gaps were filled by omens created for the purpose on the basis of some principle, like the contrast between right and left (pars familiaris - pars hostilis). In extispicy, such principles can be discerned in some detail. They were investigated, e.g., by Starr [1983] pp. 15-24, and Jeyes [1980].

As explained above, the events announced by omens were not considered inevitable by the Babylonians and Assyrians. The message of the gods always allowed one to prevent the imminent evil by appropriate actions and behavior.

The rites to avert such evil are called nam-búr-bi. This is a Sumerian word, taken over into Akkadian as *namburbû*, meaning "its (i.e., the announced evil's) loosening". These rites were believed to have been devised by the god Ea, who was also thought to send omens.[3] They were studied in detail by Maul [1994].[4] Among the numerous *namburbi*s there are also some especially concerning evil omens coming from celestial events.

Eclipses are among the most dangerous omens. It is no surprise that rituals against the disasters predictable from eclipses were devised. An example is a tablet from the Seleucid period found in Uruk (BRM 4, 6). An important part of the ritual is the playing of a bronze kettledrum. This instrument is mentioned not only in the ritual, but also in a letter from 671 B.C. (Parpola [1993] No. 347) where actions after the observation of a lunar eclipse are described. Evidence for the actual performance of eclipse rituals in the Neo-Babylonian period was discussed by Beaulieu-Britton [1994].

The validity of omens extends over a fixed period called *adannu* in the Akkadian sources. Such periods (e.g., of 100 days) are mentioned several times in the letters to the Assyrian king (see Parpola [1983] p. 122). For a discussion, see Rochberg-Halton [1988a] p. 42f.

We do not know when this belief in omens originated; by the time when texts containing omens are attested, it is already well established. That is about the last third of the third millennium B.C. Huber [1987a] proposed to see the origin of celestial omens in three lunar eclipses which are described in *Enūma Anu Enlil* and which happened (according to the chronology accepted by him) at three occasions shortly before a change of reign during the Akkade dynasty. As pointed out by Koch-Westenholz ([1995] p. 36), not only is the textual basis for these eclipses uncertain, but it is also quite unlikely that celestial omens would have been invented separately from other types of divination. In any case, no omens in Sumerian are preserved,

---

[3] He is not the only one: omens from the liver, e.g., are sent by Šamaš and Adad. For the role of Ea, see Reiner [1995] p. 81f. with previous literature.

[4] Cf. the review by N. Veldhuis, AfO 42/43 (1995/96) 145-154.

although there is evidence that ominous signs were observed and interpreted.[5]

An introduction to Babylonian celestial omens, with many examples from the texts, can be found in Koch-Westenholz [1995].

## A. SECOND MILLENNIUM B.C.

### *1. Old Babylonian*

Most celestial omens preserved from this period concern lunar eclipses (Rochberg-Halton [1988a] pp. 19-22). Four lunar eclipse tablets are known so far. Since the texts are still unpublished, we give only a short description and refer the reader to Rochberg's work.

The tablets are not exact duplicates, but seem to stem from a common source. All contain omens derived from the date of occurrence of an eclipse. Eclipses are considered not only for days 14, 15, and 16 of a synodic month, but also days 18 through 21 which are not possible lunar eclipse dates. Two of the tablets have in addition omens derived from other circumstances. These are: the time of night, indicated by means of one of three "watches"; the part of the lunar disk affected by the eclipse; the direction which the shadow takes in crossing the moon; and the duration of the eclipse (this part is badly damaged).

The tablets do not yet use the standard formulary that is found in later texts.

These four tablets of eclipse omens are clearly prototypes of the later, more organized series *Enūma Anu Enlil* (below B.1). Several of the omens contained in the older tablets recur unchanged or almost unchanged in the later series.

No Old-Babylonian tablets labelled *Enūma Anu Enlil* have turned up so far. But there is an indication that this title might have existed already. In a catalogue of literary texts[6] a Sumerian incipit $u_4$ an-né, translated into Akkadian as *i-nu* AN *ù* $^{d}$En-líl "When (the gods) Anum and Enlil" is attested; this was taken as a reference to the celestial omen series. Since it occurs in a list of Sumerian literary works, doubts remain; after all, an incipit "When Anum and Enlil" could identify more than one work of literature.

A few other Old Babylonian celestial omen texts are known to us:

---

[5] E. g., Gudea of Lagaš looks for signs from extispicy before beginning the rebuilding of the temple of his city's god, Ningirsu (H. Steible, Die neusumerischen Bau- und Weihinschriften, Teil 1 (Freiburg 1991), Gudea Statue B iii 14).

[6] S. N. Kramer, RA 55 (1961) 172 lines 49f.

Šileiko [1927][7] contains ten omens from the moon and the appearance of the sky. The terminology is different from other omen texts, and the tablet does not seem to have become a part of the later *Enūma Anu Enlil* tradition.

A tablet in private possession (Dietrich [1996]) contains 19 omens from solar eclipses. Five omens are derived from (mostly impossible) days of solar eclipses in Nisannu[8]; the other months (including intercalary Addaru) have one omen each (no day numbers); two more omens are appended, one involving the Sun's setting during the eclipse.[9]

An unpublished tablet in the British Museum is mentioned in Rochberg-Halton [1988a] p. 9 n. 5.

Little is known about the context in which celestial omens were used in Old Babylonian times. By far the most frequent source of omens in this period is the inspection of the entrails of sacrificial animals. A vivid picture of this practice can be gained from the Mari correspondence (Durand [1988]).

The Old Babylonian lunar omens found their way into neighboring countries, where they seem to have been valued.

## 2. *Areas outside Babylonia*

In this section, we consider texts not only in Akkadian, but also in other languages. Most likely, these are translations from Akkadian originals. All these texts are later than Old Babylonian and can be dated to the middle of the second millennium or later.

The Old Babylonian celestial omens which were later organized into *Enūma Anu Enlil* (see below) found widespread interest around the middle of the second millennium. While very few Old and Middle Babylonian sources are known from Babylonia or Assyria itself, many exemplars are found in so-called peripheral areas. Omens corresponding to those found in the later Tablets 17,18, 19, and 22, e.g., can be documented in Ḫattuša in the West and in Elam in the East. In Babylonia and Assyria, evidence for these particular eclipse omens is so far missing. Whether one can see in this a relative lack of interest in celestial divination, or just the accidents of excavation, is uncertain. It was noted, however, by Farber [1993] that the

---

[7] Treated again by Th. Bauer, ZA 43 (1936) 308-314.

[8] The choice of day numbers in Nisannu is not the same as in the later *(Enūma Anu Enlil)* omens from solar eclipses.

[9] Translate omen XIX: "If the god (i.e., the Sun) sets very darkened, Mars will rise and destroy the herds." Cf. the similar omen: *šumma ad-riš īrub* MUL ḪUL : $^{d}$*Ṣal-bat-a-nu* : $^{d}$UDU.IDIM *išaḫḫiṭ*(GU$_4$.UD)*-ma būla* [*uḫallaq*] "If (the Sun) sets darkly, an evil star, variant: Mars, variant: a planet, will rise and [destroy] the herds" Virolleaud [1905-12] Shamash 13:23. A similar apodosis occurs in Old Babylonian lunar eclipse omens (quoted in Rochberg-Halton [1988c] p. 326) and in liver omens (*Ṣalbatānu inappaḫamma būla uḫallaq* RA 65 73:62).

first part of *Enūma Anu Enlil* Tablet 22 exhibits Elamite graphic conventions and is very likely derived from a tradition at home in Susa. This seems to indicate that at least for parts of the material on lunar eclipses, Babylonian scribes preferred to rely for the composition of *Enūma Anu Enlil* on sources from Susa, i.e. from outside their own country, possibly because they had not enough documentation available at home. For the textual history of the lunar eclipse texts, see below.

### *2.1. Mari*

Characteristically, the one ominous celestial event attested in the huge correspondence from Mari, a lunar eclipse, could not be immediately interpreted by the diviners; they resorted to extispicy to find its meaning (Durand [1988] p. 221 No. 81). There exists however a tablet with eclipse omens, edited by Durand [1988] pp. 504-506 No. 248. The few lines preserved begin: "If an eclipse takes place in month ..." The editor considers the tablet a copy of a Babylonian original.

### *2.2. Ḫattuša*

The evidence from Ḫattuša has been discussed by Koch-Westenholz [1993].

There are many celestial omens among the divinatory material from Ḫattuša, both in Hittite and Akkadian. Güterbock [1988] edited bilingual moon omens. A parallel to these in Akkadian was found in Emar (Arnaud [1987] No. 651).

KUB 4 63+ was edited by Leibovici [1956]. There is also a Hittite translation of this text.[10] The colophon of the text reads: "Tablet 1, eclipse of the Sun". Most of its omens are indeed derived from the time and appearance of solar eclipses, but occasionally other phenomena (such as the rising) of the Sun and their circumstances are treated. Koch-Westenholz [1993] p. 231 considers this tablet as Old Babylonian or at least as copied from an Old Babylonian original.

KUB 4 64 contains omens from lunar eclipses, partly similar to what is later found in *Enūma Anu Enlil* Tablet 19, see Rochberg-Halton [1988a] p. 33.

Two fragments of lunar omens in Hurrian were also found in Ḫattuša.[11]

An important detail is the existence of a parallel in Hittite[12] to the introductory passage of the later Babylonian series *Enūma Anu Enlil*. The text is poorly preserved, but it can be seen that it continued with lunar

---

[10] Vieyra, Revue d'histoire des religions 116 (1937) 136-142.
[11] E. Laroche, Catalogue des textes hittites § 774.
[12] KUB 34 12.

omens, as did *Enūma Anu Enlil*. A fragment containing a similar passage was found in Emar (see below).

*2.3. Emar*

16 tablets with celestial omens, some very fragmentary, were discovered in Emar. Two (No. 650 and 651 in Arnaud [1987]) contain omens from the moon, taken from the moon's horns or from halos;[13] No. 650 is a longer version of No. 651.

No. 652 is an Emar version of lunar eclipse omens. It is partly paralleled by a text from Qatna and another from Alalaḫ (Wiseman [1953] No. 452) (see below). Lines 80-82 are similar to the introduction of the later series *Enūma Anu Enlil* but are placed at the end of the text. This has a parallel in one manuscript of *Enūma Anu Enlil* 22 (source E, see Rochberg-Halton [1988a] p. 270) which places a paragraph reminiscent of the introduction at the end of the lunar eclipse section. Moreover, lines 31-47 of No. 652 have parallels in Hittite and in *Enūma Anu Enlil* Tablets 21 and 22, see Koch-Westenholz [1993] pp. 241-246.

No. 653 concerns the Sun.

The remaining fragments cannot yet be confidently assigned to parallels from other places or periods; of course, there need not be any.

*2.4. Ugarit*

A fragment of mostly lunar omens in Ugaritic was found in Ras Ibn Hani near Ugarit and edited by Dietrich-Loretz [1990] pp. 165-195. The text is badly damaged so that many details are uncertain; but there is no doubt about its character.

A rather fragmentary tablet with remnants of solar eclipse omens in Akkadian was published by Arnaud [1996]. Almost nothing is left of the protases, but the text can be identified with the help of versions from Hattuša.

The short Ugaritic text KTU 1.78 has been understood by several authors —e. g., de Jong-van Soldt [1987/1988]; Dietrich-Loretz [1990] pp. 39-62 and W. C. Seitter - H. W. Duerbeck, ibid. pp. 281-286—as referring to a total solar eclipse. The text begins with a disputed word (or words) *bṯt* and then speaks of the Sun's setting on the New Moon day of (the month) Ḫiyar. Then follow the words *ṯġrh ršp*, generally translated as "her (i.e. the Sun's) gatekeeper was Rešep". The remaining sentence of the text refers to an investigation of livers. Assuming that Rešep here refers to the planet Mars (which is not at all certain, see Walker [1989]), one can assume here the astronomical situation of Mars being visible after sunset on the New Moon

13 Arnaud's statement that they are parallel to *Enūma Anu Enlil* 21 must be an error.

day of the sixth month of Ugarit. Such an observation could very well have been considered ominous[14] and be interpreted by extispicy, as seems to be implied by the last sentence of the text. The situation is reminiscent of the lunar eclipse seen in Mari, which also had to be interpreted by extispicy (see above 2.1). It has to be remarked, however, that the astronomical significance of the word "gatekeeper" is not clear. Some modern scholars felt that a Mars observation was not rare enough to be an omen and tried to find a solar eclipse in the text. The wording does not justify this interpretation.

### *2.5. Alalaḫ*

Two tablets with lunar eclipse omens were found in Alalaḫ: Wiseman [1953] Nos. 451 and 452. They are similar both to Old Babylonian eclipse omens and to (contemporary) Hittite texts (see Koch-Westenholz [1993] p. 236), but are clearly different from the later *Enūma Anu Enlil* (Rochberg-Halton [1988a] p. 32).

### *2.6. Qatna*

Bottéro [1950] edited a fragment that was excavated at Qatna in a context from the middle of the 2nd millennium. He showed that it was practically a duplicate to *Enūma Anu Enlil* 22, see now Rochberg-Halton [1988a] p. 271f.

### *2.7. Nuzi*

A tablet of earthquake omens was found at Nuzi and edited by Lacheman [1937]. It seems to contain excerpts from an already damaged original, in two sections. The first section has a parallel in the weather section of later *Enūma Anu Enlil*,[15] the second in Tablet 22, belonging to the lunar eclipse section, see Rochberg-Halton [1988a] pp. 262-269. Both sections of the Nuzi tablet are also parallel to *Iqqur īpuš* § 100-101 in Labat [1965].

### *2.8. Susa*

Not only is there a fragment of *Enūma Anu Enlil* 22 from Susa (Scheil [1917b] = Dossin [1927] No. 258), but we even have an Elamite text (Scheil [1917a]) which included lunar obscurations among other celestial omens. The first part of *Enūma Anu Enlil* 22 probably was composed using sources written in Susa, according to Farber [1993].

---

[14] Stars and planets in some relation to the sun at sunset are found in *Enūma Anu Enlil*: Virolleaud [1905-12] Shamash 16:10ff., see van Soldt [1995] p. 89f.

[15] Virolleaud [1905-12] Adad 20:35-48.

### 3. *Middle-Babylonian and Middle-Assyrian*

From Mesopotamia itself, few tablets of celestial omens come from the second half of the second millennium. One uncertain example is mentioned in Weidner [1952/1953] p. 200.[16] The lunar eclipse texts are treated in Rochberg-Halton [1988a], pp. 23-26. Only five exemplars are available. Two seem to be from a transitional stage between Old and New Babylonian. Three of them are close to the first millennium series *Enūma Anu Enlil* so that they can be used to reconstruct the relevant Tablets. Similarly, a fragmentary Middle-Babylonian tablet from Nippur (PBS 2/2 No. 123, see A. Ungnad, OLZ 1912, 446-449) contains meteorological omens attested in almost identical wording later in *Enūma Anu Enlil.* This shows that the elaboration of the *Enūma Anu Enlil* series had already begun in the 2nd half of the 2nd millennium, as is the case with other Babylonian "canonical" texts.

## B. First Half of the First Millennium B.C.

This section includes texts in both the Neo-Assyrian and the Neo-Babylonian dialect from the time of the Neo-Assyrian empire. The omens themselves are in Babylonian (literary dialect); letters and reports may be in either dialect, depending on their sender.

### 1. *Enūma Anu Enlil*

In the first millennium B.C., celestial omens are found organized in a series of tablets, called *Enūma Anu Enlil* ("When Anu (and) Enlil") after the opening words of its mythological introduction. For a description of its astronomical content, see below II A 1.

An edition by Virolleaud [1905-12], while an impressive achievement for its time, is by now both incomplete (there is no translation; many more texts have been discovered) and frequently erroneous. It is therefore rather difficult to get a clear picture of the series.

A decisive improvement was made in the articles by Weidner([1941], [1954], and [1968]) on *Enūma Anu Enlil.* He not only gave a listing of all exemplars known to him, but also contributed to the elucidation of many details. The need for a new edition was only more obvious after his articles appeared. For parts of *Enūma Anu Enlil*, such new editions are now available in Reiner-Pingree [1975], [1981], and [1998], Rochberg-Halton

[16] The tablets in the so-called library of Tiglat-pilesar I most likely come from the reign of Ninurta-apil-Ekur, who preceded him by about seventy years, see Freydank [1992] pp. 94-97.

[1988a], Al-Rawi-George [1991/1992], and van Soldt [1995]. An up-to-date and detailed overview of *Enūma Anu Enlil* is given by Koch-Westenholz [1995] pp. 74-92.[17] It is not possible to list here all the sources for the series, especially since there is much material still unpublished.

*Enūma Anu Enlil* can be divided into four sections which were named by Virolleaud after gods: Sin (lunar omens); Šamaš (solar omens); Adad (weather omens); and Ištar (omens from stars and planets). This division is made according to the contents of the sections, but is only partly based on the texts themselves. A catalogue from Uruk (Weidner [1941] pp. 186-189) groups lunar omens (Tablets 1-14, called IGI.DUḪ.A$^{me}$ *ša Sin* "appearances of the moon") in a section separate from the lunar eclipse omens (Tablets 15-22). The following section in the catalogue, on the Sun, is mostly broken so that its label is unknown. A fragmentary catalogue from Assur (Weidner [1941] pp. 184-186) still shows remnants of entries for Tablets 39 through 60; no subdivision of the series appears here, although both weather omens and omens from stars and planets are listed.

The colophons of the tablets themselves do not indicate subdivisions but merely indicate tablets of *Enūma Anu Enlil* by number. Note, however, that a commentary on Tablet 14 refers to the main text as "14th tablet of IGI.DUḪ.A[$^{me}$ *ša Sin*]," thus confirming this name of a sub-series as it is used in the catalogue from Uruk, see Al-Rawi-George [1991/1992] p. 52. In numbering the tablets, there are clearly discrepancies among the sources: tablets with identical text have different numbers. Weidner assumed that these differences were the result of different "schools" connected with different cities. However, the distribution of the differences does not correspond well enough with the provenance of the tablets, as far as it can be ascertained (Van Soldt [1995] p. 2; Koch-Westenholz [1995] p. 80f.). Nevertheless, it is possible to identify different recensions for certain Tablets of *Enūma Anu Enlil* (Rochberg-Halton [1988a] pp. 177-179).

For the library of Assurbanipal in Nineveh, tablets of *Enūma Anu Enlil* were collected from different sources. Many had been copied by the famous scribe Nabû-zuqup-kēnu from Kalaḫ who lived during the reigns of Sargon II and Sennacherib. The originals used by him came mostly from Babylonia. He may also have been the one who produced a luxury edition of the series on sixteen wax-covered ivory boards forming a polyptych. This exemplar was probably intended for Sargon's newly constructed residence in Khorsabad,[18] but was never actually brought there; it was excavated in Kalaḫ. Unfortunately, almost nothing remains of the text. Numerous tablets of *Enūma Anu Enlil* from Babylonia were among the booty brought to

---

[17] Many omens were excerpted in Gössmann [1950], mostly based on Virolleaud [1905-12].

[18] D. J. Wiseman, Iraq 17 (1955) pp. 3-13.

Nineveh in 647 B.C. after the war between Assurbanipal and his brother Šamaš-šum-ukin.[19] It is no surprise then that different recensions of the work existed at Nineveh.

The tradition of *Enūma Anu Enlil* did not end with the Assyrian empire. Tablets were copied in the Achaemenid and Hellenistic periods (e.g., von Weiher [1993] nos. 160-162; F. Thureau-Dangin, TCL 6 no. 16), and men charged with celestial observations were called "Scribe of *Enūma Anu Enlil*" as late as the 2nd century B.C. (Van der Spek [1985] pp. 548-554). This continued availability of celestial omens will have provided the source for omens in similar style preserved in Greek sources.[20]

*Contents of Enūma Anu Enlil*

Due to the bad state of preservation of most of the exemplars, and the lack of a complete text edition, a reasonably comprehensive overview of the series cannot be given. The following remarks are necessarily uneven and can only indicate some of its contents.

The mythological introduction (lines 1-8) traces the order of heaven and earth back to the gods Anu, Enlil, and Ea. It comes in a Sumerian and an Akkadian version which are slightly different from each other. The Sumerian version mentions the Moon god, the Akkadian version the Sun god, but in different functions. For a recent translation, see Koch-Westenholz [1995] p. 77f.; see also Oppenheim [1977] p. 224 n. 66. The introduction has a precursor in Hittite (see above A 2.2); this shows that some version of it very likely existed in Old Babylonian times in Babylonia proper, from where it was imported to the Hittites. It cannot be said whether the introduction was connected to celestial omens already then. A passage similar to the introduction is found on a tablet from Emar (see above A 2.3); this text however is placed not at the beginning, but rather at the end of a tablet with celestial omens.

### *1.1. The Moon*

Lunar omens fill Tablets 1 to 22. The first sections contain omens concerning unusually early or otherwise irregular (*ina lā minâtišu*) appearances of the moon. There follow possibly dark risings of the moon. Many omens are taken from the moon's horns, and from the stars which are observed next to them. Tablets 8-10 deal with lunar halos. Little is preserved

[19] S. Parpola, JNES 42 (1983) p. 6 lists 107 tablets and 6 writing-boards of *Enūma Anu Enlil* among the tablets brought to Nineveh. Some of them are said to have come from Nippur (p. 7).

[20] C. Bezold and F. Boll, Reflexe astrologischer Keilinschriften bei griechischen Schriftstellern, SHAW, phil.-hist. Kl., 1911/7.

of Tablets 11-13. Tablet 14 is made up of tables for the moon, i.e. it is largely astronomical and therefore treated below II A 1.4.

Tablets 15-22 deal with lunar eclipses. This part was edited by Rochberg-Halton [1988a], on whose introduction we rely for the following description. For the astronomical contents, see below II A 1.3.

Some of the omens are clearly related to the (still unpublished) Old Babylonian eclipse omens. This is especially evident for Tablets 17, 18, 19, and 22. The Old Babylonian antecedents of these Tablets must also have been the source for the numerous adaptations and translations in peripheral areas. On the other hand, material parallel to Tablets 15, 16, and 20 is found in a few exemplars from the 2nd half of the 2nd millennium both in Babylonia and Assyria, but not outside Mesopotamia.

The aspects of a lunar eclipse which are considered in the protases are the following:

A. Date of eclipse occurrence (month, day);
B. Time and duration;
C. Appearance of eclipse (magnitude, direction of motion of the eclipse shadow, color);
D. Phenomena associated with the eclipse (wind, weather, earthquakes etc.; stars and planets visible).

No single tablet contains all of these elements in its protases.

Ad A. Eclipses are possible in all months, and the omen tablets go through them in sequence. Of the two intercalary months used in the first millennium, only intercalary Addaru is taken into account. In order to connect a prediction with a particular country, in one of the texts (Rochberg-Halton [1988a] p. 37f.) the months are arranged in four groups of three, each of which is associated with a (traditionally named) geographical area: months I, V, IX with Akkad; months II, VI, X with Elam; months III, VII, XI with Amurru; and months IV, VIII, XII with Subartu. This scheme is however not applied consistently in the omens; on the contrary, they mostly use different associations, which in addition are not all compatible with each other. The explicit mention of such a scheme, then, shows only one attempt to formulate a principle which was by no means universally applied.

Since the Babylonian calendar has strictly lunar months, lunar eclipses are possible only around the middle of the month, on days 12 to 15. In spite of this, the "canonical" *Enūma Anu Enlil* of the Neo-Assyrian period considers eclipses mostly on days 14, 15, 16, 20, and 21; sometimes, days 17, 18, and 19 are added. In this way, not only are impossible days retained, but also the possible days 12 and 13 omitted. Days 12 and 13 do occur in "non-canonical" omens, and they are also attested in Reports (see below), as is to be expected. Tablet 22 even allows eclipses to happen "from the 1st day to the 30th day," i.e. on all days of a month. This looks like an excessive attempt to cover every possibility by some schema. But already the Old Babylonian eclipse texts have days from the 14th to the 21st (only the 17th day does not occur there). Nevertheless, the 14th day seems to have been the "normal" day for a lunar eclipse.

Ad B. Daytime and nighttime are each divided into three watches. The omens frequently take notice of the watch in which an eclipse occurs. Certain predictions are associated with particular watches, but no scheme is consistently applied.

The duration of an eclipse too is expressed by referring to the watches. Parts of a watch (1/3 or 2/3) occur, but they have no ominous significance. There are different expressions used to indicate eclipse duration, for which see Rochberg-Halton [1988a] pp. 44-47. If the moon sets during the eclipse, this is also ominous (found in Tablets 19 and 20). The rising of the moon while it is eclipsed is not unambiguously attested; the relevant omens may only be concerned with clouds obscuring the moon.

Ad C. The magnitude of an eclipse is described in the omen texts but not measured. Occasionally the remark "it left a finger" indicates an almost total eclipse. Fingerbreadths are not yet used to measure the magnitude of an eclipse in the omens. The extent of the eclipse can be gathered from the description of which quadrants of the moon are affected.

The movement of the shadow across the lunar disk was the most important element for the decision concerning which country would be affected by the evil announced by the eclipse. The disk was divided into four quadrants; each is associated with a country name. There is no system of these associations that would be valid for all of Babylonian celestial omens; for a description, see Schott-Schaumberger [1941/1942] and Rochberg-Halton [1988a] pp. 51-55. The names of the countries are traditional and do not correspond to the actual political entities of the time when *Enūma Anu Enlil* found its final recension. They are: Akkad, Subartu, Elam, and Amurru. In the interpretation of the times, Akkad is equated with Babylonia, Subartu with Assyria, and Amurru with any country in the west; Elam alone has not changed its name for more than a millennium. The country represented by the quadrant which is first darkened is expected to suffer the evil announced. Conversely, the country whose quadrant first begins to clear is not affected.

The color of the totally eclipsed moon is a dark dull red because of the refraction of sunlight by the earth's atmosphere. Red and varieties of red are therefore the most common colors occurring in eclipse omens. In the versions of *Enūma Anu Enlil* from the Neo-Assyrian period other colors are also attested. They follow a standard sequence found throughout the omen literature: white (*peṣû*), black (*ṣalmu*), red (*sāmu*), yellow-green (*arqu*), and variegated (*burrumu*). In some cases such colors may actually appear on the moon because of clouds and other atmospheric phenomena.

Ad D. The direction from which the wind is blowing during an eclipse is used to determine the country that is affected. Schemes similar to those associating countries with quadrants of the lunar disk are used.

Apart from winds, other weather phenomena such as clouds and thunder are occasionally added to the protases of eclipse omens, as are earthquakes. In general, however, the weather is the topic of the third section of *Enūma Anu Enlil.*

Stars or planets are rarely mentioned in eclipse omens. By contrast, the appearance of planets is an item of interest in the eclipse reports (see below II B 2.5).

A further development of lunar eclipse omens is found in a text published by Rochberg-Halton [1984a] and probably dating to the Achaemenid period. It combines the moon and the four planets Jupiter, Venus, Saturn, and Mars in the following pattern: if there is a lunar eclipse in zodiacal $sign_1$ and the night watch comes to an end and the ... wind blows, Jupiter (or: Venus) is not present, and Saturn or Mars stand in zodiacal $sign_2$ and zodiacal $sign_3$ (respectively). The zodiacal signs are arranged in four groups of three so that their distance is always 4 signs (corresponding to 120°). This is similar to the concept of trine aspect found in later Hellenistic astrology. Such a system presupposes the zodiac of twelve signs of equal length, which was invented in Babylonia in the 5th century B.C. Each "triplicity" is also associated with a particular wind; the correlations of these winds correspond exactly to what is found in Geminos' Isagoge. The predictions derived from these omens use the stock phrases of *Enūma Anu Enlil*, and they show no relation to the presence or absence of the benefic and malefic planets mentioned. In this respect, the Babylonian predecessor was not imitated in Hellenistic astrology.

The concept of arranging a group of twelve into four "triplicities" is attested already in a commentary tablet of *Enūma Anu Enlil* (Virolleaud [1905-12] 2nd Supp. 19:9-20) and in a compilation of sundry information on celestial divination (Weidner [1959/60], now conveniently accessible in Koch-Westenholz [1995] p. 202:244-277). There it is the twelve months of the year which are assigned to countries so that months I, V, and IX; II, VI, and X, etc., correspond to the same country. The scheme is applied by the scholars advising the Assyrian king in the 7th century. This shows the Babylonian origin of the concept of "triplicity": the system used for grouping the months was transferred to zodiacal signs. Such equivalence of months and signs can be found in several Babylonian texts from the Neo-Babylonian period, e. g., von Weiher [1983] 43, von Weiher [1988] 104 and 105, see Reiner [1995] p. 115 n. 522. Especially instructive are some texts which provide suitable times for the use of incantations: the older version gives calendar dates, while the younger version gives positions in the corresponding zodiacal signs (Reiner [1995] p. 111).

It should be noted that the text published by Rochberg-Halton [1984a] does not contain a drawing that would show the relation of the zodiacal signs occurring in the text. Such a drawing is however attested in another cuneiform text.[21] That the zodiac could be represented by a circle is supposedly shown by a tablet found in Sippar.[22]

### *1.2. The Sun*[23]

Tablets 23 (or 24) to 29 (30) concern solar omens (it should be noted that the numbers found on the actual tablets are already divergent at an earlier point in the series). Tablet 23 (24) deals with the appearance of the Sun at sunrise at the beginning of the month; all the apodoses contain eclipses of the Sun. Tablet 24 (25) mostly concerns what is called "disk" (*šamšatu*). This is not always the disk of the Sun itself, but at least in some cases must be a parhelion. "Disks" are also mentioned in connection with the moon. The following Tablets contain omens from the appearance of the Sun in relation to clouds and other atmospheric phenomena. The last topic, on Tablets 28 (29) and 29 (30), is phenomena connected with the appearance of a cloudbank near the Sun. Here, as in the preceding Tablets, colors play an important role.

The general solar omens are followed by omens from solar eclipses, as in the case of the moon; they are found on Tablets 31 to 36 according to Weidner's ([1968/69]) reconstruction.

The boundary between Sun omens and weather omens cannot be drawn exactly in the present state of our knowledge. After the omens from solar eclipses, the sequence and contents of *Enūma Anu Enlil* are particularly fragmentary and uncertain.

### *1.3. Weather*

The weather omens have received the least attention of all sections of *Enūma Anu Enlil* in recent years.

Tablet 43 deals with lightning and earthquakes. For earthquakes, there is a forerunner in a tablet from Nuzi (see above A 2.7).

A fragmentary tablet on earthquakes and thunder was published as Wiseman-Black [1996] No. 18. Colors and shapes of clouds are the topic of No. 17. Neither of them can be assigned to a particular Tablet of *Enūma Anu Enlil*.

Tablets 44 to 46 deal with thunder under various circumstances.

The last Tablet of the weather section seems to concern the wind.

---

[21] TCL 6 13, see Rochberg-Halton [1987b].

[22] Walid al-Jadir, Archeologia 224 (1987) p. 26f.

[23] This section (except the solar eclipse omens) is edited by van Soldt [1995].

*1.4. Stars and Planets (Tablets 50 to end)*

This section of *Enūma Anu Enlil* is still only partly accessible. Weidner did not continue his series of articles on the organization of *Enūma Anu Enlil* beyond Tablet 50. A new edition of the section by D. Pingree and E. Reiner is in progress; at present, it covers Tablets 50,51 and 63, as well as the Venus omens (Reiner-Pingree [1975], [1981], and [1998]). The numbers assigned to Tablets vary, as was explained above. Since some Tablets seem to be entirely missing from the fragments preserved, it is not yet possible to establish the contents of each Tablet. 50 to 53 are concerned with constellations. 56 deals with planets in general, 57 (at least in part) with constellations. Tablets 59 to 63 concern Venus; a description of their contents can be found in Pingree [1993] and Reiner-Pingree [1998]. Tablets 64 and 65 deal with Jupiter. It is likely that the other planets, at least Mars, had Tablets of their own, but until now no reconstruction is possible. Tablet 68, according to one numbering the last of the series, seems to return to "stars" in a general sense. An even higher number, 70, is mentioned in a subscript, see Reiner-Pingree [1981] p. 23.

Tablet 50, as far as it can be reconstructed, could be called a guide to stellar omens. At least part of it does not contain omens, but sentences of the pattern: "Star X is for Y," Y being some terrestrial event or topic of the omen apodoses. Unfortunately, the source material is rather uneven, containing excerpts and commented texts. In one of the commented texts, the sequence of stars seems to be dependent on the tradition of the "Astrolabes" and of MUL.APIN.

Tablet 51 is again related to the list of stars in the "Astrolabes". In its first section, after stating in which month a certain constellation rises, omens are then quoted for this constellation. Section 2 is a variant of section 1, and section 3 deals with the constellation Ikû.

Tablet 56 was edited by Largement [1957]. The topic of this tablet is *bibbu* which means some kind of sheep and is used as a term for planet. The text begins with appearances of a planet. There follows a statement about the visibility periods of the particular planet Mercury. Then the meeting of a *bibbu* with other planets and with another *bibbu* and a few fixed stars is considered. The next section concerns a red *bibbu* together with stars. There follow the encounters between a *bibbu* and constellations. Then *bibbus* "flare up" (*ṣarāru*) and meet stars or planets. More sections deal with *bibbus* and stars or constellations, also with other planets and the moon. Then comes another general statement about visibility periods of Saturn, and again about Mercury. These statements concerning Saturn and Mercury are also found, with little variation, in MUL.APIN Tablet II i 53-59 and 64f. Two more lines on *bibbu* and three more items on Mars lead up to the end of the Tablet.

Tablet 57 deals in the beginning of its preserved part with omens from the constellations of the Raven, the Eagle and the Fox. The remainder is so broken that a continuous topic cannot be restored.

Tablet 58 may contain Venus omens, because the catch-line on Tablet 57 in Virolleaud [1905-12] Ištar 23:32 refers to Venus.

Tablets 59 and 60 contain Venus omens arranged by months, 59 for the first half of the year, 60 for the second. Some of them are also found in *Iqqur īpuš* § 82-86, see Labat [1965] pp. 164-170. The contents of Tablets 61 and 62 are more difficult to establish, although there are many fragments available. An arrangement of them is found in Pingree [1993].

Wiseman-Black [1996] No. 15 is said to belong to Tablet 61; it is edited in Reiner-Pingree [1998] 236-242.

Tablet 63 contains omens from the disappearances and reappearances of Venus. From the choice of calendar dates mentioned for these phenomena it is clear that actual observations were the ultimate source. They can be associated with the reign of the Old Babylonian king Ammiṣaduqa because of a year name given in the text. The astronomical content of this Tablet is treated in detail below II A 1.1.

Apart from what could be called the main series *Enūma Anu Enlil*, we find excerpts and commentary tablets. Many of these were organized into a kind of auxiliary series, while others remain (to our knowledge) singular compilations. Weidner ([1941] p. 183f.) grouped these additional texts as follows:

Excerpts:

a) *rikis girri Enūma Anu Enlil* ("guide to *Enūma Anu Enlil*"): omens excerpted from the main series and following its sequence exactly.
b) sundry excerpts containing a small number of omens from one or several Tablets of the main series. These excerpts sometimes deal with a special topic, and may have been made for a particular occasion.

Commentaries:

c) *mukallimtu Enūma Anu Enlil* "explanation of *Enūma Anu Enlil*"
d) *ṣâtu u šūt pî*.
e) The commentary series *Šumma Sin ina tāmartišu* "If the moon at its appearance".

For the commentaries, see below 3.

## 2. *Non-canonical*

In addition to the groups just mentioned, we find the so-called *aḫû* omens. Literally, this means "other, strange, extraneous". It was therefore assumed that such omens were not part of the main series *Enūma Anu Enlil*. The actual situation is not quite so clear. The scholars at the Assyrian court (see below 5.) occasionally quote omens which they call *aḫû*. They consider such material just as authoritative as other omens. The catalogue from Assur (see above) had a list of 29 *aḫû* tablets appended to *Enūma Anu Enlil*,[24] and from this and other references it appears that the "extraneous" material is itself arranged in a series like *Enūma Anu Enlil*. But (at least on some tablets) occasionally omens called *aḫû* are even incorporated into the main series, see Koch-Westenholz [1995] p. 90. According to Lieberman [1990] p. 308, it is therefore more appropriate to think of *aḫû* omens as additions, not as something excluded from *Enūma Anu Enlil*. The only *aḫû* tablet identified so far is the 29th, edited by Rochberg-Halton [1987a]. It concerns lunar eclipses, as do Tablets 15-22 of the main series. A comparison is therefore possible, which shows that the *aḫû* tablet provided additional material and used different phrases to express the variable circumstances of an eclipse. Why and when some omens were not included in *Enūma Anu Enlil* and then considered "extraneous", is difficult to say because we do not know the history of these traditions (Rochberg-Halton [1984b]).

### *3. Commentaries and explanatory texts*

Three types of commentaries on *Enūma Anu Enlil* can be listed according to the classification used by Weidner:

*mukallimtu Enūma Anu Enlil* "explanation of *Enūma Anu Enlil*": comments on the protases of a Tablet of the main series. The comments can deal with words or content. For instance, stars are said to be equivalent to one of the planets; in this way an omen is made applicable to a situation not previously intended (see below 5).

*ṣâtu u šūt pî*: comments on words, their reading and meaning, in a Tablet of the main series. Such comments existed also for other kinds of omens, such as extispicy.

Finally, there are excerpts with comments, both on content and on single words. These excerpts are organised as a commentary series which is called *Šumma Sin ina tāmartišu* "If the moon at its appearance". An example is Borger [1973] where he edits an exemplar of Tablet 2 of the series. Tablet 4 is attested by Virolleaud [1905-12] 2 Supp 19.

In general, there is little in *Enūma Anu Enlil* that corresponds to the later mathematical astronomy. The few planetary periods mentioned in Tablet 56 are very crude. The periodic phenomena which are the topic of mathematical astronomy are not ominous. Stationary points, e.g., are in themselves not of interest; only the movements of the planets in and out of constellations are

[24] Unfortunately, only the incipits of the last two are preserved.

considered by the omens. For a detailed explication of the astronomical content of *Enūma Anu Enlil*, see below II A 1.

There are other celestial omen texts which show no obvious relation to *Enūma Anu Enlil*:

A small group of tablets published by Gadd [1967] and Hunger [1969] used numbers as a kind of cryptography for omen protases. Since five elements (represented by the numbers 3, 4, 7, 11.30, 12) vary cyclically, it is likely that they concern the five planets. The phenomenon connecting the planets is given by numbers as well, so it can only be guessed that it may be "conjunction" or the like.

A mixture of star-list and comments of different sorts was called by Weidner (Weidner [1959/1960]; Koch-Westenholz [1995] pp. 187-205) "astrologischer Sammeltext" or "Great Star List". Its preserved part begins with explanations of one star or constellation by another, or by god names. Occasionally, omen-like statements are included. There are also keys to the interpretation of omens inserted: the parts of the moon corresponding to certain countries, or the months related to those same countries. The text includes three groups of twelve stars each, summarized as stars of Elam, Akkade, and Amurru. These are followed by several groups of seven, e.g., seven names of Mars or seven planets (including Sun and Moon). We find a list of Elamite month names, and the famous statement about the three heavens, each made of a different stone. After a description of the changing shape of the Moon, more mythological explanations follow. The end, like the beginning, is not preserved.

Partly concerned with astral matters is a text with the incipit I.NAM.GIŠ.ḪUR.AN.KI.A, edited by Livingstone [1986] pp. 17-52. Among mythological comments, it gives explanations to words for the moon's changing shape in the course of the month. There is also a passage similar to MUL.APIN (Tablet II ii 14ff.) and a table for the visibility of the moon which is attested as part of *Enūma Anu Enlil* Tablet 14 as well (see below II A 1.4).

### *4. Diviner's Manual*

This text was composed in the 7th century at the latest (the earliest exemplars come from Assurbanipal's library). A. L. Oppenheim, who edited it (Oppenheim [1974]), called it a "Diviner's Manual". It begins with a list of 13 incipits, all of them omen protases, which are summed up[25] as "tablets with signs occurring on earth". Then follows another list of 11 such incipits, called in a summary "signs occurring in the sky". After each list, we find general statements about omens: "The signs on earth just as those in the sky

[25] The number 14 in the summary is an error.

give us signals; sky and earth both produce portents, though appearing separately they are not separate (because) sky and earth are related." The connection between celestial and terrestrial signs is stressed more than once in the text. Then follows the instructional part. The question of the validity of an omen is addressed: sometimes a sign can be cancelled by another one of opposite portent. What however will happen if no annulment takes place? How can the evil announced by the sign be dispelled? To this the text proposes a solution not attested elsewhere. First, the calendar date has to be established beyond doubt. For this, allusions are made to the possible ways of determining whether a year needs an intercalary month[26], and to find out the lengths of the months, i.e. whether they have 29 or 30 days. Once the date is certain, it may turn out that the bad omen was not really applicable because it occurred at a different time. Also, hemerologies can be consulted to see if a day is propitious for what had been planned when the omen occurred. For this purpose, a table is appended which shows which months are propitious for a few examples of actions. This last part seems to be much abbreviated; there are also differences between the exemplars.

This text was not the only one of its kind because fragments exist which have the instructions for the diviner on one side and omens on the other side which are different from the ones in the main text.

In spite of the fact that several copies of this text existed, it still looks very much like a singular attempt at solving a recurring problem in divination. It also has to be noted that almost all of the incipits listed in the text (see above) are not attested anywhere else in the omen literature. This makes one wonder whether the Diviner's Manual is based on sources which were rarely used among the professionals known to us.

### *5. Letters and Reports*

From the last century of the Neo-Assyrian empire survive remnants of the royal archives which were found in Nineveh. Among the thousands of tablets belonging to these archives are letters and letter-like messages (called Reports) sent to the Assyrian kings by experts in divination, specifically in celestial divination. The latest editions are Parpola [1993] and Hunger [1992]; for an introduction and commentary, see also Parpola [1983]. Almost all the preserved tablets were written during the reigns of Esarhaddon (-679 to -668) and Assurbanipal (-667 to -626); practically no

[26] For intercalation, the calendar date of the conjunction (lit., "balancing") of the Moon and Pleiades is observed (the translation "opposition" is inappropriate here). This is the so-called "Plejaden-Schaltregel" found in MUL.APIN II Gap A 8f. and discussed already by Schaumberger [1935] pp. 340-344.

letters were found to be later than -647, and the majority comes from the time between -677 and -665.

The senders of the Letters and Reports were residing in cities all over Assyria and Babylonia. Since in their messages they cite omens from both *Enūma Anu Enlil* and other sources considered equally authoritative, they must have had at their disposal reasonably comprehensive collections of such omens, even if they may have known many of the omens by heart. We know from the excavations that such a collection was available in Nineveh, where the most influential scribes were living. It is possible that Assurbanipal himself, who boasts of his ability to write and read, consulted such tablets. Lieberman [1990] p. 328f. assumes that Assurbanipal wanted to check on the information presented to him by the scribes in order to control them.

When one compares the omens quoted in the Letters and Reports with *Enūma Anu Enlil*, it appears that the senders were quite selective in their choice of omens. Some omens are quoted again and again, particularly good omens from the Moon which happen frequently. Also, some of the frequently quoted omens have not been found yet in what remains of *Enūma Anu Enlil.*

In a letter (Parpola [1993] No. 100, dateable to -656) to the Assyrian king Assurbanipal, Akkullānu quotes an omen interpreting a lack of rain as favorable to the king. He says that it came from a Report which was sent by Ea-mušallim to the Babylonian king Marduk-nādin-aḫḫē (who ruled at the beginning of the 11th century B.C.). Apparently, records of their Babylonian predecessors' activities were available to Assyrian experts in Nineveh more than 400 years later.

The correspondence can be subdivided into Letters and Reports. Letters address the king, politely of course, and describe the actions that the senders propose to be taken in consequence of omens observed in the sky (and elsewhere[27]). As we said above, predictions in omens are not inevitable fate, and almost everything could be averted by the proper measures. Paramount among these measures were the so-called *namburbi* rites; for a detailed description, see above p. 6. We frequently read in the letters that such rites were recommended by the experts and performed with the participation of the king. Balasî writes after an earthquake: "(The god Ea) who caused the earthquake has also created the apotropaic ritual against it" (Parpola [1993] No. 56; similarly No. 10). Undesirable omens from planets can also be counteracted in this way (Parpola [1993] No. 67, 206, 362, 381). The Babylonian Bēl-ušēzib considers a *namburbi* rite advantageous in the case of a possible eclipse, although he explains to the king that the gods, by hiding

[27] Omens not from the sky will not be discussed here; they do not figure prominently, however, in this correspondence.

the moon behind clouds, had indicated that the eclipse (if there was one) was not intended as a sign for the king (Parpola [1993] No. 114). If an eclipse did not threaten the Assyrian king, nevertheless apotropaic rites were advisable (Parpola [1993] No. 347). Sometimes the king is told in detail what he has to expect: "It is said in the relevant *namburbi* as follows: 'He (the king) sits seven days in a reed hut, and purification rites are performed upon him; he is treated like a sick person. During the seven days, 'hand-lifting' prayers before the gods of the night and the *namburbi* against evil of any kind are performed as well. For the seven days he sits in the reed hut, he recites benedictions to his god and his goddess'". After this, the sender of the letter politely asks the king to agree to his proposal (Parpola [1993] No. 277).

A rather strange way of avoiding the dangerous consequences indicated by certain eclipses was the substitute king rite.[28] If an eclipse implied the death of the king of Assyria, some man was chosen to be put in his place, at least for all appearances. Usually someone whose life was not considered important, like a condemned criminal, seems to have been used for the purpose. He was clad like a king and made to sit on the throne, but of course he had no influence on government. In order to make it clear to everyone who was to suffer the impending evil, the bad portents were recited to the substitute king. The true king, in the meantime, had to behave as inconspicuously as possible, avoid being seen outside the palace, and undergo extensive purifying rites. In letters written to him during such a period, the king was to be addressed as "farmer" in order to avoid any association with kingship. It was expected that the dire fate announced by the omen would fall on the substitute king. The assumed time of validity of such an omen was 100 days. If additional unfavorable portents were expected (e.g., other eclipses), the substitute would remain enthroned for most of this time. Otherwise, his "reign" could be rather short; it was neither convenient nor necessary to extend it. In any case, the substitute king had to die. It is unknown how his death was brought about, but it was the decision of the true king: in the letters, the advisers ask the "farmer" on which day the substitute king "should go to his fate". He was then buried and mourned like a king (Parpola [1993] No. 352).

According to literary tradition, a substitute king was enthroned during the reign of Erra-imittī of Isin in the early part of the 2nd millennium; this case was atypical insofar as the true king died while the substitute sat on the throne, and so the reign passed to the latter.[29] Since omens announcing the death of the king are known from the Old Babylonian period, such an early

---

[28] A study based on all the available material is found in Parpola [1983] pp. xxii-xxxii.

[29] Grayson [1975] p. 155.

date for the rite is not unlikely. The fact that substitute kings are known from Hittite texts[30] (although no actual performance is attested) also indicates that the idea goes back to Old Babylonian times.

There are many other topics dealt with in the Letters: Esarhaddon frequently received medical advice, but there are also reports about the health of members of the royal family. Matters of cult and ritual are much discussed, especially when the king's participation is required; an overview is found in Parpola [1983] pp. 472-477.

The Reports differ from the Letters inasmuch as they do not normally address the king. Their shape too differs from the Letters: in Reports, the writing is parallel to the longer side of the tablet, whereas in Letters it is parallel to the shorter side.

Reports usually contain only observations and the omens derived from them. In most cases, an observation will not be reported directly, but instead only the pertinent omens are quoted. Strictly speaking, this is enough, because the protasis of an omen always implies an observation. When the actual observation is described in addition to the omen, it is often quite different from what the omen's protasis literally said. Sometimes explanations are given of why a quoted omen is considered applicable to the observation made by the reporter. Frequently, a fixed star or constellation mentioned in the omen is considered equivalent to a planet. By this method traditional omens could be applied to actual observations much more frequently than if taken literally. For the king's correspondents, this was a regular procedure.

The interpretation of constellation names as planets has been investigated by Bezold (in Boll [1916]). His theory that it was mostly the color of stars and planets which provided the association was recently questioned by Koch-Westenholz [1995], pp. 131f. This whole topic needs re-investigation; unfortunately, many sources for planetary omens are still unpublished.

## C. SECOND HALF OF THE FIRST MILLENNIUM B.C.

### *1. Proto-horoscopes*

At the end of the 5th century B.C., the earliest examples[31] of what has been called Babylonian horoscopes are attested. A few of them were edited by Sachs [1952]; Rochberg-Halton [1989] gives an overview of the 32 horoscopes known so far. Her edition of them appeared recently (Rochberg

[30] Kümmel [1967].
[31] Dateable to -409.

[1998]). Strictly speaking, they should not be called horoscopes; see Pingree [1997] p. 20. They begin with the date on which a child was born. Rarely is the name of the child mentioned. Then follow the positions of the planets, in the sequence Moon, Sun, Jupiter, Venus, Mercury, Saturn, and Mars. Their positions are more often given by zodiacal sign alone, less often by degree within a sign. Apart from these positions, other astronomical data are included in the horoscope. These can be more or less distant in time from the date of birth, but were probably considered as possibly significant. Such are the length of the month (whether 29 or 30 days), the time interval between sunrise and moonset just after full moon, and the time between moonrise and sunrise towards the end of the month. Further events listed are eclipses, including those that were not visible in Babylonia, equinoxes and solstices, and conjunctions of the moon with reference stars.[32]

The positions of the planets at any given time cannot be easily found by means of the procedures contained in the mathematical astronomical texts known to us. These are rather designed to compute the dates and positions of the synodic phenomena of the planets. It is possible to find intermediate positions between the synodic phenomena by interpolation, but this seems to have been done rarely.[33] Nevertheless the positions of the planets in horoscopes are apparently computed. The most likely source for these computations are the so-called Almanacs (for the term see Sachs [1948] p. 277, and for a discussion see below II B 4). The astronomical data listed in the horoscopes are found in the Almanacs too (Rochberg-Halton [1989] pp. 107f.). It should be noted though that one of the two horoscopes for -409 (Rochberg [1998] Text 1) is closer to the Astronomical Diaries (see below II B 2) because it lists the synodic phenomena of the planets in the course of the year, rather than their positions in the zodiac.

A horoscope-like text gives positions of planets required at the time of conception of a child in order that he will become king (Rochberg-Halton [1989] p. 107 n. 21).

Most of the horoscopes do not give any predictions about the future life of the child. Such predictions were probably to be found on different tablets. There exist a number of nativity omen texts which could have served this purpose (see below). Occasionally, such nativity omens are quoted in horoscopes. One could see in a horoscope a listing of the "signs" available for the date of birth, a kind of omen protasis, for which the apodosis was to be found in the omen literature. Seen in this way (Rochberg-Halton [1989] p. 110; Rochberg [1998] p. 16), the horoscopes would be an expansion of the traditional omen procedures, and not a radical departure from them.

32 For these reference stars, termed "Normal Stars", see below II B 2.3.
33 See Huber [1957].

A text from the reign of a king Artaritassu (possibly Artaxerxes II) reports on stars seen by a person in a dream.[34] Another tablet from the same reign contains records of the places of the planets at the birth of several named persons.[35]

## 2. *Hypsomata*

The Babylonian term *ašar niṣirti* or *bīt niṣirti*, literally "place of secret", occurs in texts referring to the planets' positions. Each planet has its specific *ašar niṣirti*. These places seem to correspond to the Greek term hypsoma, which is the place in the ecliptic where a planet exerts its strongest influence (see Rochberg-Halton [1988b] pp. 56f.). The underlying concept in Babylonia is different, however.

The earliest attestation of this concept is found in the inscriptions of Esarhaddon of Assyria (see below II B 1.2). There it is said of Jupiter that he reached his *ašar niṣirti* in the month Pēt-bābi, which corresponds to Tešrītu (see Hunger-Pingree [1989] pp. 146f.; Koch [1999] p. 22). The place of the *ašar niṣirti* of Jupiter is between Cancer and Leo, as can be seen both from computation of the celestial events described in Esarhaddon's inscriptions and from a passage in the so-called GU Text (see below II A 6). From these two texts the *ašar niṣirti* of Jupiter seems to be at a longitude of 91°. This differs, though not much, from Jupiter's hypsoma in Greek astrology which is situated at 15° of Cancer, i.e. 105°. It has to be remembered, however, that the concept of longitude measured in degrees was not present in Assyrian times.

Another passage of Esarhaddon (Borger [1956] § 2 i 39 - ii 5) mentions the *ašar niṣirti* of Venus (see below II B 1.2). Several omens derived from Venus' *ašar niṣirti* do not suffice to locate it precisely.

The *bīt niṣirti* of Mercury is placed in Virgo by a text from the middle of the 3rd century B.C. (Weidner [1967] p. 11).

In the "horoscopes" the term *ašar/bīt niṣirti* occurs several times. Unfortunately, the positions do not agree with what can be derived from other sources, be they Babylonian or Greek, see Rochberg-Halton [1988b] p. 56.

Weidner [1967] published a group of texts[36] which had become famous because of the pictures of constellations engraved on some of the exemplars. They use older material which probably was not yet provided with astral connections (Weidner [1967] pp. 39f.; more fragments are still unpublished). The texts seem to have formed a kind of series of twelve sections (some

---

[34] T. G. Pinches, Recueil de travaux 19 (1897) p. 101-103.

[35] It will be edited by H. Hunger.

[36] The tablets had been described and excerpts published already in Weidner [1919]. A further fragment is published in von Weiher [1993] No. 167.

taking up a whole tablet, some half of one), each of which was devoted to one sign of the zodiac. Of the preserved sections, each begins with a description of a lunar eclipse; sometimes a drawing of the constellation concerned is found. Weidner proposed that the drawings represent the planets in their hypsomata. Then there follow twelve columns which subdivide the zodiacal sign into a "micro-zodiac" of twelve signs of 2½° each. The 2½° subdivisions are named like the zodiacal signs, beginning with the name of the sign that was subdivided, and continuing in their order. The columns then contain associations with a temple or city, and with trees, plants, and stones. Then follow notes of omens from the planets and prescriptions in the style of hemerologies.

The same eclipse omens are also found separately, probably arranged in "triplicities", see Rochberg-Halton [1984a] (above I B 1).

### 3. *Innovations*

A compilation text from Uruk dateable to the 3rd century[37] is TCL 6 (= TU) 11. It contains both omen-related and astronomical material[38]. Only part of it can be understood. First come sections dealing with eclipses and omens derived from them. Then we find rules for predicting whether a month will be full or hollow. The text also contains rules for computing the interval between moonrise and sunrise towards the end of the month (discussed by van der Waerden [1950] pp. 307f.) and for other intervals related to the moon's movement.

The general idea of trying to find periodicities in celestial events was extended even to weather predictions. We have two tablets from Uruk (TCL 6 19 and 20, edited by Hunger [1976b]) which derive the weather from planetary conjunctions and other phenomena; they also seem to connect the return of the same weather phenomena to periods of the planets. Similarly, a tablet written around -320 makes predictions for business from the positions of the planets (Hunger [1976a] No. 93).

A group of texts called "Kalendertexte" (calendar texts) by Weidner ([1967] pp. 41-49; additional texts are listed by Reiner [1995] pp. 114-118) uses enumerations of trees, plants and stones, as well as hemerological notes. They associate each of these with two points in the zodiac which are computed by the following rule: one point runs through the degrees of the zodiac in sequence. Both the sign and the degree within the sign are expressed by numbers, so there are pairs 1 1, 1 2, etc., with a difference of

---

[37] There is no date on the tablet, but its scribe wrote another tablet in year 97 of the Seleucid era.

[38] It was mentioned in Neugebauer [1947] as going back to the 8th century, on the basis of an investigation by Schaumberger, which however never appeared in print. An edition by L. Brack-Bernsen and H. Hunger is in preparation.

1° from line to line. The other point begins at 10 7 and has a difference of 277° from line to line, so that the next line is 7 14 (the first number, representing the sign, is 10 + 9 = 19 = 7 modulo 12), and so on. One could equivalently state that the second point is found by multiplying the distance between the first and the beginning of the zodiac by 277. Since there is no astronomical phenomenon which moves by 277° each day, the whole computation can only have an astrological purpose. Equivalently, the numbers could be understood as pairs of dates in the schematic calendar, but this too does not relate to anything astronomical.

The scheme (as explained by van der Waerden in Weidner [1967] p. 52) is analogous to the Dodekatemoria (where 13 is the distance instead of 277). Just numbers computed according to both schemes are listed side by side on a Neo-Babylonian tablet (Hunger [1996]).

For the "micro-zodiac", see above C 2.

### *4. Genethlialogy*[39]

In general, celestial omens concern the country as a whole or the king as its representative; this is especially true of *Enūma Anu Enlil.* There are, however, a few "personal" celestial omens from the late period, i.e. from about the same time as the horoscopes (see above C 1). A collection of them is found on a tablet from Uruk (TCL 6 14, edited by Sachs [1952a] pp. 65-75). They describe positions and phenomena of one or more of the planets at the time of a child's birth and make predictions about his future life. There is also a section using the appearance of certain fixed stars, the so-called *ziqpu* stars (see below II A 5). Nativity omens occur in TCL 6 13, for which see Rochberg-Halton [1987b].

Such personal omens are not a new invention. They are represented in earlier periods by the physiognomic omens, which refer to the fate of the person on whose body some special traits are observed. Personal omens are also frequent in *Šumma ālu*, the series dealing with ominous happenings in everyday life. Similar omens can be found in the series *Iqqur īpuš*, which includes a section making predictions from the month in which a child is born (Labat [1965] § 64). A close parallel to this section is found already in Hittite sources, i.e. from the second half of the second millennium.[40]

In this literature, we also find the idea of benefic and malefic planets. Jupiter and Venus (and occasionally Mercury) were considered benefic, Mars and Saturn malefic. It is likely that the sequence of the planets in

---

[39] An overview of the roots of this kind of astrology is given in Pingree [1997] pp. 21-29.

[40] KUB 8, 35:1-10, mentioned in K. K. Riemschneider, Babylonische Geburtsomina in hethitischer Übersetzung, Wiesbaden 1970, p. 44 fn. 39.

Hellenistic cuneiform texts is based on this attribute, as was argued by Rochberg-Halton [1988c].

### 5. *Egyptian borrowings*

Omens in Babylonian style found their way into Egypt, in the Achaemenid period or later, as attested by a Demotic papyrus edited by Parker [1959]. A collection of omens in Greek similar to *Enūma Anu Enlil* comes under the authorship of Pseudo-Petosiris, see Pingree [1997] pp. 25-26.

### 6. *Indian borrowings*

Babylonian omens as well as astronomical knowledge were transmitted to India, beginning probably around the middle of the first millennium B.C.; for an overview, see Pingree [1997] pp. 31-33. The transmission of divinatory works included not only celestial omens, but also terrestrial ones of the type found in the cuneiform collection *Šumma ālu* (ominous events in everyday life). Pingree [1987a] gave a detailed comparison of an Indian text on Venus omens from about the 1st century A.D. with possible Babylonian predecessors; these are now included in the edition of Reiner-Pingree [1998] pp. 29-34. For Indian astronomical works influenced by Babylonia, see below II A 3, II A 5, II B 5, II C 2.3, II C 4.2b.

CHAPTER TWO

# ASTRONOMY

## A. The Early Period

### 1. *Astronomy in Enūma Anu Enlil*

#### *1.1. The Venus Tablet of Ammiṣaduqa*

This text constitutes Tablet 63 of *Enūma Anu Enlil*, which is the last Tablet of the section on Venus. Twenty sources were available for the edition by Reiner and Pingree [1975], indicated by the letters from A to V (omitting I and S); but it was suggested that A (K 2321 + 3032) and M (K 3105) may be parts of the same Neo-Babylonian copy; F (BM 37010) and H (BM 36758 + 37496) of another Neo-Babylonian copy; L (K 12344 + 12758), P (K 7072), and Q (Sm 174) of a third; and T (K 5963 + Rm. 134) and U (K 12186) of a fourth. If all these associations were correct, the number of independent sources would be reduced to fifteen. Walker [1984] further added to the combination of L, P, and Q both G (Rm 2, 531) and R (K 7090) to form a single, but incomplete, Late Assyrian tablet; and he added to F and H sources J (BM 36395) and O (BM 37121 + 37432). These associations, assuming that they are correct, reduce the number of known independent sources to eleven. Of these A (+ M) was copied by Nergal-uballiṭ, who identifies it as Tablet 63; and B (W 1924.802), which was found at Kish in 1924, was copied from a tablet written at Babylon by Nergal-ēpuš while Sargon was king of Assyria, between -720 and -704. B is the oldest datable copy of Tablet 63 that we have. According to Walker the latest sources—E (BM 41498), K (BM 34227 + 42033), and N (BM 41688), which are Late-Babylonian (i.e., early sixth century B.C.)—represent a late edition in which additional material was inserted between sections III and IV.

The four sections of the Tablet are constituted as follows:

Section I: omens 1 - 21, where each omen (except omen 10, which is a dated report of a disappearance of Venus in the East) consists of four parts: the date (month and day) of a disappearance of Venus in the East or the West; a statement expressed in (months and) days of its period of invisibility; the date (month and day) of an appearance in the West or the East; and an apodosis.

Section II: omens 22 - 33, which have the structure: the date (month and day) of an appearance of Venus in the East or the West; an apodosis; a statement that it remains present in the East or the West until a given date (month and day, eight months and four days later); the date (month and day, one day later than the preceding date) of its disappearance; the period of its invisibility (always three months at superior conjunction, 7 days at inferior conjunction); the date (month and day) of its appearance in the West or the East; and a second apodosis.

Section III: omens 34 - 37, structured like those in section I.

Section IV: omens 38 - 59, in which omens 1 - 21 and 34 - 37 (with the omission of omens 10, 11, and 34) are rearranged in the order of the dates of the first phenomenon (disappearance) in each.

In source B section IV is omitted, but section III is followed by a subscript, omen 60 (which is a correction of omen 17), and another subscript; the same subscript, omen 60, and second subscript, followed by a colophon, are the sole surviving lines of source R (K 7090). Instead of omen 17 section III contains, in omen 50, a copy of omen 60. The correction of omen 17 may have been made independently by the compiler of section III, or he may have known the source of B and R.

This introduces the problem of the sources of the data in Tablet 63. This was discussed at length in Reiner-Pingree [1975] 21-25. Their conclusions may be summarized as follows:

Sections I and III constitute a single source, designated β, into which section II was inserted sometime before the reign of Sargon II. Section IV had a source, designated γ, that rearranged the same omens with the exception of omen 10 (which is a report), omen 17 (for which γ has the same correction as does omen 60), omen 34 (in which uniquely the first phenomenon occurred in an intercalary month), and omen 11 (which contains an egregious error). It is noted that omen 21, the last omen in section I, is in fact a corrected form of omen 11; and that omen 21 appears in γ as omen 59. These facts lead to the conclusion that there was a source α from which β and γ were independently derived. It may not, therefore, be without significance that B omits γ and that T+U contains only γ.

The common source of β and γ, designated α, had three sources for the omens found in sections I, III, and IV. One contained in almost completely correct form omens 1 to 10 in β, which cover the first eight years of the reign of Ammiṣaduqa and is the main check (not determinant) on the chronology of the First Dynasty of Babylon provided by this text; the one error in β (a month VII for a month IV in omen 4) is not present in γ (omen 43, which had month IV). The second group, omens 11 to 21 in β, represents the second eight-year period in the reign of Ammiṣaduqa, with omen 21 being a correction of omen 11. But the text of β is extremely corrupt (about half of its entries are astronomically impossible if they are regarded as a continuation of omens 1 to 10). Some of the errors in this part were already present in the α text because they are also found in γ. These omens do not preserve trustworthy data, and necessitate the rejection of data that is not compatible with any given chronology. Such rejection of data that contradicts a theory obviously undermines the persuasiveness of the theory, which cannot explain the given data. Even more dubious are the data in omens 34 to 37 of β. Those who wish to see this group as a continuation of omens 1 to 21 of β must assume that the entries for year 18 of Ammiṣaduqa were missing in α and must also deal with the fact that there are astronomical and textual problems with three of the four omens. It is likely that these omens came from some other source than the records of the reign of Ammiṣaduqa.

But α cannot represent the form in which the observations of appearances and disappearances of Venus were originally recorded. This is clear from the fact that the intervals of invisibility are computed; their computation is established by the fact that the months in these intervals always contain 30 days. Thus, in omen 3, VI 23 + 20 days gives VII 13; in omen 6, VIII 28 + 3 days gives IX 1; in omen 8, V 21 + 2 months 11 days gives VIII 2; in omen 9, IV 25 + 7 days gives V 2; in omen 13, VI 26 + 11 days gives $VI_2$ 7; and in omen 18, V 20 + 2 months 15 days gives VIII 5. And, when the dates in an omen are wrong, the interval is computed from the erroneous values. Moreover, it seems likely that section II (omens 22 to 33) was a part of α, and, as we shall see, this section was certainly composed long after the original observations. The form of the reports that are the basis of the omens in α was presumably that of omen 10: In Addaru the 25th day Venus set in the East: Year of the Golden Throne. Fortunately we know that the Year of the Golden Throne was year 8 of the reign of Ammiṣaduqa.

Table of Data in β

| Year | Omen | Last visibility | Interval of invisibility | First visibility |
|---|---|---|---|---|
| [1] | 1 | Ω XI 15 (B) | 3d (AB) | Γ XI 18 (B) |
| | | Ω (A) | | Γ (A) |
| [2] | 2 | Σ VIII 11 (B) | 2m 7d (B) | Ξ X 19 (B) |
| | | Σ (A) | 2m 8d (A) | Ξ (A) |
| [3] | 3 | Ω VI 23 (B) | 20d (AB) | Γ VII 13 (B) |
| | | Ω (A) | | Γ (A) |
| [4] | 4 | Σ VII! 2 (B) | 2m 1d (AB) | Ξ VI 3 (B) |
| | | Σ (A) | | VI 3 (J) |
| | | | | Ξ (A) |
| [5] | 5 | Ω II 2 (B) | 18d (B) | Γ II 18 (B) |
| | | Ω (A) | 15d (A) | II (AJ) |
| | 6 | Σ IX 25 (B) | 2m 4d (AB) | Ξ XI 29 (B) |
| | | Σ IX 12 (A) | | Ξ XI 16 (A) |
| | | | | XI 28 (J) |
| [6] | 7 | Ω VIII 18! (B) | 3d (AB) | Γ IX 1 (B) |
| | | VIII 20+x (A) | | IX 1 (A) |
| | | | | IX (J) |
| [7] | 8 | Σ V 21 (A) | 2m 11d (B) | Ξ VIII 2 (A) |
| | | Σ (BC) | xm x+1d (A) | VIII 2 (C) |
| | | | | VIII (J) |
| [8] | 9 | Ω IV 25 (AC) | 7d (BCD) | Γ V 2 (ACJ) |
| | | | | Γ (D) |
| | 10 | Σ XII 25 (AC) | | |
| [9] | 11 | Ω III 11 (AC) | 9m! 4d (CD) | Γ XII 15 (AC) |
| | | | 9m! xd (A) | Γ (D) |
| | | | xm! 5d (F+J) | XII 16 (F+J) |
| [10] | 12 | Σ VIII 10 (AC) | 2m 6d (C) | Ξ X 16 (ACF+J) |
| | | | xm 6d (D) | |
| | | | 2m 16?d (F+J) | |
| [11] | 13 | Ω VI 26 (C) | 11d (CF+J) | Γ $VI_2$ 7 (CF+J) |
| [12] | 14 | Σ I! 9 (C) | 5!m 16d (CF+J) | Ξ VI 25 (CF+J) |
| [13] | 15 | Ω II! 5 (CG) | 7d (CFG) | Γ (CF+J) |
| | | | | 12 (G) |
| | 16 | Σ X 20 (C) | 15d! (C) | Ξ XI ⌜21⌝ (C) |
| | | XI 21 (G) | | Ξ XI 11 (G) |
| | | x+1 (F+J) | | Ξ (F+J) |
| [14] | 17 | Σ! VII 10 (C) | 1m! 16d (C) | Ξ! VIII 26 (CG) |
| | | VII 10 (G) | | |

| | | | | |
|---|---|---|---|---|
| [15] | 18 | Σ V 20 (C) | 2m 15d (C) | Ξ VIII 5 (C) |
| | | Σ V 21 (G) | | Ξ IX 5 (G) |
| [16] | 19 | Ω V$^!$ 5 (C) | 15d (C) | Γ IV 20 (G) |
| | | Ω VIII$^!$ 5 (G) | | Ξ$^!$ V$^!$ 20 (C) |
| | 20 | Σ XII 15 (CG) | 3m$^!$ 9d (C) | Ξ III 25 (C) |
| | | Σ (H) | 2m 7d (H) | Ξ (G) |
| | 21 | XII 10 (C) | 4d (CH) | Γ XII 14 (C) |
| | (see 11) | | | |
| | 34 | Ω $VI_2$ 1 (C) | 15d (C) | Γ $VI_2$ 17 (C) |
| | | | 16d (M) | $VI_2$ (M) |
| | 35 | Σ III$^!$ 25 (C) | 2m 6d (C) | Ξ VI 24 (C) |
| | | | 2m 16d (M) | Ξ (M) |
| | 36 | Ω I 27 (C) | 7d (C) | II 3 (C) |
| | 37 | Σ (C) | | XII 28 (C) |

Table of Data in γ

| Omen | Last visibility | Interval of invisibility | First visibility |
|---|---|---|---|
| 38 (= 14) | Σ I 8 (K) | 5m$^!$ 18d (K) | VI x (K) |
| | I 8 (T) | 5m$^!$ 17d (T) | VI 25 (T) |
| | I (P) | | VI 24 (P) |
| 39 (= 36) | I 26 (PT) | 6d (T) | Γ II 3 (P) |
| | Σ$^!$ I 27 (K) | | Γ (T) |
| | | | Ξ$^!$ II 3 (K) |
| 40 (= 5) | Ω II 2 (KP) | x d (P) | Γ II 28 or 18 (K) |
| | II 2 (T) | | Γ (T) |
| 41 (= 15) | Ω II 5 (K) | 7d (T) | Γ (KT) |
| | 5 (T) | 6d (V) | III$^!$ (V) |
| 42 (= 35) | III (V) | 1$^?$ m$^!$ 9d (V) | Ξ (V) |
| | | | x+5 (K) |
| 43 (= 4) | [IV] (V) | | |
| 44 (= 9) | IV (V) | | Ξ$^!$ (Q) |
| | | | IV (V) |
| 45 (= 19) | Ω (Q) | | |
| 46 (= 8) | Σ (Q) | | Ξ (Q) |
| 47 (= 18) | Σ (Q) | 1+xd(O) | Ξ (A) |
| 48 (= 3) | Ω VI 23 (A) | 20d (A) | Γ VII 13 (A) |
| | Ω (Q) | | |
| 49 (= 13) | Ω VI 26 (A) | 12d (A) | Γ $VI_2$ 8 (A) |
| 50 (= 60; cf. 17) | Ω VII 11 (A) | 1m 17d (A) | Γ VIII 28 (A) |

| | | | |
|---|---|---|---|
| 51 (= 7) | Ω VIII 28 (A) | 5d (A) | Γ IX (A) |
| 52 (= 2) | Σ (A) | 2m 8d (A) | Ξ X 19 (A) |
| 53 (= 12) | Σ (O) | 2m 8d (A) | Ξ X 16 (A) |
| | | 2m xd (O) | Ξ (O) |
| 54 (= 6) | Σ x+1 (A) | 2m xd (AO) | Ξ (AO) |
| | Σ 12 (O) | | |
| 55 (= 13) | Ω! 24 (O) | 1m xd (O) | Γ! XI 28 (A) |
| | Ω! (A) | xm 4d (A) | Γ! (O) |
| 56 (= 37) | Σ 28 (O) | 2m 0d (A) | Ξ (OU) |
| | Σ (A) | xm 0d (J) | |
| 57 (= 1) | Ω (AU) | 3d (AJ) | Γ XI 18 (J) |
| | | | XI 18 (U) |
| | | | XI 28 (A) |
| 58 (= 20) | Σ (AUV) | 2m 7d (AJ) | Ξ (A) |
| | | | III 4 (J) |
| 59 (= 21; cf. 11) | Ω (A) | 4d (AJ) | Γ XII 14 (J) |
| | Σ! (V) | | Γ (A) |
| 60 (= 17; cf. 50) | Ω 11 (N) | 1m 7d (R) | VIII 28 (R) |
| | Ω 3 (R) | | VIII 27 (B) |
| | VII (B) | | |

Many scholars have been tempted by the possibility of dating this sequence of observations even though they faced the difficulties of translating dates in a Babylonian calendar wherein intercalation occurred in an arbitrary fashion into Gregorian equivalents, of computing phenomena of first visibility and first invisibility when the appropriate *arcus visionis* was not accurately known, and of choosing the best data when the dates of Ξ, Ω, Γ, and Σ of Venus recur on given days of synodic months at intervals of 56 or 64 years since the synodic period of Venus is less than 99 months (eight years) by 4 days, which means that it slips back by 28 days in 56 years, by 32 days in 64. Moreover, any solution using the data in omens 11 to 21 and 34 to 37 must drastically emend the text.

Nevertheless, the search for a solution has been made repeatedly. Two fragmentary tablets, A and C, were studied by Schiaparelli [1906/1907]. He assumed that the observations must date from the seventh, eighth, or ninth century B.C., and so computed that possible dates for the beginning of the series were -656 (or -664), -811, or -867 (or -875). Kugler [1912] pp. 257-306, relying on the identification of the Year of the Golden Throne with the eighth year of Ammiṣaduqa and setting -2060 and -1800 as the limits of the period in which Ammiṣaduqa reigned, computed that Ammiṣaduqa 1 corresponds to -1976/1975. Weidner [1914b] accepted the date -2000 without any astronomical argument, but in Weidner [1917] opted for -1808/1807.

Kugler [1924] pp. 563-571 and 622-627 recomputed the first year of Ammiṣaduqa to be -1800/1799. Fotheringham in Langdon [1923] vol. 2, p. iii, arrived at -1920/1919; this was supported by Schnabel [1925b]. Schoch [1925] placed the first year of Ammiṣaduqa in -1856/1855. Langdon-Fotheringham-Schoch [1928] gave astronomical arguments to reject -1808/1807 and -1800/1799, and non-astronomical (economic) arguments as well as the argument from full and hollow months to favor -1920/1919. Neugebauer [1929], reviewing Langdon-Fotheringham-Schoch, demonstrated the impossibility of using only the Venus Tablet to date the First Dynasty of Babylon.

A decade later, theorizing that the First Dynasty of Babylon must be dated later than had hitherto been accepted, Sewell in Smith [1940] pp. 26-27 and 50-52 claimed that -1645/1644 agreed with the data as well as did -1920/1919; Sidersky [1940] chose -1701/1700; and Ungnad [1940] elected -1659/1658. Neugebauer [1941a], reviewing Smith [1940], again pointed out that astronomy alone cannot determine the date of Ammiṣaduqa. Cornelius [1942] p. 7, fn. 2, claimed to have found that -1581/1580 is a possible date for the first year of Ammiṣaduqa. Using the new tables he had prepared (van der Waerden [1943a]), van der Waerden [1943b] and [1945-1948] supported Cornelius' solution, as he continued to do in van der Waerden [1965] pp. 34-47. Weir [1972] favored Sewell's dating, -1645/1644. Reiner-Pingree [1975], accepting Neugebauer's arguments, noted that omens 1 to 10 could be used as a check on dates for the first eight years of Ammiṣaduqa's reign, which must be shown to be in reasonable agreement with them, but that the data in these omens by themselves are insufficient to establish a date.

Huber [1982] combined the data of the Venus Tablet, which he purified by excluding 18 of the 49 preserved dates (one in omen 5, two in omen 9, two in omen 14, two in omen 15, one in omen 16, one in omen 17, one in omen 20, and all eight dates in omens 34 to 37), with attested 30-day (full) months and attested intercalations; statistical analyses of these combined data seemed best to fit the hypothesis that the first year of Ammiṣaduqa was -1701/1700, the date favored by Sidersky [1940].

But the most interesting part of the Venus Tablet for the historian of mathematical astronomy is section II, designated source δ. It is likely that δ was already included in α, and it is probable that α was compiled within a century or two of -1000; in any case, δ once existed independently since it was excerpted in at least monthly recensions of *Iqqur īpuš* (Labat [1965] pp. 205-239) as the last omens for Nisannu, Simānu, Kislimu, and Ṭebētu. *Iqqur īpuš* also contained a full transcription of δ; see Reiner-Pingree [1975] p. 63 (omens 22 to 27). The original of δ was simply a list of omens in which the phenomena of the protases were the first visibilities of Venus on I 2, II 3, III 4, IV 5, V 6, VI 7, VII 8, VIII 9, IX 10, X 11, XI 12, and XII 13; each protasis was followed by an appropriate apodosis. To these simple omens were later attached supplemental protases and apodoses. The supplemental protases were based on the following mean periods of visibility and invisibility:

| | Visibility | Invisibility |
|---|---|---|
| East | 8 months 5 days | 3 months |
| West | 8 months 5 days | 7 days. |

One synodic period in this scheme consists of 19 months 17 days. If each month contained 30 days the synodic period would be 587 days; this appears a better approximation to reality than the 567½ days that would result from taking a month to be 29½ days long. This, the first attested approximation to a mean synodic period in Babylonian astronomy, must have been devised in the late second millennium B.C., and testifies to the fact that the periodicity of Venus' phenomena was already recognized by about -1000. The Babylonian theory is comparable to that of the Mayas (Thompson [1972] p. 66), for whom:

| | Visibility | Invisibility |
|---|---|---|
| East | 236 days | 90 days |
| West | 250 days | 8 days. |

The Mayan scheme yields a mean synodic period of 584 days.

Table of Data in δ

| Omen | Appearance | Disappearance | Invisibility | Appearance |
|---|---|---|---|---|
| 22 | Γ I 2 | Σ IX 7 | 3 mo. | Ξ XII 8 (read 7) |
| 23 | Ξ II 3 | Ω X 7 (read 8) | 7 days | Γ X 15 |
| 24 | Γ III 4 | Σ XI 8 (read 9) | [3] mo. | Ξ II 9 |
| 25 | Ξ IV 5 | Ω XII 10 | 7 days | Γ XII 17 |
| 26 | Γ V 6 | Σ I 11 | 3 mo. | Ξ IV 11 |
| 27 | Ξ VI 7 | Ω II 12 | 7 days | Γ II 19 |
| 28 | Γ VII 8 | Σ III 13 | 3 mo. | Ξ VI 13 |
| 29 | Ξ VIII 9 | Ω IV 14 | 7 days | Γ IV 21 |
| 30 | Γ IX 10 | Σ V 15 | 3 mo. | Ξ VIII 15 |
| 31 | Ξ X 11 | Ω VI 16 | 7 days | Γ VI 23 |
| 32 | Γ XI 12 | Σ VII 17 | 3 mo. | Ξ X 17 |
| 33 | Ξ XII 13 | Ω VIII 17 (read 18) | 7 days | Γ VIII 25 |

### *1.2. Planetary Theory in Enūma Anu Enlil*

Most of the omens in *Enūma Anu Enlil* that involve planets utilize as protases distortions of the planets' appearances as their reflected light passes through the atmosphere (see Reiner-Pingree [1981] pp. 16-22), or their conjunctions or other relationships to each other or to the constellations, or the points along the eastern horizon above which they rise, or the months and days of the month in which they rise or set. Still, there are scattered among the tablets that have been investigated passages that indicate an awareness of the predictability (or periodicity) of planetary phenomena and, in the latest strata, some attempts to measure intervals between phenomena. Undoubtedly more such passages will be discovered as the remaining fragments of the last section, "Ištar", of *Enūma Anu Enlil* are brought to light.

Two commentaries on tablets related to Tablet 50 (III 19a and 20a and IV 12a in Reiner-Pingree [1981] pp. 42-43 and 46-47) refer to the planets passing by their specified times (UD.SUR or *adannu*) and not rising promptly, or to the planets not completing their days and setting promptly. Both tablets come from Assurbanipal's library; as we shall see in II B 1.1, the diviners of the reigns of Esarhaddon and Assurbanipal were quite familiar with the predictions implied by the terms UD.SUR and *adannu*; such predictions could have been made on the basis of the periods found in MUL.APIN (see II A 3) or, for Venus, of those in section II of Tablet 63.

However, the oldest level of the Venus omens not found on Tablet 63 (Group A in Reiner-Pingree [1998]) preserves somewhat different periods than does the δ text. Two successive omens—105 and 106—of VAT 10218 read (p. 51):

> If Venus stands in the West and sets—on the seventh day [she rises?] in the East.
>
> If Venus for 9 months in the East, 9 in the West changes, variant: turns back, her position (KI.GUB).

The first of these omens refers to the interval of 7 days between Ω and Γ also found in the δ text, while the second replaces the latter's symmetrical periods of visibility in the East and the West, 8 months 5 days, with 9 months. The effect of two 9-month periods of visibility in Venus' synodic period would be to deny sufficient time for the period of invisibility at superior conjunction; the sum of 9 months of visibility between Ξ and Ω, 7 days of invisibility between Ω and Γ, and another 9 months of visibility between Γ and Σ is approximately 538 days; this leaves only about one and a half months for the period of invisibility between Σ and Ξ. We expect something more than two months. The number of months in omen 106 of

VAT 10218, therefore, is not part of a scheme for the mean periods of Venus, but an extreme. A commentary on Tablet 59's omen 3 of month V, K 2907:35, also mentions the period of 9 months (p. 133).

But Tablet 59 itself, in omen 8 of month VI, gives a period of 8 months, from VI to $XII_2$ inclusive (pp. 123 and 129), while omen 10 of the same month gives a period of invisibility from VI to VII (pp. 123 and 129); this could at best represent two months. The omen is repeated with the substitution of Elamite month names in K 229:30 (pp. 172 and 181). The corresponding section of Tablet 60's month IX is lost, but can be restored from K 229:37 which, again using Elamite month names, refers to Venus' setting at the New Moon of month IX and rising in month XII (pp. 173 and 182); unfortunately, this period of four months is too long for the interval of invisibility of Venus at superior conjunction.

Another group of late texts, from Group F, contains omens or comments on omens that imply the use of UŠ and *bēru* in the measurement of arcs along the circle of the horizon; these tablets should not be dated much before -700, if that early. Thus K 3601 r. 8 has (pp. 216 and 223): "ascends by 2/3 *bēru*" (that is, by 20 UŠ); K 3601 r. 22: "every day she goes higher by one UŠ"; and K 3601 r. 24 (pp. 218 and 223): "she completes x *bēru*". These statements all refer to the "motion" of Venus' KI.GUB (the point on the horizon above which it rises) north or south along the eastern horizon. An arc of 20° is the approximate measure of the arc between the East-point and the northern or southern boundary of the Path of Anu, while the 2 *bēru* or 60° mentioned in DT 47:20-26 approximates the arc over which the KI.GUB of the Sun ranges at a terrestrial latitude of 36°; see Pingree [1993] 269-271. The rate of 1 UŠ per day would be close to the maximum "velocity" of Venus' KI.GUB. This evidence, then, suggests that a circle was divided into 12 *bēru* and 360 UŠ in the late Neo-Assyrian period, though the use of UŠ for indicating degrees in the equator (whether or not that was regarded as a circle) is only attested from texts written several centuries later.

The theory presented in Weidner [1913c] that the Babylonians knew the phases of Venus and Mars because their "horns" are mentioned in the omens is without merit.

### *1.3. Lunar and Solar Theory*

Aside from eclipses the omens involving the Moon utilize primarily aspects of its appearance and the appearance of its horns at New Moon, and the stars and planets with which it comes into contact. But Tablets 15 to 22 indicate that the compilers of *Enūma Anu Enlil* carefully observed various aspects of lunar eclipses that later were useful in constructing eclipse theory. A brief survey of this material, derived from Rochberg-Halton [1988a], may be useful here.

The Old Babylonian predecessors of the lunar eclipse section of *Enūma Anu Enlil* (pp. 19-23) are concerned with the month and the day within the month (14, 15, 16, 18, 19, 20, or 21) on which the eclipse occurs; the watch of the night—evening, middle, or morning—in which it happens; the part of the Moon—right side, middle, (or left side)—that is eclipsed; the direction in which the eclipse begins and clears; and whether the Moon rises while already eclipsed or sets while still eclipsed.

Tablet 15 (pp. 67-81) schematically presents the consequences of the lunar eclipse beginning in each of the four cardinal directions and clearing in each of the same four, thus providing 16 variations of which seven never occur (for the eclipse can neither begin in the West nor clear in the East). One must imagine the quadrants as being determined somewhat as is indicated in the diagrams and table in Rochberg-Halton [1988a] pp. 53-55.

The shadow, centered on the ecliptic, is approached by a point on the eastern rim of the Moon, which extends into the northern and southern quadrants, and is left by a point on the western rim. The points of impact and of clearing depend on the latitude of the Moon at the times of those two phases.

A commentary in K 778 (pp. 80-81) interprets the omen "If the Moon is early" to mean: "an eclipse occurs not according to its expected time (*adannu*), on the 12th or 13th day". Clearly the *adannu* of a lunar eclipse began on the 14th day of the lunar month.

Tablet 16 (pp. 82-111) goes through the twelve months and intercalary Addaru indicating the effects of lunar eclipses occurring on certain days of the month (14th and 15th, with the sporadic addition of the 16th, 19th, 20th, 21st, 24th, and 25th) with, as further variables, the color of the eclipsed Moon, the direction of the eclipse's beginning and clearing, the watch in which it occurs (infrequently), and whether it lasts through the watch, goes until daylight, or "faces the Sun".

Tablet 17 (pp. 112-137; months I to VI) and 18 (pp. 138-155; months VII to $XII_2$) include the same phenomena, but always specify the watch in which the eclipse occurs. They are also concerned with eclipses that happen "not at the calculated time" (*ina* NU ŠID.MEŠ, *ina lā minâti*); there is no indication on these two tablets of how the calculation was made.

Tablet 19 (pp. 156-173) is concerned with the watch in which the eclipse begins and that in which it ends; the magnitude of the eclipse (a third of the Moon or totality); and eclipses on the 14th, 15th, 16th, 20th, and 21st of each month that have not cleared when the Moon sets.

In Tablet 20 (pp. 174-229) the main phenomena are eclipses on the 14th of each month beginning in a given direction and clearing in another (often the direction is further qualified by the adjectives "above" or "below"); to this simple structure is sometimes added that Venus enters the Moon's *šurinnu* (this seems to signify its crescent on the next to last day of the month), and the eclipse's occurrence at a not calculated time (*ina lā minâti*). These additional phenomena are connected to the rules for predicting future eclipses normally appended to each month's omens in both recensions of Tablet 20:

Nisannu. Recension A (p. 180): "In Kislīmu (month IX) the 28th (or) 29th day, observe his last visibility ... and Venus entered within him ... The day of last visibility will show you the eclipse."

Recension B (p. 182): "On [the 28th?] of Nisannu observe his *šurinnu* ... and Venus entered within him ... On the 28th of Kislīmu observe his last visibility, and [you will predict] an eclipse (whose effects will last) for 100 days."

Recension B, composite text (p. 186): "[at last visibility the right horn] protrudes into the sky; predict an eclipse for the 14th of Nisannu."

Aiaru (p. 188): "[observe] the last [visibility] on the 28th of Nisannu; on the 14th of Aiaru [predict] an eclipse."

Simānu. Recension A (p. 189) (cf. Recension B (p. 191)): "observe the last visibility on the 28th [of Nisannu] and in the Eagle[41] he becomes dark and low—on the 14th of Simānu predict an eclipse."

Duʾūzu. Recension B (p. 194) (cf. Recension A (p. 193)): "Observe on the 28th of Nisannu the last visibility ... and [predict an eclipse] on the 14th of Duʾūzu."

Abu. Recension B (p. 198) (cf. Recension A (p. 196)): "In Duʾūzu the *šurinnu* ... predict an eclipse on the 14th of Abu."

Ulūlu (pp. 199 and 201): "[Observe] the last visibility on the 28th of Abu; [predict] an eclipse for the 14th of Ulūlu."

Though it is not possible to glean very much from these statements, it is clear that certain characteristics of the last visibility of the Moon, such as Venus' entering into its crescent, were believed to signify that a lunar eclipse would occur in the next month or several months later.

Tablets 21 (pp. 230-250) and 22 (pp. 251-272) add nothing of significance to what we have already found in the preceding Tablets. One other tablet published in Rochberg-Halton [1988a], BM 82-5-22,501 (pp. 283-284), though not part of *Enūma Anu Enlil*, is relevant to the history of Mesopotamian eclipse theory since it seems to indicate that for both lunar and solar eclipses there was expected to be an eclipse six months or some multiple (2 or 3) of six months after its predecessor; it is likely that six months is the period indicated by the word *adannu* in this text.

---

[41] The longitude of the Eagle was about 260° in -700, in opposition to 80° which is about where the Sun would be on the 14th of Simānu.

Eventually, the mathematical astronomy of the Seleucid period allowed Babylonian astronomers to predict the times and magnitudes of lunar eclipses; in other cultures, such as the Indian and its successors, if not in Babylon itself, the directions of impact and clearing were regularly computed and illustrated, and the colors of the eclipses were related to their magnitudes.

Little of solar theory can be derived from *Enūma Anu Enlil*. The principal and astonishing element is the claim in Tablet 23 (24) that certain features of the Sun's appearance on the 1st day of a month allow one to predict a solar eclipse at the end of the month: see van Soldt [1995] pp. 5 (Nisannu), 6 (Simānu), 10 (Tašrītu), 11-12 (Kislīmu), 13 (Ṭebētu), and 14 (Šabāṭu). In some of these cases the same phenomena are used to predict an eclipse in the middle of the month, i.e. a lunar eclipse. Compare also the eclipse predictions in Tablet 25 (26) III 45-48 (pp. 59-60), Tablet 26 (27) I 11 (p. 71), Tablet 27 (28) II 4-7 (p. 87), and Tablet 28 (29) 8 and 12 (p. 94) and 27 (p. 98).

Otherwise, we are informed in Tablet 24 (25) of the Sun, a disk, or two disks rising, at an unexpected time (*ina lā simāne* in II e (p. 22); *ina lā* AN.NI in IV 8 and 8a (p. 39); *ina lā minâti* in III 35 (p. 30), and *ina* NU UD.BA in III 9 (p. 23) and IV 5 and 6 (p. 39)), and the Sun rising late for its expected time (*ana* UD.BA-*šú*). We are, of course, not told how the expected time of sunrise was determined; the texts may simply refer to the fact that clouds or mist made the Sun invisible at dawn, but they could also mean that the Sun did not appear precisely at the time indicated by the water-clock.

### *1.4. Tablet 14 and related texts*

The earliest tablet of this complex to be investigated was K 90, which was found in Nineveh but is a copy of a tablet from Babylon (Sayce-Bosanquet [1879/80]). This tablet, together with 80-7-19,273 and K 6427 (both belonging to Tablet 14 of *Enūma Anu Enlil*), were next discussed by Kugler [1909/1910] pp. 45-53. The related material in K 2164+2195+3510 (I.NAM.GIŠ.ḪUR.AN.KI.A; see below II A 4) was published in Weidner [1912c]. Kugler [1913] pp. 88-106 criticizes Weidner's interpretation and adds to the discussion passages from MUL.APIN (see II A 3) and BE 13918, as well as K 6427 and 80-7-19,273, which are fragments of Tablet 14. Weidner [1914] pp. 82-91 responds to Kugler, adducing BM 45821, which is a commentary to Tablet 14. Many of these same texts, and others were again discussed in Neugebauer [1947b], van der Waerden [1950] pp. 299-301 and 306-307, and [1954] pp. 22-24. A recent presentation is in Al-Rawi-George [1991/1992], who edit the four tables of Tablet 14 from five sources, the commentary from BM 45821, and, in addition, K 90 and BM

37127. A tablet from the Nabû temple at Kalḫu which employs for some numbers Sumerian words is edited by Hunger [1998].

Table A records the duration, expressed in UŠ, of the time between sunset and moonset on the first 15 days of an ideal equinoctial month, and the duration of the time between moonrise and sunrise and between sunset and moonrise on days 16 to 30 of such a month. The equinox, indicated by a night-time of 3,0 UŠ = 6 *bēru*, on day 15, is assumed to fall on the fifteenth of the equinoctial month as it does, e.g., in the "Astrolabes" and in MUL.APIN. The colophon states that this table was copied from an exemplar from Nippur. We "translate" the table in order to facilitate its understanding.

| Day of month | Sunset to moonset | Sunset to moonrise |
|---|---|---|
| 1 | 3;45 | |
| 2 | 7;30 | |
| 3 | 15 | |
| 4 | 30 | |
| 5 | 1,0 | |
| 6 | 1,12 | |
| 7 | 1,24 | |
| 8 | 1,36 | |
| 9 | 1,48 | |
| 10 | 2,0 | |
| 11 | 2,12 | |
| 12 | 2,24 | |
| 13 | 2,36 | |
| 14 | 2,48 | |
| 15 | 3,0 | |
| | Moonrise to sunrise | |
| 16 | 2,48 | 12 |
| 17 | 2,36 | 24 |
| 18 | 2,24 | 3[6] |
| 19 | 2,12 | 4[8] |
| 20 | 2,0 | 1,0 |
| [21] | 1,48 | 1,[12] |
| [22 | 1],36 | 1,[24] |
| [23 | 1,24 | 1,36] |
| [24 | 1,12 | 1,48] |
| [25 | 1,0 | 2,0] |
| [26 | 30 | 2,30] |
| [27 | 15 | 2,4]5 |
| [28 | 7;30] | 2,52;30 |
| [29 | 3;]45 | 2,56;15 |
| [30] | the god during the day stands | |

The entries for column 2, nights 5 to 25, form a linear zigzag function in which:

m = 1,0 (2 *bēru*)
M = 3,0 (6 *bēru*)
d = 12 (0;24 *bēru*)

$$\mu = \frac{M + m}{2} = 2,0 \qquad (4\ b\bar{e}ru)$$

$$P = \frac{2(M - m)}{d} = 20 \text{ nights}.$$

Note that d, 12 UŠ, is precisely a thirtieth of a day, 48 minutes of our time. This measure of time occurs again in the *ziqpu*-star texts (see II A 5), and is the equivalent and probably the source of the Indian division of the nychthemeron into 30 muhūrtas, and then into 60 ghaṭikās or nāḍikās, a unit of time measurement that first appears, with many other Mesopotamian features, in the *Jyotiṣavedāṅga* composed by Lagadha in ca. -400.

The entries for column 2, days 1 to 5 and 25 to 29, form geometrical progressions where each entry from 1 to 5 is double its predecessor, each from 26 to 29 half of its predecessor. Since the sum of the time from sunset to moonrise and from moonrise to sunrise on days 16 to 29 must be approximately one equinoctial night, the table correctly has the entries in column 3 equal 3,0 UŠ minus the entries in column 2.

Table B gives the duration of the time from sunset to moonset on days 1 to 15 of an ideal equinoctial month and that of the time from sunset to moonrise on days 16 to 30 of the same month. The durations are measured by the minas and šiqlus of water that will flow out of a water-clock in those times. Again, the equinox presumably occurs on day 15 when the amount of water is 3 minas. The table is said in the colophon to have been copied from an exemplar from Babylon.

The two halves of the column of water-weights are identical as the second, for days 16 to 29, represents the inverse of the descending slope of a linear zigzag function of which the first half, for days 1 to 15, is the ascending slope. The parameters are:

m = 12 šiqlu (0;12 mina)
M = 3 mina
d = 12 šiqlu
$\mu$ = 1 mina 36 šiqlu
P = 28 days.

This text proves that 1 mina of water in the water-clock measured a third of an equinoctial night, that is, 2 *bēru* or 1,0 UŠ.

Table C gives the length of daylight and of nighttime, measured in mina and šiqlu, for the 15th and 30th day of each month in the ideal calendar beginning with Nisannu. In the ideal calendar, a solar year contains 12 months and each month contains 30 days. The table begins with Nisannu 15, but the equinoxes, when the length of both daytime and nighttime is measured by 3 minas, occur on Addaru 15 and Ulūlu 15, the summer solstice when the daytime is measured by 4 minas on Simānu 15, and the winter solstice when the daytime is measured by 2 minas on Kislīmu 15. A similar table with the vernal equinox at Nisannu 15 is found in MUL.APIN II ii 43 - iii 15. The lengths of daylight in table C form a linear zigzag function:

m = 2 minas (= 4 *bēru*)
M = 4 minas (= 8 *bēru*)
d = 10 šiqlu (= 0;10 minas = 0;20 *bēru*)
$\mu$ = 3 minas (= 6 *bēru*)
P = 24 half-months.

The sum of the entries for daylight and nighttime for each date must be 6 minas (= 12 *bēru*). The ratio 2 : 1 of the longest to the shortest daylight is incorrect for Mesopotamia; it implies a terrestrial latitude of about 48° if daylight meant the time from sunrise to sunset. Since every daylight is longer than such a period because of morning and evening twilight the consequences of the ratio should probably be modified somewhat for these early texts, but the error will not be greatly diminished.

Table D, finally, records in UŠ and NINDA the duration of the time between sunset and moonset on the first day of every month beginning with Nisannu and that of the time between sunset and moonrise on the fifteenth day of every month. As they are put together in chronological order (i.e., Nisannu 15 immediately follows Nisannu 1, etc.), the two functions form a single linear zigzag function with minimum at Simānu 15 and maximum at Kislīmu 15; the mean value occurs at Addaru 15 and Ulūlu 15, the equinoxes (note that the manuscripts' numbers are all correct except that ⸢12⸣ at Šabāṭu 15 should be 13;20).

m = 8 UŠ
M = 16 UŠ
d = 40 NINDA (= 0;40 UŠ)
$\mu$ = 12 UŠ
P = 24 half-months.

Since 1 *bēru* = 30 UŠ, m = 8 UŠ = $\frac{4}{15}$ *bēru* and M = 16 UŠ = $\frac{8}{15}$ *bēru*. This indicates that each entry in the table is $\frac{1}{15}$ of the nighttime. A similar scheme is found in MUL.APIN II ii 43 - iii 15, where it is further noted that, since d = 40 NINDA over 15 days, the difference for one day will be $\frac{40 \text{ NINDA}}{15} = \frac{40 \text{ NINDA}}{60} \times 4 = 2;40$ NINDA. Again, the MUL.APIN scheme shifts the vernal equinox to Nisannu 15. Since in the MUL.APIN passage the ratio of 2 : 1 for the longest nighttime to the shortest is expressed both in minas of water and in UŠ of time, it is clear that the suggestion in Neugebauer [1947b] must be rejected, that 4 minas to 2 minas will approximately give a ratio of longest to shortest nighttime of 3 : 2 ($\sqrt{9} : \sqrt{4}$), since the rate of outflow of water from a cylindrical water-clock is not proportional to the weight of the water but to the square-root of the weight. Time was measured in Mesopotamia on the assumption that the measurement depended directly on the weight of water in the clepsydra.

An unfortunate consequence of Table D of Tablet 14 is that the night of Full Moon is shifted from day 15, where it had been in Tables A and B, to day 14; for on day 15 the period from sunset to moonrise is already $\frac{1}{15}$ of the nighttime. This means that a linear zigzag function could not be constructed to fit the whole month of 30 days because the two halves of the month have become unequal. Moreover, in making the interval between sunset and moonset on the first night of the month $\frac{1}{15}$ of the length of that nighttime, Table D has ignored the geometrical progression for days 1 to 5 found in Tables A and B. However, we know that linear zigzag functions based on a daily difference of $\frac{1}{15}$ of the nighttime were used to approximate the periods from sunset to moonset because they are extant in Greek and Latin sources. Vettius Valens in *Anthologies* I 12, written in the middle of the second century A.D., presents a linear progression in the periods of visibility of the Moon for days 1 to 15 of a month. He uses the following parameters:

m = 0 hours (night 30)
M = 12 hours
d = ½ ¼ $\frac{1}{20}$ hour (= 0;48 = $\frac{12}{15}$ hours)

If the hours are seasonal, the scheme will work for every synodic month; if they are equinoctial, only for the equinoctial months. Pliny in *Natural History* II 14 (58) gives as the value of d a dodrans semuncia of an hour, that is, $\frac{3}{4}$ $\frac{1}{24}$ hours = 0;47,30 hours, a close approximation to 0;48 necessitated by Pliny's choice to express the fractions in words rather than in numbers. The unintended implication is that the Moon would be visible on night 15 for only 11;52,30 hours rather than 12. In *Natural History* XVIII 32 (324-325) Pliny gives the value of d as a dextans siculus, that is, $\frac{5}{6}$ $\frac{1}{48}$ = 0;51,15 hours. This would give a visibility of the moon lasting 11;57,30 hours on night 15 if the moon were visible a negligible time on night 1. This second scheme of Pliny is clearly a Roman invention based on the previous crude adaptation of the Babylonian linear zigzag function.

A further degradation of the Roman idea in Pliny II 58 is found in Cassianus Bassus' *Geoponica* I 7, where it is attributed to Zoroaster. This gives the hour of moonset and that of moonrise on every nychthemeron of a synodic month; since the number of the hour for moonset is always identical with that for moonrise, but in the other half of the nychthemeron, we can compare the hours of moonset at night (on nights 1 to 15) and the hours of moonrise at night (on nights 16 to 30) to the intervals from sunset to moonset on nights 1 to 14 and from sunset to moonrise on days 17 to 30 in a hypothetical Babylonian table, with a constant difference of 0;24 *bēru* = 0;48 hours; cf. Table B, where the measurements are in minas. Following Bilfinger [1884] and Bidez-Cumont [1938] vol. 2, pp. 174-178, we add the reconstructed table of Pliny II 58, from which the "Zoroaster" text clearly has been derived, and, for completeness' sake, the reconstructed table of Pliny XVIII 324-325.

| | Babylonian (Valens) | Pliny II 58 | *Geoponica* nights 1 to 15 | *Geoponica* nights 30 to 16 | Pliny XVIII 324-325 |
|---|---|---|---|---|---|
| 1 | 0;48 | 0;47,30 | 0;30 | 0;30 | 0 |
| 2 | 1;36 | 1;35 | 1;30 | 1;35(?) | 0;51,15 |
| 3 | 2;24 | 2;22,30 | 2;15 | 2;22,30 | 1;42,30 |
| 4 | 3;12 | 3;10 | 3;20 | 3;15 | 2;33,45 |
| 5 | 4;0 | 3;57,30 | 3;30 | 3;30(?) | 3;25 |
| 6 | 4;48 | 4;45 | 4;45 | 4;45 | 4;16,15 |
| 7 | 5;36 | 5;32,30 | 5;32,30 | 5;32,30 | 5;7,30 |
| 8 | 6;24 | 6;20 | 6;20 | 6;20 | 5;58,45 |
| 9 | 7;12 | 7;7,30 | 7;57,30 | 7;57,30 | 6;50 |
| 10 | 8;0 | 7;55 | 7;55 | 7;55 | 7;41,15 |
| 11 | 8;48 | 8;42,30 | 8;42,30 | 8;42,30 | 8;32,30 |
| 12 | 9;36 | 9;30 | 9;30 | 9;47,50 | 9;23,45 |
| 13 | 10;24 | 10;17,30 | 10;30(?) | 10;7,30 | 10;15 |
| 14 | 11;12 | 11;5 | 11;15 | 11;15 | 11;6,15 |
| 15 | 12;0 | 11;52,30 | 0 | 0 | 11;57,30 |

The tablet K 90 is given in a new edition in Al-Rawi-George [1991/1992] pp. 66-68. It is constructed as is Table A, with a geometrical progression from nights 1 to 5 and 25 to 29 and a linear zigzag function from nights 5 to 25. The latter is based on the following parameters:

m = 1,20 UŠ (nights 5 and 25)
M = 4,0 UŠ (night 15)
d = 16 UŠ
$\mu$ = 2,45 UŠ
P = 20 nights

The difference of 16 and the length of nighttime, 4,0 UŠ (= 8 *bēru*), indicate that the table is meant for the ideal month in whose middle occurs the winter solstice. There must have been a similar table for the month in whose middle occurs the summer solstice, in which M = 2,0 UŠ (= 4 *bēru*) and d = 8 UŠ.

The geometrical progression from night 5 to night 1 and from night 25 to night 29 is: 1,20 - 40 - 20 - 10 - 5. But the column of durations of time from sunset to moonrise on nights 16 to 29 is a straight arithmetical progression in which d = 16 and the entry for night 16 is 16 UŠ; therefore, the sums of the entries in columns 2 and 3 for nights 26 to 29 are not 4,0 UŠ as they should be, but 3,36 - 3,32 - 3,38 - 3,49.

It is clear from this review of elements of astronomy found in *Enūma Anu Enlil* that by the end of the second millennium and beginning of the first B.C. the inhabitants of Mesopotamia recognized the periodicity of many celestial phenomena and had devised methods to predict them. Some of these, as in the case of eclipses, were not mathematical, but others were. In particular, some periodic deviations from mean time-intervals were being described by linear zigzag functions, some of which (e.g., Table A of Tablet 14 and K 90) were modified by the intrusion of non-linear elements.

But Tablet 14 does not represent the earliest known application of a linear zigzag function to the problem of a periodic variation in time. The Old Babylonian tablet BM 17175+17284, published in Hunger-Pingree [1989] pp. 163-164, gives a scheme for the lengths of daylight and nighttime measured in minas of water on the equinoxes and solstices. The data in the tablet may be represented as follows:

| Colure | Date | Daylight | Nighttime |
|---|---|---|---|
| Vernal equinox | Addaru 15 | 3 mina | 3 mina |
| Summer solstice | Simānu 15 | 4 mina | 2 mina |
| Fall equinox | Ulūlu 15 | 3 mina | 3 mina |
| Winter solstice | Kislīmu 15 | 2 mina | 4 mina |

This is precisely the scheme found in Table C of Tablet 14.

### 2. *"Astrolabes" or "Three Stars Each" Texts*

Pinches [1900] 572-576 combines information from four tablets—Sm 162 (copy in CT 33, 11), K 14943 + 81-7-27,94, and 83-1-18,608 (copy in CT 33, 12), and two additional star-lists—to form a composite list known as "Pinches' Astrolabe". We will use this name to designate all tablets that conform to its type. For each of the twelve months of an unintercalated (i.e., "ideal") year it lists three stars—allegedly one from each of the three Paths

—and, normally, the weight of water to be poured into a water-clock to measure the lengths of daylight, half of daylight, and a quarter of daylight in each ideal month. At least some copies of "Pinches' Astrolabe" are circular. The following table lists the stars named in "Pinches' Astrolabe".

| Month | "Path of Ea" | "Path of Anu" | "Path of Enlil" |
|---|---|---|---|
| I | Field | Venus | Plow |
| II | Stars | Old Man | Anunītu |
| III | True Shepherd of Anu | Lion | Crab |
| IV | Arrow | Twins | Jupiter |
| V | Bow | Great Twins | Wagon |
| VI | Kidney | Raven | ŠU.PA |
| VII | Ninmaḫ | Scales | Ḫabaṣirānu |
| VIII | Mad Dog | Scorpion | King |
| IX | Mars | Panther | She-goat |
| X | GU.LA | Crab | Eagle |
| XI | Numušda | Swallow | Pig |
| XII | Fish | Fox | Marduk |

Clearly, this list does *not* identify stars that rise in each of the months; four of them are planets, and two (Wagon and Fox) never rise since they are circumpolar. Nor are the stars in each Path properly positioned; from MUL.APIN we know that Field, Stars, True Shepherd of Anu, Arrow, and Bow are Anu rather than Ea stars; that Old Man, Twins, Great Twins, Panther, and Crab are Enlil rather than Anu stars; that Anunītu and Eagle are Anu rather than Enlil stars; and that Ḫabaṣirānu is an Ea rather than an Enlil star. There were presumably mythological as well as astronomical reasons for the choice of the three stars of each month.

"Astrolabe B", on the other hand, contains three sections: the first (transliteration in Reiner-Pingree [1981] pp. 81-82) is a bilingual listing of, for each month, a constellation, a god, and "mythological" notes; the second (transliteration in Weidner [1915] pp. 76-79 and 145; summarized in Reiner-Pingree [1981] p. 5) names twelve stars in each of the three Paths, often with remarks on their locations relative to each other; and the third (transliteration in Weidner [1915] 66-67; summarized in Reiner-Pingree [1981] p. 4) in lines 1-12 names one constellation from each Path for each of the twelve months (these are generally identical with the constellations named on the "Pinches' Astrolabe"), and in lines 13-36 states that the three constellations of each month rise in that month and that the constellations in the seventh month from it set in that same month. This is not correct astronomically.

The tablet containing "Astrolabe B", VAT 9416 (KAV 218), was copied probably in the reign of Ninurta-apil-Ekur (-1190 to -1178). Section B of this text, as indicated above, lists twelve constellations in each of the three Paths, or 36 constellations in all. The assumed Tablet 51 of *Enūma Anu Enlil* closely follows this list; see Reiner-Pingree [1981] pp. 52-53.

| Path of Ea | Path of Anu | Path of Enlil |
|---|---|---|
| Field | Venus | Plow |
| Stars | Scorpion | Anunītu |
| Jaw of the Bull | Scales | Snake |
| True Shepherd of Anu | Panther | Wagon |
| Arrow | Old Man | [ŠU.PA?] |
| Bow | Swallow | She-goat |
| NUN.KI | [Lion] | Wolf |
| Ninmaḫ | [Twins] | Eagle |
| Mad Dog | Great Twins | Pig |
| Mars | Crab | Jupiter |
| Ḫabaṣirānu | Raven | Fox |
| Fish | Nēberu | Southern Yoke |

With phraseology copied from both Section B of "Astrolabe B" and the assumed Tablets 50 and 51 of *Enūma Anu Enlil* are two tablets, HS 1897, presumably from Nippur, and BM 55502, from Babylon, published by Oelsner-Horowitz [1998]. Neither of these tablets is complete, nor is BM 55502 a direct descendant of the much earlier HS 1897, though the overlap is considerable. HS 1897 lists the first 10 constellations of Ea (with Bull of Heaven in place of Jaw of the Bull and Kidney in place of NUN.KI), the first 10 constellations of Anu (with a planet, <Saturn>, in place of Scales, and with Crab and Raven omitted), and the first constellation, Plow, of Enlil, before breaking off. Of BM 55502's copy of this catalogue all that is left are the 10 constellations of Ea and the 10 constellations of Anu. If differs from HS 1897 by retaining Jaw of the Bull in the Path of Ea, by stating that Scales is Saturn, and by collapsing Twins and Great Twins into one entry: [Twins ... Lugalg]irra and Mesl[amtaea].

Kugler [1907] pp. 228-258 studied six monthly star-lists, one of which is that from "Pinches' Astrolabe". Here and in Kugler [1914] pp. 201-206 there is no discussion of the weights of water. Weidner [1915] pp. 64-66 presents a transliteration of the "Pinches' Astrolabe," and in the following pages a quite unconvincing discussion of the date of its original. Schaumberger [1935] pp. 323-330 again tried to use the "Astrolabe" texts to identify the constellations. Van der Waerden [1949] wished to connect the "three stars each" with the Egyptian decans, a connection that is not at all convincing; he discusses the numbers in "Pinches' Astrolabe" on pp. 17-19, and again in van der Waerden [1954] pp. 21-22. Sachs [1955] suggested that

BM 34713 (LBAT 1499; copy on p. 233) was a major source for Pinches in constructing his "Astrolabe". Weidner [1967] p. 19 fn. 60, on the basis of correspondence between Pinches and F. Hommel, asserts that Pinches used the tablet 85-4-30,15 and an unnumbered text containing explanations (see presently), but *not* LBAT 1499, which evokes "kaum lösbare Fragen". He then mentions duplicates to the omens contained in LBAT 1499. Unfortunately, 85-4-30,15 is not a duplicate of the "Astrolabe" texts, so that Weidner must have been mistaken here. LBAT 1499 is a Late Babylonian copy beginning with a list which suffers from a major mistake: the second and third columns of numbers and constellation-names are shifted by one line so that they do not fit the entries in the first column. The reverse, lines 10-30, gives the rising-times of the dodecatemoria of Aries and part of Taurus; see II C 1.1d. Walker-Hunger [1977] edit a tablet, BM 82923, probably copied in the second half of the first millennium B.C., that Pinches seems to have used for the "explanations" quoted by him. In Reiner-Pingree [1981] p. 3 it was pointed out that the associations of constellations and planets (for they are included in the "Astrolabe" lists) with the ideal months of the year and with the Paths of Enlil, Anu, and Ea are influenced by mythological as much as by astronomical considerations; attempts to identify these constellation and planet names with stars that have their heliacal rising in the ideal months are, therefore, devoid of persuasive force.

Thus, the "Pinches' Astrolabe" tablets seem to follow "Astrolabe B" which dates from ca. -1100 if not before; and the "Pinches' Astrolabe" is older than -700 since Sm 162 was copied by Nabû-zuqup-kēnu, a scribe who was active during the reigns of Sargon II (-720 to -704) and Sennacherib (-703 to -680). For the constellation names see the list in the Appendix. The weights of water form three linear zigzag functions; the values in the middle ring of the circular form (Path of Anu) are half of those in the outer ring (Path of Enlil), those in the inner ring (Path of Ea) are half of those in the middle ring. The parameters of the outer ring are:

$m$ = 2 mina (Kislīmu)
$M$ = 4 mina (Simānu)
$d$ = 0;20 mina
$\mu$ = 3 mina (Addaru and Ulūlu)
$P$ = 12 months.

This scheme is identical with that in the Old Babylonian text mentioned at the end of II A 1.4. That it continued to be copied, even mistakenly, until the Late Babylonian period attests to the extreme conservatism of some Mesopotamian scribes.

The reverse of Sm 162 seems to be related to the often discussed "Hilprecht Text" (formerly HS 229, now HS 245); see Horowitz [1993]. The

"Hilprecht Text" is a Middle Babylonian tablet from Nippur, which may be copied from an Old Babylonian original. It gives a list of numbers representing the distances between the Moon and seven stars:

| | |
|---|---|
| 19 | Moon to Stars |
| 17 | Stars to True Shepherd of Anu |
| 14 | True Shepherd of Anu to Arrow |
| 11 | Arrow to Yoke |
| 9 | Yoke to ŠU.PA |
| 7 | ŠU.PA to Scorpion |
| 4 | Scorpion to AN.TA.GUB |

The sum, as the scribe remarks, is 1,21, which corresponds to 2,0 *bēru* (= 1,0,0,0 NINDA). The problem (for this is a mathematical problem text, not an astronomical text) is to find the number of NINDA in each of the seven given intervals. However, for many years after it was first partially published in 1908, it evoked elaborate speculations on the Babylonian measurements of the universe; see Kugler [1909/10] pp. 93-94; Kugler [1912] pp. 312-320; Kugler [1913] pp. 73-87; Weidner [1914b] pp. 1-28; Weidner [1915] pp. 128-131; and Thureau-Dangin [1931]; the last recognized that the text is speculative. The remainder of the text was published in Neugebauer [1936], since which time its purely mathematical character has been recognized. See further van der Waerden [1949] pp. 6-7 and Rochberg-Halton [1983], who records the suggestion by Pingree that the Moon and the seven stars will rise in the order given in the text during the night of the autumnal equinox on the 15th of the ideal month Tešrītu, when the moon is in opposition to the Sun. The intervals between their risings, of course, are not proportional to those indicated on the "Hilprecht Text".

Koch in Donbaz-Koch [1995] p. 71 n. 25, points out that, if one identifies Scorpion with $\beta$ or $\delta$ Scorpii and AN.TA.GUB with $\lambda$ or $\varkappa$ Scorpii, these two stars rise after sunrise at Nippur ($\varphi$ = 32;6°). The situation would be different further north—say, where $\varphi$ = 36°—but the text is an exercise in arithmetic, not a record of an observation. The interpretation of the text depends on two lines from MUL.APIN: I iii 13-14 ("The Stars rise and the Scorpion sets. The Scorpion rises and the Stars set"). Clearly, while the Stars are of relatively small longitudinal extension, the Scorpion is a very large constellation. Since we do not know its boundary, we cannot exclude the possibility of its including more northerly stars or stars included in our (and their?) Libra. Nor is there any certainty concerning the identification of AN.TA.GUB. But, however Scorpion and AN.TA.GUB are interpreted, the intention of the author of the text was to name the Moon and the seven stars in the order in which they rose on the night near the vernal equinox when the Moon was opposite the Sun (see MUL.APIN II i 14-15: "On the 15th of

Tešritu the Sun rises in the Scales in the East, and the Moon stands in front of the Stars behind the Hired Man").

The reverse of Sm 162, which is broken, names the Arrow, states "14 from Arrow to Yoke" (the "Hilprecht Text" has 11), gives the sum, 1,21, and then lists the Moon and Stars, Bull(!), Arrow, Yoke, Scorpion, and Ḫabaṣirānu (which does not fit into the sequence since it rises before Scorpion does). Clearly this is a different problem from that of the "Hilprecht Text", though closely related to it.

What Koch calls a "third generation" of Astrolabes is represented by a text found in Nineveh, Nv. 10, published by Donbaz-Koch [1995]. On what remains of the obverse are sections for months VIII to XII; in each month are listed three constellations, one from each path, and, in the bottom line, two numbers followed by MUL AN.TA.GUB.BA ("the star standing above") height of Swallow (there is not a MUL preceding "Swallow").

| Month | Star | Path |
|---|---|---|
| VIII | Lion | Enlil |
| | Mars | Anu |
| | Ninmaḫ | Ea |
| IX | Eru | Enlil |
| | Furrow | Anu |
| | ŠU.PA | Ea |
| X | DI | Enlil |
| | Furrow (and) Circle | Anu |
| | Šarur (and) Šargaz | Ea |
| | 4 4 | |
| XI | Dog | Enlil |
| | Sitting Gods | Anu |
| | Pabilsag | Ea |
| | 4 6 | |
| XII | Lammu | Enlil |
| | Eagle | Anu |
| | Panther | Ea |
| | 4 4 | |

The assignation of constellations to Paths is far from perfect (e.g., ŠU.PA, Sitting Gods, and Panther belong to Enlil), and the inclusion of Mars is difficult to explain.

The reverse contains entries for months I to VI, breaking off at the beginning of this last month. In the left column for each month are listed four constellations, and in the right column for each a constellation called "the star standing above" and another called MUL AN.TA.ŠÚ.UR.RA, which Koch translates as "the star flashing above".

| Month | Stars | AN.TA.GUB.BA | AN.TA.ŠÚ.UR.RA |
|---|---|---|---|
| I | (broken) | (broken) | (broken) |
| II | (broken) | Field | Raven |
| III | ŠU.PA<br>Swallow<br>She-goat<br><LI$_{9}$.>SI$_{4}^{?}$ (and) King | Anunītu | Furrow |
| IV | Old Man<br>Fish<br>ŠU.PA<br>Šarur (and) Šargaz | Stars | Scales |
| V | Crook<br>AN.Ú.GI.E<br>Dog<br>Pabilsag | True Shepherd of Anu | Sitting Gods |

The choices of four constellations for each month makes no sense to us. The times of their heliacal risings, for instance, seem never to be in the months they are associated with, and the members of each tetrad have, to our knowledge, no connections with each other. Koch (pp. 74-75) considers them to be also AN.TA.GUB.BA and AN.TA.ŠÚ.UR.RA stars, though the text does not imply this.

Koch (p. 72-75) cleverly hypothesizes that the AN.TA.ŠÚ.UR.RA stars are those above the western horizon just after the Sun sets, and that the AN.TA.GUB.BA stars are those above the eastern horizon just before the Sun rises. This is certainly possible for the constellations named in the right-hand column. See the end of II A 6.

On pp. 77-84 Koch suggests that the obverse represents an intercalation rule. The "height of the Swallow" he interprets as the altitude of ε Pegasi on the meridian, and the numbers "4 4", "4 6", and "4 4" as the altitudes of the AN.TAB.GUB.BA stars measured in cubits and fingers. For month X he chooses DI (which he identifies as $\alpha^2$ Librae) as the AN.TA.GUB.BA star, for month XI Sitting Gods (which he identifies with υ Virginis), and for month XII Eagle (α Aquilae). He then claims that when the true altitude of the AN.TA.GUB.BA stars is below the altitude of ε Pegasi on the meridian, the year in which this occurs is an intercalated year. The scanty evidence appears to us incapable of bearing such an interpretation.

## *3. MUL.APIN*

The first part of MUL.APIN to be published was an almost complete copy of Tablet I, BM 86378, by L. W. King in CT 33, plates 1-8 (London 1912). This tablet was probably copied in about -500. King [1913] draws attention to the importance of this text. Immediately a host of articles and books appeared attempting to utilize the information of this text, often in combination with material from other sources, to identify the Babylonian constellations; we mention here Kugler [1913] pp. 1-72 and 88-106; Weidner [1913a]; Bezold-Kopff-Boll [1913]; Weidner [1914b] pp. 42-56; Kugler [1914] pp. 141-181 and 207-224; Weidner [1915] pp. 35-51; Schaumberger [1935] pp. 330-350; van der Waerden [1949] 13-21; Papke [1978]; Reiner-Pingree [1981] pp. 6-9; Hunger-Pingree [1989] pp. 137-146; Koch [1989]; Neumann [1991/1992a]; and Koch [1991/1992]. For our views concerning the identifications of the Babylonian constellations see the Appendix.

Parts of Tablet II (II i 1-8, on the path of the Sun and the planets; II i 9-13, on the Sun's risings on the eastern horizon on the days of the solstices and equinoxes; II ii 21-40, the shadow table; and II iii 22-39, on astral omens) were published from VAT 9412 and AO 7540 and discussed by Weidner [1924]. There were further studies of these published parts of MUL.APIN by Weidner [1931/1932]; Neugebauer-Weidner [1931/1932]; Schaumberger [1935] pp. 321-322 and 340-344; Neugebauer [1947b] 38 and 40-41; van der Waerden [1949], [1950], and [1952/1953]; and Neugebauer [1975] vol. 1, pp. 544-545. The complete text was edited from forty tablets, with a translation and commentary, in Hunger-Pingree [1989].

One tablet, VAT 9412+11279, is dated in its colophon -686, and a number of others come from Assurbanipal's library. Moreover, it is pointed out in Hunger [1982] that two copies, VAT 9412 (dated -686) and VAT 9527, were made in Assur. These tablets guarantee that the text is at least as old as the early first millennium B.C. MUL.APIN was still being copied in the late Babylonian period; see, e.g., Horowitz [1989/1990]. Its final composition may have occurred as late as -700, but its component parts date sometimes centuries earlier.

A fragment of MUL.APIN, Rm 4, 337, was discussed by Kugler [1907] p. 230; it contains the beginnings of I ii 42 - iii 4 on the obverse, and the middles of iii 31-39 on the reverse.

The contents of MUL.APIN should be discussed under four headings: stars, planets, intercalation schemes, and time-keeping.

### *3.1. Stars*

There are six lists of stars in Tablet I: a general list of 60 constellations in the Paths of Enlil, Anu, and Ea plus 6 circumpolar constellations (inserted into the Path of Enlil) and the 5 star-planets (Jupiter at the end of the Path of Enlil, Venus, Mars, Saturn, and Mercury at the end of the Path of Anu; cf. sections 1 and 2 of "Astrolabe B", above p. 51); a sequential list of dates in the ideal calendar on which certain constellations rise heliacally; a list of simultaneously rising and setting constellations (cf. lines 13-36 of section 3 of "Astrolabe B"); a list of the time-intervals in the ideal calendar between the dates of the heliacal risings of pairs of constellations, based on the second list; a list of *ziqpu* constellations and a list of *ziqpu* stars with the calendar dates on which they culminate simultaneously with the risings of certain constellations; and a list of 17 constellations in the path of the Moon. We believe that the first, second, third, and fifth lists were composed in Assyria (latitude ca. 36°) in about -1000.

The first list is based in part upon the mythological association of constellations with gods assumed in *Enūma eliš* V and detailed in "Astrolabe B"; cf. the "Great Star List", for which see Weidner [1915] pp. 6-20, Weidner [1959/1960], and most recently, Koch-Westenholz [1995] pp. 187-205. We attempt to tabulate these relations in the following, where for MUL.APIN we list the constellation name followed, where given, by the name of the associated god. We also include information on the constellations of the Paths in the third section of "Astrolabe B" (= "Pinches' Astrolabe"), with the numbers of the months, and in the second section of "Astrolabe B", with their numbers within the Path and their gods.

| MUL.APIN | | "Astrolabe B" | | |
|---|---|---|---|---|
| Constellation | Gods | first section | second section | third section |
| Path of Enlil | | | | |
| 1. Plow | Enlil | | Enlil 1 | I 3 |
| 2. Wolf | | | Enlil 7 | |
| 3. Old Man | Enmešarra | | Anu 5. Enmešarra | II 2 |
| 4. Crook | Gamlu | | | |
| 5. Great Twins | Lugalgirra Meslamtaea | | Anu 9. Nabû, LUGAL | V 2 |
| 6. Little Twins | Alammuš, Nin-EZEN×GUD | | [Anu 8] Lugalgirra, Meslamtaea | IV 2 |
| 7. Crab | seat of Anu | | Anu 10. Anu | |
| 8. Lion | Latarak | | [Anu 7] | III 2 |
| 9. King | | | | VIII 3 |
| 10. Tail of Lion | | | | |
| 11. Frond of Eru | Zarpanītu | | | |
| 12. ŠU.PA | Enlil | Enlil | [Enlil 5. Enlil] | VI 3 |
| 13. Abundant One | messenger of Ninlil | | | |
| 14. Dignity | messenger of Tišpak | | | |
| Circumpolar | | | | |
| 15. Wagon | Ninlil | | Enlil 4. Ninlil | V 3 |
| 16. Fox | Erra | | Enlil 11 | XII 3 |
| 17. Ewe | Aya | | | |
| 18. Hitched Yoke | Anu | | | |
| 19. Wagon of Heaven | Damkianna | | | |
| 20. Heir of the Sublime Temple | first-ranking son of Anu | | | |
| 21. Standing Gods of Ekur | | | | |
| 22. Sitting Gods of Ekur | | | | |
| 23. She-Goat | Gula | | Enlil 6. Gula | IX 3 |
| 24. Dog | | | | |

| | | | | |
|---|---|---|---|---|
| 25. Lamma | messenger of Baba | | | |
| 26. two stars | Nin-SAR, Erragal | | | |
| 27. Panther | Nergal | | Anu 4. Nergal | IX 2 |
| 28. Pig | Damu | | Enlil 9 | XI 3 |
| 29. Horse | | | | |
| 30. Stag | messenger of Stars | | | |
| 31. Ḫarriru | Rainbow | | | |
| 32. Deleter | | | | |
| 33. Jupiter | | | Enlil 10. | IV 3, XII 2 |
| Path of Anu | | | | |
| 34. Field | seat of Ea | seat of Ea | Ea 1 | I 1 |
| 35. Swallow | | | Anu 6 | XI 2 |
| 36. Anunītu | | | Enlil 2 | II 3 |
| 37. Hired Man | Dumuzi | | | |
| 38. Stars | Great Gods | Great Gods | Ea 2. Great Gods | II 1 |
| 39. Bull of Heaven | | | | |
| 40. Jaw of Bull | crown of Anu | crown of Anu | Ea 3. crown of Anu | |
| 41. True Shepherd of Anu | Papsukal, messenger of Anu and Ištar | Papsukal, messenger of Anu and Ištar | Ea 4. Papsukal, messenger of Anu and Ištar | III 1 |
| 42. twin stars | Lulal, Latarak | | | |
| 43. Rooster | | | | |
| 44. Arrow | arrow of Ninurta | arrow of Ninurta | Ea 5. Ninurta | IV 1 |
| 45. Bow | Elamite Ištar | Elamite Ištar | Ea 6. Elamite Ištar | V 1 |
| 46. Snake | Ningizzida | | Enlil 3. Ningizzida | III 3 |
| 47. Raven | Adad | | Anu 11 | VI 2 |
| 48. Furrow | Šala | | | |
| 49. Scales | horn of the Scorpion | | Anu 3 | VII 2 |
| 50. star | Zababa | | | |
| 51. Eagle | | | Enlil 8 | X 3 |
| 52. Dead Man | | | | |
| 53. Venus | | | Anu 1 | I 2 |
| 54. Mars | | | Ea 10 | IX 1 |
| 55. Saturn | | | | |
| 56. Mercury | | | | |

| | | | |
|---|---|---|---|
| Path of Ea | | | |
| 57. Fish | Ea | Ea 12. Ea | XII 1 |
| 58. Great One | Ea | | X 1 |
| 59. Eridu | Ea | [Ea 7] | |
| 60. Ninmaḫ | | Ea 8 | VII 1 |
| 61. Ḫabaṣirānu | Ningirsu | Ea 11. Ningirsu | VII 3 |
| 62. Harrow | weapon of Marbīti | | |
| 63. Šullat and Ḫaniš | Šamaš and Adad | | |
| 64. Numušda | Adad | | XI 1 |
| 65. Mad Dog | Kusu | Ea 9 | VIII 1 |
| 66. Scorpion | Išḫara | Anu 2. Išḫara | VIII 2 |
| 67. Lisi | Nabû | | |
| 68. sting of Scorpion | Šarur and Šargaz | | |
| 69. Pabilsag | | | |
| 70. Bark | | | |
| 71. Goat-Fish | | | |

From this table it is easy to see that the compiler of the first list of constellations in MUL.APIN and the author of the first and second sections of "Astrolabe B" drew upon the same source for at least some of their associations of gods with constellations. It is also clear that the Paths of Enlil, Anu, and Ea meant something different to the author of the second section of "Astrolabe B" and the compiler of the first list in MUL.APIN. For the latter as for the author of the late Venus omens in *Enūma Anu Enlil* (Group F in Pingree [1993] and Reiner-Pingree [1998]) the Paths are defined by arcs along the eastern horizon (and the western) over which the stars and planets rise (or set); see also the "late" commentary III 24b in Reiner-Pingree [1981] pp. 42-43 and the comment on that, *ibid.*, pp. 17-18. The Path of Anu seems to occupy the arc of the horizon over which stars with declinations between ca. 15° N and 15° S rise, the Path of Enlil the arc on the horizon north of this, the Path of Ea the arc south of it. One expression of this concept is the statement found in MUL.APIN II A 1-7 that the Sun stays in each Path for the following months of the ideal calendar:

| | |
|---|---|
| Anu | months XII - II |
| Enlil | months III - V |
| Anu | months VI - VIII |
| Ea | months IX - XI |

The equinoxes occur on I 15 and VII 15, the solstices on IV 15 and X 15. It is also indicated by the fact that the declinations of the constellations in the

Path of Enlil in MUL.APIN are north of ca. 13° N and those of the constellations in the Path of Ea are south of ca. 11° S.

The constellations, of course, were conceived of before the introduction of this definition of the Paths, and therefore do no fit nicely into it at all. The astronomical concept of the Paths as being delimited, at least approximately, by arcs on the horizon must be posterior to the description in *Enūma eliš* V 1-10 of Marduk's assigning to the stars gates along the horizon through which they pass in rising, and his placing the Ford (Nebiru) between Enlil and Ea (i.e., at the East point, "halving the sky") to open the gates (at the appointed times). This same myth is alluded to in "Astrolabe B" (see CAD s.v. nēberu mng. 3a; see also Weidner [1915] p. 79): "the red star which, halving the sky, stands in the direction of the south, after the gods of the night have finished (their course), this star is the Ford, Marduk", and as well in MUL.APIN I i 37: "One big star—its light is dim—divides the sky in half and stands there: (that is) the star of Marduk, the Ford." These reflections make it likely that *Enūma eliš* was composed earlier than the reign of Nebukadnezar I (-1123 to -1103)—earlier enough so that it might be quoted in a text copied in Assyria probably some time before the reign of Nebukadnezar's contemporary, Tiglath-pileser I.[42]

*Enūma eliš* also refers to the already existing system of assigning three constellations to each month, which we find in the third section of "Astrolabe B", from which "Pinches' Astrolabe" seems to be derived. Since "Astrolabe B", which is a compilation from older sources, was copied in about the 12th century B.C. and the composition of *Enūma eliš* probably occurred at about the same time, we would date the origin of the Paths in "Astrolabe B" earlier than -1100, those more strictly defined, as in MUL.APIN and *Enūma Anu Enlil*, after -1100. For the constellations in the second section of "Astrolabe B", twelve to each Path, do not fit into astronomically defined Paths; rather, they are associated together because of their contiguity sometimes, and because of mythological factors at other times. Many of the same constellations appear in section three of "Astrolabe B" (and in "Pinches' Astrolabe"), where there are three constellations (apparently intended to be in the order of the Paths of Ea, Anu, and Enlil) in each month. Thirty-two constellations are certain in the second section, including three planets (Mars, Venus, and Jupiter), and the Ford, Marduk, who is the last in the Path of Anu; many of the remaining twenty-eight constellations appear in different Paths in the first list of MUL.APIN. Either the same names refer to different constellations in the two texts, or the criteria for including a constellation in one or another of the Paths were different in the two cases. We find it hard to believe that the names of the constellations changed, especially since the constellations involved in the

[42] According to Freydank [1992] pp. 94-97.

change of Paths include the Old Man, the Great Twins, [the Little Twins], the Crab, [the Lion], the Panther, the Field, Anunītu, the Stars, the Jaw of the Bull, the True Shepherd of Anu, the Arrow, the Bow, the Snake, the Eagle, and the Scorpion. One would be contemplating a total revolution in the star names! The "Astrolabe" lists provide no information useful for identifying the constellations because we do not know the principles of their categorizations.

Section three of "Astrolabe B" includes twenty-eight of the thirty-two constellations that were certainly in section two, including the three planets (Venus, Jupiter, and Mars) and Marduk in place of the Ford. But the table in Reiner-Pingree [1981] p. 4 demonstrates that, while thirteen of these constellations are said in the second list of MUL.APIN to rise (heliacally) in the same ideal months, thirteen do not, three (the planets) can rise in any month, four, while they do have heliacal risings, are not included in the second list of MUL.APIN, and two (the Wagon and the Fox) are circumpolar; and Marduk is not a star! This list is meaningless as an astronomical document (it is basically mythological), as is also the list at the end of section three of "Astrolabe B" where this list is mechanically converted into one in which three constellations rise in a month and three set. The person who composed this totally misunderstood the nature of his source (already in -1100!) and unfortunately misled several scholars of this century.

The Mesopotamian association of gods with constellations in the late second millennium B.C. probably gave the idea to the Vedic Indians to associate one or a set of their gods to each of the twenty-eight nakṣatras or constellations alleged to be in the path of the Moon; see Pingree [1989b] p. 442. Note that the ideal calendar mentioned in the next paragraph was also utilized by the Vedic Indians; see ibid. p. 441.

An unpublished tablet, MLC 1866, was kindly made available to us by P.-A. Beaulieu and J. P. Britton. It was copied by [Ina-qibīt-Ani], son of Anu-aḫa-ušabši, a descendant of Ekur-zākir of Uruk, in SE 97 = -214/3, and consists to a large extent of citations from the first star-catalogue of MUL.APIN followed by comments and elaborations drawn from several sources. The order in which the quotations from MUL.APIN are given is somewhat irregular. The lines that we can identify and the constellations to which they refer are as follows:

| MUL.APIN I | | Constellation |
|---|---|---|
| i 40 | | Field |
| i 41 | | Swallow |
| i 42 | | Anunītu |
| i 43 | | Hired man |
| i 28 | | Panther |
| | (break) | |
| i 6 | | Little Twins |
| ii 2 | | True Shepherd of Anu |
| ii 3-4 | | Lulal and Latarak |
| ii 5 | | Rooster |
| i 7 | | Crab |
| ii 6 | | Arrow |
| ii 7 | | Bow |
| | (break) | |
| ii 21 | | Ninmaḫ |
| ii 22 | | Ḫabaṣirānu |
| ii 23-24 | | Harrow |
| i 15 | | Wagon |
| i 16-17 | | Fox |
| i 18 | | Ewe |
| i 19 | | Hitched Yoke |
| | (break) | |
| ii 19 | | Fish |
| i 44 | | Stars |
| i 21 | | Heir of the Sublime Temple |
| i 5 | | Great Twins |
| i 6 | | Little Twins |
| ii 2 | | True Shepherd of Anu |
| | (break) | |

For some of these constellations an image is described. These and similar images associated with constellations are also described in VAT 9428, published by Weidner [1927], a much older tablet from Assur. The constellations whose images it describes include Old Man, Great Twins, Little Twins, Crab, Lion, Frond, Wagon, Wagon of Heaven, Dog, and the She-goat, all constellations in the Path of Enlil.

The second star-list in MUL.APIN gives dates in the ideal calendar (in which 1 month = 30 days and 1 year = 12 months = 360 days) of the heliacal risings of thirty-five constellations. The choice of dates is restricted to days 1 (= 0), 5, 10, 15, 20, and 25 of a month, so that the list gives only the order of the heliacal risings and the approximate intervals between them. These data are not very useful for determining the boundaries of the constellations, but only their general locations. Since the fourth list is derived from the second, we combine their data in the following table.

| List II | | List IV | |
|---|---|---|---|
| Constellations | Dates of visibility | (Daycount) | Intervals |
| Hired Man | 1 Nisannu | (0) | 35 days |
| Crook | 20 Nisannu | (20) | |
| Stars | 1 Ajjaru | (30) | 10 days |
| Jaw of the Bull | 20 Ajjaru | (50) | 20 days |
| True Shepherd of Anu | 10 Simānu | (70) | 20 days |
| Little Twins, Crab | 5 Duʾūzu | (95) | 35 days |
| Arrow, Snake, Lion | 15 Duʾūzu | (105) | |
| Bow, King | 5 Abu | (125) | 55 days |
| — | 1 Ulūlu | (150) | 60 days |
| Eridu, Raven | 10 Ulūlu | (160) | |
| ŠU.PA | 15 Ulūlu | (165) | |
| Furrow | 25 Ulūlu | (175) | 10 days |
| Scales, Mad Dog, Ḫabaṣirānu, Dog | 15 Tešrītu | (195) | 20 days |
| Scorpion | 5 Araḫsamnu | (215) | 30 days |
| She-goat, breast of Scorpion | 15 Araḫsamnu | (225) | 30 days |
| Panther, Eagle, Pabilsag | 15 Kislīmu | (255) | 30 days |
| Swallow | 15 Ṭebētu | (285) | |
| (Arrow evening rising | 15 Ṭebētu) | | 20 days |
| Great One, Field, Stag | 5 Šabāṭu | (305) | |
| Anunītu | 25 Šabāṭu | (325) | 40 days |
| Fish, Old Man | 15 Addaru | (345) | |

It is notable that beginning with the rising of the Little Twins and the Crab the constellations on each date are normally multiple and that all the dates except for two end with a 5 while the first five entries have single constellations rising on dates ending in 0 or its equivalent, 1. Van der Waerden [1949] 19 suggests that a "stellar" year began on 15 Duʾūzu with the heliacal rising of Arrow (i.e., of Sirius), as is reflected in List IV, and a "solar" (we would say, rather, "civil") year with 1 Nisannu. He then supposes that the "solar" year was used for the first three months, the "stellar" year for the remaining nine months. If this hypothesis is true, the days in each ideal month in which a given constellation rises become even less precise, because the intervals between them become, in general, multiples of 10 days rather than of 5.

In the ideal year of List II, supported by statements in MUL.APIN II 1 9-24 and II ii 21-42, the date of the rising of the Arrow, IV 15, is also the date of the summer solstice; accordingly, the date of the winter solstice is X 15 and the dates of the equinoxes are I 15 and VII 15. List II remarks at Duʾūzu 15 that 4 minas measure the daytime, 2 minas the nighttime, at Tešrītu 15 that 3 minas measure both the daytime and the nighttime, and at Ṭebētu 15 (the night of the symmetrical evening rising of the Arrow) 2 minas measure the daytime, 4 the nighttime. The absence of an entry for the vernal equinox on Nisannu 15 tends to confirm van der Waerden's conjecture about the dual origin of List II.

Thus, the dates of the equinoxes and solstices have been arbitrarily shifted from their positions according to the older tradition in which the vernal equinox occurred in (the middle of) the month preceding Nisannu to positions in a new tradition in which it occurs in (the middle of) Nisannu itself. This positioning of the colures on the 15th continued into early Greek astronomy, when Eudoxus, using zodiacal signs instead of months, set the equinoxes in the middle (τὰ μέσα) of Aries and of Libra, the summer solstice in the middle of Cancer, and the winter solstice in the middle of Capricorn (fragments 65, 69, and 72 in Lasserre [1966] pp. 52-54). But in Mesopotamia the ratio of the longest to the shortest day, 2 : 1, remained the same, as did the linear zigzag function describing it.

The third list in MUL.APIN is of the simultaneous risings and settings of constellations. Whereas the parallel passage at the end of section three of "Astrolabe B" was simply a meaningless rearrangement of the preceding list of three constellations in each month, List III of MUL.APIN is independent of the schematic dates of rising in List II and the simultaneously setting constellations are clearly determined by observation. This list is the foundation of our identifications of the constellations and of our determination of the time (ca. -1000) and place (ca. 36° N) of the observations. The data are as follows:

| Rising | Setting |
|---|---|
| Stars | Scorpion |
| Scorpion | Stars |
| Bull of Heaven | ŠU.PA |
| True Shepherd of Anu | Pabilsag |
| Arrow, Snake, Lion | Great One, Eagle |
| Bow, King | She-goat |
| Eridu, Raven | Panther |
| ŠU.PA | Field |
| Ninmaḫ | Anunītu |
| Scales, Mad Dog, Ḫabaṣirānu | Hired Man |
| Scorpion, Dog | Eridu, Stars |
| Breast of Scorpion, She-goat | Old Man, True Shepherd of Anu |
| Pabilsag, Zababa, Standing Gods | Arrow, Bow, Crook |
| Panther, Eagle | Great Twins, Little Twins |
| Field, Great One, Stag | Lion, Snake, Ḫabaṣirānu |
| Fish, Old Man | Furrow, Mad Dog |

That this list is independent of List II is shown by its placing the rising of the Dog with that of the Scorpion rather than with those of the preceding trio, which in List II occurred 20 days earlier, and with its placing the rising of Pabilsag before the risings of Panther and Eagle. Of the intervals between the risings and settings of the same stars that can be derived from List III a comparison can be made with the intervals according to Ptolemy for a latitude of 36° N; the two lists agree remarkably well (see Hunger-Pingree [1989] p. 145).

It is also noteworthy that List III begins (omitting the Scorpion) with the Stars, the Bull of Heaven, and the True Shepherd of Anu (that is, with the Pleiades, Taurus, and Orion), while in List II, starting with 1 Ajjaru, the successively rising constellations are the Stars, the Jaw of the Bull (α Tauri and the Hyades), and the True Shepherd of Anu. List VI, the constellations in the path of the Moon, as we shall see, also begins with the Stars, the Bull of Heaven, and the True Shepherd of Anu. The sequence of the Stars, the Jaw of the Bull, and the True Shepherd of Anu also appears in "Astrolabe B" and in Reiner-Pingree [1981] IX 2-4. With this compare Homer, *Iliad* 18, 483-489, describing the shield of Achilles:

> Ἐν μὲν γαῖαν ἔτευξ', ἐν δ'οὐρανόν, ἐν δὲ θάλασσαν,
> Ἠέλιόν τ'ἀκάμαντα Σελήνην τε πλήθουσαν,
> ἐν δὲ τὰ τείρεα πάντα, τὰ τ'οὐρανὸς ἐστεφάνωται,
> Πληϊάδας θ'Ὑάδας τε τό τε σθένος Ὠρίωνος
> Ἄρκτον θ'ἣν καὶ Ἄμαξαν ἐπίκλησιν καλέουσιν,
> ἥ τ'αὐτοῦ στρέφεται καί τ'Ὠρίωνα δοκεύει,
> οἴη δ'ἄμμορός ἐστι λοετρῶν Ὠκεανοῖο.

In this passage, not only does the line "the Pleiades, the Hyades, and the strength of Orion" correspond to the sequence in the Akkadian star lists, but "the earth, the heaven, and the sea" represent the domains of Enlil, Anu, and Ea respectively; "the unwearying Sun and the full Moon" represent the situation on the 14th of the month when "the Moon and the Sun" are "in opposition," "in balance," or "seen together"; "the Wagon" is Greek for MAR.GÍD.DA, the Mesopotamian name for Ἄρκτος; and the phrase "which (Bear) circles around it (i.e., heaven)" translates the Akkadian mulMAR.GÍD.DA *kal* MU DU-*az ma-a i-lam-ma-a*, "The Wagon stands all year, namely, it circles round" (BPO 2 Text III 28c). Note that, according to Callimachus as cited by Diogenes Laertius (I, 23), Thales "measured the stars of the Wagon by which the Phoenicians sail." Diogenes identifies the Wagon with Ursa Minor—MAR.GÍD.DA.AN.NA.

Lines 487-489 are repeated in *Odyssey* 5, 273-275, where they are preceded by:

Πληϊάδας τ'ἐσορῶντι καὶ ὀψὲ δύοντα Βοώτην.

The ὀψὲ, "afterwards," makes this correspond to entries 1 and 3 in List III.

List IV of MUL.APIN, which gives the intervals between the dates in the ideal calendar of the risings of 15 constellations, is entirely dependent on the data in List II, as is shown in Hunger-Pingree [1989] 141. The list begins (twice) with the Arrow, Sirius, and includes the Arrow again in its main sequence. This indicates that the rising of Sirius, on Duʾūzu 15 in the ideal calendar, the date also of the summer solstice, was already a crucial phenomenon for establishing the dates of the colures; this importance is reflected in the later scheme of ca. -600 (Neugebauer-Sachs [1967] 183-190) and that of the Seleucid period (Sachs [1952]).

The next section of Tablet I of MUL.APIN contains two lists of *ziqpu* stars, which are described as follows (I iv 1-3; cf. I iv 7-9): "The *ziqpu* stars which stand in the Path of Enlil in the middle of the sky (*ina* $MURUB_4$ AN-*e*) opposite the breast of the observer of the sky, and by means of which he observes the rising and setting of the stars at night." In fact, in general these stars had northern declinations such that they passed over his head if he stood at Nineveh ($\varphi$ = 36;30° N), Nimrud ($\varphi$ = 36;4° N), or Assur ($\varphi$ = 35;17° N) in -1000. The first list (List Va) includes 14 constellations; in the following table we add a bright star from each with its right ascension and declination in -1000.

| Constellation | *ziqpu* star | RA | δ |
|---|---|---|---|
| 1. ŠU.PA | α Boötis | 179° | +35;2° |
| 2. Dignity | α Coronae Borealis | 201;52° | +39;50° |
| 3. Standing Gods | β Herculis(?) | 215;49° | +31;47° |
| 4. Dog | ϑ Herculis | 243;54° | +41;11° |
| 5. She-Goat | α Lyrae | 254;15° | +39;47° |
| 6. Panther | α Cygni | 285° | +37;33° |
| 7. Stag | γ Andromedae | 350;29° | +26;7° |
| 8. Old Man | α Persei | 5;29° | +35;16° |
| 9. Crook | α Aurigae | 28;59° | +36;12° |
| 10. Great Twins | α Geminorum | 65;8° | +31;43° |
| 11. Crab | ε Cancri | 85;22° | +24;36° |
| 12. Lion | ε Leonis | 100;40° | +32;57° |
| 13. Frond | γ Comae Berenices | 145;31° | +44;17° |
| 14. Abundant One | β Comae Berenices | ca. 158° | ca. +43° |

Where there was a choice of stars (especially for the Abundant One) the identification is based on the notion that the sequence of the constellations is that of their crossing the meridian. The first constellation, ŠU.PA, will be on the local meridian at midnight of the day of the vernal equinox. The two constellations that are the most southerly in this list—the Stag and the Crab—are omitted from the next list of *ziqpu* stars, List Vb.

That List Vb gives thirteen dates on which a given star is on the meridian while another is rising; the observation of these simultaneously rising and culminating stars is made in the morning, as the text indicates (I iv 10-12): "If you are to observe the *ziqpu*, you stand in the morning before sunrise, West to your right, East to your left, your face directed towards South."

| Dates | *ziqpu* stars | Rising constellations |
|---|---|---|
| 1. I 20 | Shoulder of the Panther | Crook |
| 2. II 1 | Breast of the Panther | Stars |
| 3. II 20 | Knee of the Panther | Jaw of the Bull |
| 4. III 10 | Heel of the Panther | True Shepherd of Anu |
| 5. IV 15 | bright star of the Old Man | Arrow |
| 6. V 15 | dusky stars of the Old Man | Bow |
| 7. VI 15 | Great Twins | ŠU.PA, Eridu |
| 8. VII 15 | Lion | Scales |
| 9. VIII 15 | Frond | She-goat |
| 10. IX 15 | ŠU.PA | Panther |
| 11. X 15 | Standing Gods | Swallow |
| 12. XI 15 | Dog | Field |
| 13. XII 15 | She-goat | Fish |

The first four dates are those of the risings of the named rising constellations in the ideal calendar according to List II; so are the dates for nos. 5, 7, 8, 9, 10, 11, and 13. But the original aim of List Vb was to give a culminating star on the 15th of each month, so that the dates of the risings of Bow (5 Abu) and of Field (5 Šabāṭu) were arbitrarily changed to fit the scheme. For reasons unknown to us the same was not done for nos. 1, 3, and 4. No. 2, the rising of Stars with the culmination of the Breast of the Panther on 1 Ajjaru, was added presumably because of the importance of the Pleiades in the intercalation rules that keep the dates of the colures within the desired months.

Since the dates of the equinoxes and solstices in the ideal calendar are given, we can establish approximate longitudes of the Sun for each of the dates in the list; if we regard these longitudes as those of the ascendent, it is easy to compute the right ascensions of the meridian on the sunrises of each of the dates at a latitude of 36° North. In the following table the right ascensions and declinations of the stars are computed for -1000.

| λ | RA of meridian | Star | RA | δ |
|---|---|---|---|---|
| 5° | 274;35° | γ Cygni | 278;52° | +33;59° |
| 15° | 283;47° | α Cygni | 285° | +37;33° |
| 35° | 302;41° | α Lacertae | 308;47° | +37° |
| 55° | 322;37° | λ Andromedae | 320;57° | +31;12° |
| 90° | 0° | α Persei | 5;29° | +35;16° |
| 120° | 32;12° | h Persei | ca. 353° | ca. +42° |
| 150° | 62;7° | α Geminorum | 65;8° | +31;34° |
| 180° | 90° | ε Leonis | 100;4° | +32;43° |
| 210° | 117;53° | γ Comae Berenices | 145;31° | +44;17° |
| 240° | 147;48° | α Boötis | 179° | +35;2° |
| 270° | 180° | β Herculis | 215;49° | +31;47° |
| 300° | 212;12° | ϑ Herculis | 243;54° | +41;11° |
| 330° | 242;7° | α Lyrae | 254;15° | +39;47° |

There are clearly serious errors in this list. The dusky stars of the Old Man (h Persei) have a right ascension less than that of α Persei; one expects as *ziqpu*-star here Crook, α Aurigae (RA = 28;59°, δ = +36;12°). The positions of the Frond, ŠU.PA, the Standing Gods, and the Dog must all be lowered by one; the *ziqpu*-star for VIII 15 should probably be δ Leonis (RA = 125;19°, δ = +34;22°). Finally, the She-goat is the *ziqpu* star for I 1 (at dawn the RA of the meridian is 256;12°). This list was compiled incorrectly from a list of the ideal dates of the heliacal risings of constellations and a list of *ziqpu* stars.

The final star list in Tablet I, List VI, is of 17 constellations in the path of the Moon (in II i 1-6 we are told that the Sun and each of the five star-planets also follow the path of the Moon). For each of the text's constellations we have provided an exemplary star whose latitude is small enough (generally 5° or below) that it actually could be "touched" by the Moon; it may not, however, have been the objective of the author of this list to meet that criterion. The coordinates are those of -1000.[43]

| Constellation | Exemplary star | Longitude | Latitude |
|---|---|---|---|
| 1. Stars | η Tauri | 18;22° | +3;45° |
| 2. Bull of Heaven | α Tauri | 28;7° | -5;40° |
| 3. True Shepherd of Anu | ζ Tauri(?) | 43;9° | -2;35° |
| 4. Old Man | ψ Tauri(?) | 23;46° | +7;34° |
| 5. Crook | γ Aurigae | 40;56° | +5;8° |
| 6. Great Twins | δ Geminorum | 66;54° | -0;33° |
| 7. Crab | δ Cancri | 87;3° | -0;3° |
| 8. Lion | α Leonis | 108;23° | +0;19° |
| 9. Furrow | α Virginis | 162;15° | -1;52° |
| 10. Scales | α Librae | 183;31° | +0;41° |
| 11. Scorpion | α Scorpii | 208;9° | -4;11° |
| 12. Pabilsag | θ Ophiuchi | 219;46° | -1;26° |
| 13. Goat-Fish | β Capricorni | 262;24° | +4;56° |
| 14. Great One | λ Aquarii | 299;55° | -0;16° |
| 15. Tails of the Swallow | λ Piscium | 315;8° | +3;32° |
| 16. Anunītu | ζ Piscium | 338;9° | -0;15° |
| 17. Hired Man | δ Arietis | 9;7° | +1;34° |

The main difficulties lie in nos. 3 and 4; the True Shepherd of Anu and the Old Man were larger than our Orion and Perseus, and apparently included some of the stars that we assign to Taurus. In the later Normal Stars, whose latitudes can be as much as 10°, β Tauri (=γ Aurigae) is called "the northern rein of the Chariot" and ζ Tauri is "the southern rein of the Chariot," so that they belong to a single constellation; and no star is included that is said to be part of the True Shepherd of Anu or of the Old Man. We have used ϑ Ophiuchi to represent Pabilsag (no. 12) since its name as a Normal Star is "the bright star on the tip of Pabilsag's arrow," but have rejected such Normal Stars as α and β Arietis, and α, β, and γ Geminorum because of their high latitudes.

[43] The coordinates were computed with the program "Uraniastar" (written by W. Vollmann and M. Pietschnig of the Vienna Planetarium).

The Indian lists of nakṣatras composed during the early centuries of the last millennium B.C. show striking resemblances to the Mesopotamian constellations, and to MUL.APIN's List VI in particular, though the Indians, wishing to have one nakṣatra for the Moon to spend each night of a sidereal month with, have split some Mesopotamian constellations in two and have added a number of nakṣatras with very high latitudes, far beyond the observed path of the Moon (see Pingree [1989] 439-442). The following list of Indian nakṣatras with their yogatārās (junction-stars) as of the fifth century A.D. will illustrate the relationship more clearly (the identifications of the yogatārās are those of Pingree-Morrissey [1989]).

| Nakṣatra | Yogatārā | Babylonian constellation |
|---|---|---|
| 1. Kṛttikā | η Tauri | Stars |
| 2. Rohiṇī | α Tauri | Bull of Heaven |
| 3. Mṛgaśiras | λ Orionis | True Shepherd of Anu |
| 4. Ārdrā | α Orionis | True Shepherd of Anu |
| 5. Punarvasu | β Geminorum | Great Twins |
| 6. Puṣya | δ Cancri | Crab |
| 7. Āśleṣā | ϰ Cancri(?) | Crab |
| 8. Maghā | α Leonis | King |
| 9. Pūrva Phalgunī | δ Leonis | Lion |
| 10. Uttara Phalgunī | β Leonis | Lion |
| 11. Hasta | δ Corvi | Raven |
| 12. Citrā | α Virginis | Furrow |
| 13. Svāti | α Boötis | ŠU.PA |
| 14. Viśākhā | ι Librae | Scales |
| 15. Anurādhā | δ Scorpii | Scorpion |
| 16. Jyeṣṭhā | α Scorpii | Scorpion |
| 17. Mūla | 45 Ophiuchi | |
| 18. Pūrva Aṣādhā | δ Sagittarii | Pabilsag |
| 19. Uttara Aṣādhā | ζ Sagittarii | Pabilsag |
| 20. Abhijit | α Lyrae | She-goat |
| 21. Śravaṇa | α Aquilae | Eagle |
| 22. Dhaniṣṭhā | α Delphini | |
| 23. Śatabhiṣaj | λ Aquarii | Great One |
| 24. Pūrva Bhadrapadā | α Pegasi | Field |
| 25. Uttara Bhadrapadā | α Andromedae | Field |
| 26. Revatī | ζ Piscium | Anunītu |
| 27. Aśvinī | β Arietis | Hired Man |
| 28. Bharaṇī | 35 Arietis | Hired Man |

The yogatārās whose latitudes lie outside of the path of the Moon are nos. 11, 13, 17, 20, 21, 22, 24, and 25. It is noteworthy that in the three cases where the Indians have divided a single constellation into pūrva (prior) and uttara (posterior) halves the single constellation corresponds to a single Mesopotamian constellation.

### *3.2. Planets*

The phrase used to differentiate the planets from the fixed stars in MUL.APIN is "keeps changing its position and crosses the sky" (I i 38 and I ii 13-15); compare "keep changing their positions and their glow [and] touch [the stars of the sky]" said of the five star-planets (II i 40-41), and "together six gods who have the same positions, (and) who touch the stars of the sky and keep changing their positions" said of the five star-planets with the Sun (II i 7-8). The first phrase refers to the fact that the points on the eastern and western horizon above which the planets rise and set keep changing as their declinations change; "touch the stars of the sky" means that they move with respect to the fixed stars; and "have the same positions" refers to the fact that the horizon positions of rising and setting are the same for the planets and the Sun as they are for the Moon in whose path they move. In the case of Mercury in I ii 16-17, where one expects the first phrase, one finds instead an addendum to the section on planetary theory (II i 66-67), which is an abbreviated form of II i 54-55; the reason may be that the scribe understood "crosses the sky" to refer to a motion from the eastern horizon to the western in one night, which is inappropriate when applied to Mercury, though its inappropriateness did not deter him from applying it to Venus.

MUL.APIN's planetary theory occupies a section (II i 38-67) that has been inserted in a section on the risings of constellations in Ulūlu and Araḫsamnu, flanking the month of the fall equinox, and in Addaru, the month before that of the vernal equinox (II i 25-37) and on the constellations as indicators of the directions from which the wind blows (II i 68-71). The introduction to the planetary section (II i 38-43) contains elements derived from the first phrase distinguishing planets from fixed stars noted above and from the description of activities to be undertaken at the risings of the constellations according to the immediately preceding section. This introduction, then, is an artificial construct composed by the compiler of MUL.APIN, while the rest of the planetary section is also a compilation consisting of a section on the periods of invisibility of Venus, Jupiter, Mars, and Saturn (II i 44-53) and another on the same planets' and Mercury's periods of visibility (II i 60-67, where 66-67 equals I ii 16-17 as remarked above). Between these two parts of a single section is inserted from some other source (quoted in a commentary on *Enūma Anu Enlil* 56; see Largement [1957] 252-253 and Hunger-Pingree [1989] 148) a special section on Mercury (II i 54-59) in which occur omens reminiscent in their phraseology of the passage immediately preceding the planetary section. It is not known whether the second theory of Mercury was paralleled elsewhere for the other star-planets.

The planetary phenomena "predicted" in MUL.APIN are the planets' heliacal risings and settings as in section δ of the Venus Tablet of Ammiṣaduqa; this situation, of course, reflects the fact that these phenomena and the circumstances that accompany them are ominous in *Enūma Anu Enlil*. The theory does not attempt to provide a precise prediction of the date of any phenomenon, but simply gives periods of visibility and of invisibility from which, once a heliacal rising or setting occurs, the approximate time at which the following setting or rising will occur is known.

For the superior planets the sum of the periods of visibility (Γ→Ω) and of invisibility (Ω→Γ) should be approximations of the planets' synodic periods. This is essentially true for Saturn whose synodic period is 378 days, and for Jupiter, whose synodic period is 398 days; MUL.APIN gives 1 year and 20 days (= 380 days) and 1 year and 20 or 30 days (= 380 or 390) respectively. For Mars, however, we find four possibilities for the period of visibility: 1 year (= 360 days), 1 year and 6 months (= 540 days), 1 year and 10 months (= 660 days), and 2 years (= 720 days); the last is the closest to one acceptable value, about 720 days. For the period of invisibility three values are given: 2 months (= 60 days), 3 months and 10 days (= 100 days), and 6 months and 20 days (= 200 days). The first is the best; in combination with 2 years for the period of visibility it gives a correct synodic period of 780 days. The origins of the alternative values are not recoverable.

For the inferior planets a synodic period contains two periods of visibility (in the East and in the West) and two periods of invisibility (at superior conjunction and at inferior conjunction). The older scheme for Venus, in Tablet 63 of *Enūma Anu Enlil*, has been considerably modified in MUL.APIN. In place of 8 months and 5 days (=245 days) as the period of visibility in both East and West the later text has 9 months (= 270 days), which is in general too large. For the periods of invisibility MUL.APIN offers a choice: at superior conjunction 1 month (= 30 days), 1 month and 15 days (= 45 days), and 2 months (= 60 days), and at inferior conjunction 1 day, 3 days, 7 days, and 14 days. The periods of invisibility at inferior conjunction fall into the expected range, but of the values listed for superior conjunction the first is not possible. Though there is a combination of periods that produces Venus' synodic period of 584 days (270 + 30 + 270 + 14 days) it is unreal. The fault lies in the excessively long periods of visibility; they repeat the data, not intended to be normative but ominous, found in *Enūma Anu Enlil*.

Of the original theory of Mercury there remains only the period of visibility, "within a month," valid in both East and West. The alternative theory of the insert simply names five periods without specifying where in Mercury's synodic period they are to be applied. The five periods are: 7 days, 14 days, 21 days, 1 month (= 30 days), and 1 month and 15 days (= 45 days). The periods of visibility for Mercury in the later ACT material vary between 14 and 48 tithis, which leaves the short 7 days for a period of invisibility at inferior conjunction. The synodic period of Mercury is 116 days, which could be approximated by an arbitrary combination of four of the five given periods; but such a reconstruction would make little sense. In general, MUL.APIN shows an awareness that the periods of visibility and invisibility of the planets fall within certain bounds, but the setting of these bounds is not precise. The text provides a basis for "predictions" such as one finds in the *Reports* and *Letters*, but only of the approximate dates of first and last visibilities.

### *3.3. Intercalation Schemes*

MUL.APIN presents two intercalation schemes, each of which is naturally based on a comparison of the solar year with synodic months. In the first scheme (II i 9-24) the solar year (beginning with the summer solstice) is defined by the solstices and the equinoxes; these in turn are defined in several ways:

1) The solstices are determined by the rising of Arrow on 15 Duʾūzu in the morning and the rising of Arrow on 15 Ṭebētu in the evening; the dates of both phenomena are derived from List II of the constellations.

2) Both solstices and equinoxes are determined by the minas of water poured into the water-clock to measure their days and nights; 4 minas for the longest day (summer solstice) and the longest night (winter solstice), 2 minas for the shortest day and the shortest night, and 3 minas for the equinoctial day and the equinoctial night, so that this scheme exactly parallels that in List II though it omitted the data for the vernal equinox. The daily change in the weight of the water is given as 40 NINDA or 0;0,40 minas; in 90 days this equals 1 mina, the difference of the linear zigzag function.

3) All four colures are identified by the position of the Sun at its turning point on the eastern horizon, the direction of the change in position of the rising Sun along the eastern horizon between the colures, the constellations that rise or set with the Sun, and, for the equinoxes, the position of the Moon among the constellations; since each colure occurs on the 15th of a month in the ideal lunar calendar, the situation of the Sun and the Moon at the vernal equinox is simply the reverse of that at the fall equinox. The data are summarized in the following table, where the longitudes and latitudes of the stars are computed for -1000.

| Dates | Phenomena | Stars | Longitude | Latitude |
|---|---|---|---|---|
| IV 15 (SS) | Sun rises to North with head of Lion | ε Leonis | 99;3° | +9;28° |
| VII 15 (FE) | Sun rises in Scales in East | α Librae | 183;31° | +0;41° |
| | Moon before Stars | η Tauri | 18;22° | +3;45° |
| | Moon after Hired Man | β Arietis | 352;21° | +8;23° |
| X 15 (WS) | Sun rises to South with head of Great One | ν Aquarii | 274;43° | +5;5° |
| I 15 (VE) | Moon in Scales in East | | | |
| | Sun before Stars in West | | | |
| | Sun after Hired Man in West | | | |

The Mesopotamian observations of the rising-points on the horizon of the Sun at the solstices and equinoxes and of the "motion" of its rising-points North and South along the horizon seem to lie behind a passage in an Indian liturgical text of the middle of the first half of the last millennium B.C., the *Kauṣītakibrāhmaṇa* (19, 3). "On the New Moon of Māgha he (i.e., the Sun) rests, being about to turn northwards . . . He goes North for six months . . . Having gone North for six months he stands still, being about to turn southwards . . . He goes South for six months . . . Having gone South for six months he stands still, being about to turn North." It is demonstrated in Pingree [1989] 444 that the months here begin with New Moon as in Mesopotamia (an earlier Vedic tradition put the beginning of the month at Full Moon); that the *Kauṣītakibrāhmaṇa*'s year is the Mesopotamian ideal year of 360 days (indicated in this passage in MUL.APIN by the daily difference of 0;0,40 NINDA of water, which amounts to exactly 4 minas in 360 days); and that its reference to the New Moon of Māgha as the date of the winter solstice implies, in -1000, as correct a date for the solstices as is found in MUL.APIN, and one that is determined by dates in an ideal calendar and the positions of the Sun and the Moon relative to the fixed stars.

Also related to this passage in MUL.APIN are two lines in Homer's *Odyssey* (15, 403-404):

νῆσός τις Συρίη κικλήσκεται, εἴ που ἀκούεις,
Ὀρτυγίης καθυπέρθεν, ὅθι τροπαὶ Ἠελίοιο.

The "turnings of the Sun" are clearly its rising-points on the eastern horizon at the solstices. According to a scholiast, who identifies Syriē with the island Syra, one of the Cyclades, observations of the solstices were made from a cave on the island: ἔνθα φασὶν εἶναι Ἡλίου σπήλαιον, δι' οὗ σημειοῦνται τὰς τοῦ Ἡλίου τροπάς (Dindorf [1855] vol. 2, p. 617). This activity of observing the solstices on the Greek islands appears to be imitative of Mesopotamian practice. With this tradition is probably to be connected the statement made by Diogenes Laertius concerning Pherecydes (I 119): σώζεται δὲ καὶ ἡλιοτρόπιον (i.e., of Pherecydes) ἐν Σύρῳ τῇ νήσῳ.

The words with which MUL.APIN instructs one concerned with determining the need for intercalation are only suggestive of a procedure (II i 22-24): "On the 15th of Nisannu, on the 15th of Duʾūzu, on the 15th of Tešrītu, on the 15th of Ṭebētu, you observe the risings of the Sun, the visibility time of the Moon, the appearance of the Arrow, and you will find how many days are in excess." The "risings of the Sun" with the stars mentioned for IV 15, VII 15, and X 15 will determine the actual day of a lunar month on which the summer solstice, fall equinox, and winter solstice occur; if that date in the real calendar is one month more than the given dates in the ideal calendar, intercalation is needed. If those sunrises cannot be observed because, e.g., of clouds in the East, one can observe the setting of the Moon in the West just before sunrise of the fall equinox and of the vernal equinox. If the Moon is situated with respect to the fixed stars as is indicated in the text on the date of the equinox in the ideal calendar, the year is normal; if it is in the right position only three days later, on the eighteenth of the real month, intercalation is needed. The first visibility of the Arrow on IV 15 in the morning and on X 15 in the evening in the ideal calendar could also be used as criteria, but not as an accurate one since Sirius' first visibility in the evening occurs about 5 rather than 6 months after its first visibility in the morning. But the first visibility of Arrow on X 15 of the ideal calendar appears again, as a criterion of the need for intercalation, in the second intercalation scheme.

Immediately after the first intercalation scheme comes the section on constellations into which the planetary theory is inserted. The second part of this is on the relation of constellations to directions (phrased as the directions from which the winds blow, a manner of designating the directions adopted also in Greek and Latin) (II i 68-71).

| | |
|---|---|
| North | Wagon (the first circumpolar constellation in I i 15) |
| South | Fish (the first star of the Path of Ea in I ii 19) |
| West | Scorpion (sets when Stars rise in I iii 13) |
| East | Old Man and Stars (Old Man rises with Fish and Stars rise when Scorpion sets in I iii 12-13). |

Koch [1995/96] argues that the author of these lines intended to refer to the North, South, West, and East points on the horizon—i.e., to azimuths of 180°, 0°, 90°, and 270°. After computing the azimuths of many stars, he concludes that this author refers to the night of 1 August -1300, for example, between 21:56 h and 23:53 h at Babylon. His best fits are:

| Direction | Constellation | Chosen star | Azimuth |
|---|---|---|---|
| North | Wagon | α Ursae Maioris | 180° |
| South | Fish | α Piscis Austrini | 359;59° |
| West | Scorpion | β Librae | 90;2° |
| East | Stars | η Tauri | 270;2° |

He necessarily includes Scales as a part of Scorpion, which is quite plausible. We feel, however, that this kind of "precision" is out of place in early astronomy—especially when it limits the text's general statement to very specific moments in time. Rather we believe that the author intended nothing so exact and so bereft of usefulness. Necessarily, then, we also are not convinced of either Koch's date or his observation site.

In Vedic India, in the *Śatapathabrāhmaṇa* (2, 1, 2, 4), it is stated: "The Saptarṣis (= Ursa Maior, the Wagon) rise in the North, these (the Kṛttikās = the Pleiades, Stars) in the East" (see Pingree [1989] 444-445). And according to the Pahlavī *Bundahishn* (see Henning [1942] 231) the "generals" of the directions are:

| | | |
|---|---|---|
| North | Haftōreng | (= Ursa Maior) |
| South | Sadwēs | (= α Scorpii) |
| West | Wanard | (= α Lyrae) |
| East | Tishtrya | (= Sirius) |

At most the Mesopotamian idea of associating stars with directions is reflected in these Iranian and Indian texts; the possibility of influence is made more plausible by the other cases of Mesopotamian influence on the astral lore of these two cultures.

In the second intercalation scheme in MUL.APIN (II A 1 - ii 20) the solar year is defined by the statement that the Sun spends three months in each of the Paths of Enlil and Ea and two periods of three months each in the Path of Anu; the months are so chosen that the equinoxes occur on Nisannu 15 and Tešrītu 15 in the ideal calendar; see above p. 61f. The intercalation criteria are phenomena which "normally" occur on or near fixed dates in the ideal calendar, but which often occur on later dates in the real calendar. When the dates in the real calendar indicate that a month of difference has accumulated, a month must be intercalated.

One type of phenomenon involves the Moon; it occurs in two varieties. In the first the Moon and the Stars are normally "in conjunction" (*šitqulū*) on 1 Nisannu in the ideal calendar, when the longitude of the Sun is ca. 345°, that of the Moon ca. 357°, and, in -1000, that of the Pleiades ca. 18°; the order of setting was Sun, Hired Man, Moon, and Pleiades. But if this sequence occurred only on 3 Nisannu in the real calendar, then the longitude of the Moon on 1 Nisannu was ca. 330° and that of the Sun ca. 318°; a month must be intercalated so that the vernal equinox should occur in the middle of Nisannu. In the second variety, the Moon and Stars are in conjunction (*šitqulū*) on 15 Araḫsamnu in the ideal calendar, when the longitude of the Sun is ca. 210°; if that phenomenon occurs a month later in the real calendar, on 15 Kislīmu, a month must be intercalated.

The other phenomena are the first visibilities of constellations:

| First visibilities | Dates | |
|---|---|---|
| | Ideal | Real |
| Stars in the morning | II 1 | III 1 |
| Arrow in the morning | IV 15 | V 15 |
| ŠU.PA in the morning | VI 15 | VII 15 |
| Arrow in the evening | X 15 | XI 15 |
| Fish and Old Man in the morning | XII 15 | I 15 |

When the second lunar phenomenon is included, there is one phenomenon for each pair of months. Failure to observe one phenomenon because of bad weather or adverse circumstances can be remedied within two months.

### *3.4. Shadow Table and Water Clock*

The shadow table in MUL.APIN (II ii 21-40) records a schematic sequence of lengths of a shadow in integer numbers of cubits at times since sunrise measured in *bēru* on the days of the equinoxes and solstices.

| Shadow | Summer Solstice | Equinox | Winter Solstice |
|---|---|---|---|
| 1 | 2 *bēru* | 2;30 *bēru* | 3 *bēru* |
| 2 | 1 *bēru* | 1;15 *bēru* | 1;30 *bēru* |
| 3 | 0;40 *bēru* | 0;50 *bēru* | 1 *bēru* |
| 4 | 0;30 *bēru* | | 0;45 *bēru* |
| 5 | 0;24 *bēru* | | 0;36 *bēru* |
| 6 | 0;20 *bēru* | | 0;30 *bēru* |
| 8 | 0;15 *bēru* | | 0;22,30 *bēru* |
| 9 | 0;13,20 *bēru* | | 0;20 *bēru* |
| 10 | 0;12 *bēru* | | 0;18 *bēru* |

As Neugebauer [1975], vol. 1, pp. 544-545 noted, each of these columns is like a table of reciprocals, that is, the number of *bēru* in each column of times since sunrise multiplied by the number of cubits in the corresponding shadow always equals 2 *bēru* for the summer solstice, 2;30 *bēru* for an equinox, and 3 *bēru* for the winter solstice. This circumstance explains why there is no entry for shadows whose length is 7 cubits; for neither $\frac{2}{7}$ nor $\frac{3}{7}$ *bēru* can be expressed with a finite sexagesimal fraction. A consequence of this reciprocal structure is that the ratio of the entries in the winter solstice column to those in the summer solstice column is 3 : 2; and, with those two columns as maximum and minimum and the column for the equinoxes as the mean, we have simple linear zigzag functions, at least for shadows of lengths 1, 2, and 3 cubits.

As Neugebauer also points out, a shadow of 1 cubit is reached 2 *bēru* after sunrise in the column for summer solstice, 3 *bēru* in that for winter solstice; however, the inverse should hold true. Indeed, if the ratio of longest to shortest day were 2 : 1 as elsewhere in MUL.APIN, the winter solstice day is only 4 *bēru* long, so that the shadow would keep diminishing till mid-afternoon; if it were 3 : 2, the winter solstice day would be 4;48 *bēru* long and the shadow would still keep diminishing till 0;36 *bēru* after noon. Therefore, the column labeled as for summer solstice is that for winter solstice and vice versa. For this reason and others the solution offered in Bremner [1993] is unacceptable, and we must regard the table as based on mathematical manipulation rather than on observation. Equally Gleßmer [1996], who, without knowledge of Al-Rawi-George [1991/92] pp. 59-60, believes that 18 mina of water measure a nychthemeron rather than 6, can not be accepted. But Neugebauer also shows that, using the reciprocal scheme and finding the time after sunrise at which a shadow of 0;50 cubits occurs, we have

$$0;50 = \frac{3}{3;36} = \frac{2;30}{3} = \frac{2}{2;24}$$

The ratio of the longest to shortest day, then, is 3;36 : 2;24 = 3 : 2—the earliest attested occurrence of this parameter in a cuneiform source. Since the longest day is 7;12 *bēru* and the shortest 4;48 *bēru* when the ratio is 3 : 2, this implies that the noon shadow is 0;50 cubits on every day of the year —a clear impossibility! Note that in BM 29371 (see Britton-Walker [1996] p. 47) a shadow of 1 cubit is stated to occur 1;40 *bēru* after sunrise on every day of the year, so that this type of error is not unknown in Mesopotamia. Again, though the ratio 3 : 2 underlies the scheme in MUL.APIN, the scheme itself is completely artificial.

But we also learn from this table, as from LBAT 1494 and 1495 and elsewhere, that the Babylonians employed gnomons to tell the time of day. From Mesopotamia the use of the gnomon spread to the Greeks, among whom either Anaximander or his pupil Anaximenes is alleged to have been the first to set a gnomon up. The Greek indebtedness to Mesopotamia is acknowledged by Herodotus (II 109, 3): πόλον μὲν γὰρ καὶ γνώμονα καὶ τὰ δυώδεκα μέρεα τῆς ἡμέρης παρὰ Βαβυλωνίων ἔμαθον οἱ Ἕλληνες. In India the gnomon or śaṅku of 12 digits appears first in texts that can be dated to the first few centuries of the Christian era: the *Arthaśāstra* ascribed to Kauṭilya (II 20, 39-42) and the *Vasiṣṭhasiddhānta* as reported in Varāhamihira's *Pañcasiddhāntikā* (2, 9-10); in these texts a Mesopotamian-style linear zigzag function is used to generate the noon shadow when the Sun is in each zodiacal sign, or, equivalently, for every month. Such a scheme is found in a tablet from Uruk (von Weiher [1993] p. 191 no. 172 col. ix bottom), where the parameters seem to be:

M = 1,45
m = 15
μ = 1,0
d = 15
P = 12

The maximum occurs at the winter solstice, the minimum at the summer solstice. The parameters in the Indian tables are:

M = 12 digits
m = 0 digits
μ = 6 digits
d = 2 digits
P = 12

The choice of 0 for the noon shadow on the day of the summer solstice places the latitude of the locality at about 24° (e.g., Ujjayinī), which puts the origin of this particular scheme no earlier than the middle of the second century A.D., when Ujjayinī became a center for Greco-Indian astronomy; but the ratio of the longest to shortest day remains 3 : 2, which is appropriate for a latitude of about 35°. This discrepancy shows that the ratio was not tested in India, and probably was not generated in India, but was borrowed from the Mesopotamian tradition along with the linear zigzag function. Since Megasthenes and Daimachus, the Seleucid ambassadors to the Mauryan court in the third century B.C., and other Greeks such as Baeton are known to have made gnomon observations in India (see Pingree [1963] 232) we cannot be sure from which source, the Achaemenids or the Greeks (both of whom learned to use gnomons from Mesopotamia), the Indians developed their use of this instrument.

With the shadow table in MUL.APIN, which we have shown to be based on a ratio of 3 : 2 of the longest to the shortest day of the year, is incongruously combined the usual scheme of weights of water in a water-clock based on the ratio of 2 : 1 (4 minas for the day of the summer solstice and 2 for the day of the winter solstice). That Neugebauer's attempt [1947] to explain that the prescribed weights of water would flow out of the water-clock in the ratio of 3 : 2 is wrong has been demonstrated by Høyrup [1997/1998]. This collocation of contradictory parameters serves to emphasize the unreality of the shadow table.

In the ideal calendar, wherein the time between successive colures is a constant 90 days, and the difference between the weights of water is 1 mina, the difference for one day is $\frac{1}{90}$ = 0;0,40 minas; this is the number "for the difference for daytime and nighttime" given in II ii 41-42. This number is to be multiplied by 0;7,30—i.e., divided by 8—to produce 0;0,5 minas, which is "the difference for 1 cubit of shadow." The rationale for this procedure is the following. If the ratio of the longest to the shortest day is 2 : 1, then 1 mina of water measures 2 *bēru* of time (4 minas measure 8 *bēru* and 2 minas measure 4 *bēru*). The difference between shadows of 2 cubits and of 1 cubit in length equals 1;30 *bēru* on the day of the summer solstice, 1;15 *bēru* on the days of the equinoxes, and 1 *bēru* on the day of the winter solstice; these times are measured by 0;45 minas, 0;37,30 minas, and 0;30 minas respectively. Therefore, the difference for 90 days is 0;7,30 minas, and the difference for one day is 0;0,5 minas.

Following the shadow table in MUL.APIN is a table similar to table D of Tablet 14 of *Enūma Anu Enlil*, save that the colures are located at I 15, IV 15, VII 15, and X 15, instead of XII 15, III 15, VI 15, and IX 15 (II ii 43 - iii 12); MUL.APIN also adds the minas of water that measure each night.

| Date | Nighttime | Sunset to Moonset | Date | Nighttime | Sunset to Moonrise |
|---|---|---|---|---|---|
| I 1 | 3;10 minas | 12;40 UŠ | I 15 | 3 minas | 12 UŠ |
| II 1 | 2;50 | 11;20 | II 15 | 2;40 | 10;40 |
| III 1 | 2;30 | 10 | III 15 | 2;20 | 9;20 |
| IV 1 | 2;10 | 8;40 | IV 15 | 2 | 8 |
| V 1 | 2;10 | [8;40] | V 15 | 2;20 | 9;[20] |
| VI 1 | 2;30 | 10 | VI 15 | 2;40 | 10;40 |
| VII 1 | 2;50 | 11;20 | VII 15 | 3 | 12 |
| VIII 1 | 3;10 | 12;40 | VIII 15 | 3;20 | 13;20 |
| IX 1 | 3;30 | 14 | IX 15 | 3;40 | 14;40 |
| X 1 | 3;50 | 15;20 | X 15 | 4 | 16 |
| XI 1 | 3;50 | 15;20 | XI 15 | 3;40 | 14;40 |
| XII 1 | 3;30 | 14 | XII 15 | 3;20 | 13;20 |

As the three lines that follow this table indicate, the entries for the periods from sunset to either moonset or moonrise equal the corresponding lengths of the night multiplied by 4—i.e., divided by 15. Moreover, since the daily difference of nighttime is 0;0,40 minas, the daily difference between sunset and either moonset or moonrise is 0;0,40 × 4 = 0;2,40 UŠ.

The activities of someone steeped in the astral lore of MUL.APIN and of *Enūma Anu Enlil* are accurately described by Apuleius in *Florida* 18 (31), where he is speaking of Thales: maximas res parvis lineis repperit: temporum ambitus, ventorum flatus, stellarum meatus, tonitruum sonora miracula, siderum obliqua curricula, solis annua reverticula, i<ti>dem lunae vel nascentis incrementa vel senescentis dispendia vel deli<n>quentis obstacula. Similarly, an inscription in Hieroglyphs on a statuette found at Tell Faraoun to the West of the present Suez Canal records the activities of an Egyptian astronomer, Harkhebi; the object is dated to the early third century B.C.

(Neugebauer-Parker [1969], vol. 1, pp. 214-216): "Hereditary prince and count, sole companion, wise in the sacred writings, who observes everything observable in heaven and earth, clear-eyed in observing the stars, among which there is no erring; who announces rising and setting at their times, with the gods who foretell the future, for which he purified himself in their days when (the decan) Akh rose heliacally beside Venus from earth, and he contented the lands with his utterances; who observes the culmination of every star in the sky, who knows the heliacal risings of every . . . in a good year, and who foretells the heliacal rising of Sothis (Sirius) at the beginning of the year. He observes her (Sothis) on the day of her first festival, knowledgeable in her course at the times of designating therein, observing what she does daily, all she has foretold is in his charge; knowing the northing and southing of the sun, announcing all its wonders (omina?) and appointing for them a ⌜time⌝, he declares when they have occurred, coming at their times; who divides the hours for the two times (day and night) without going into error at night . . . ; knowledgeable in everything which is seen in the sky, for which he has waited, skilled with respect to their conjunction(s) and their regular movement(s); who does not disclose (anything) at all concerning his report after judgement, discreet with all he has seen." Not only are the activities reminiscent of those referred to in MUL.APIN, but the decan Akh is in Pisces, the *ašar niṣirti* of Venus.

### *4. I.NAM.GIŠ.ḪUR.AN.KI.A*

The earliest publication of part of this curious text is in Weidner [1912c], which discusses K 2164+2195+3510; this tablet was discussed again in Kugler [1913] pp. 88-106. The matter was taken up again in Weidner [1914b] 82-91. It is next mentioned in Schott [1934], where it is regarded as a source for MUL.APIN. The text is included in the survey in van der Waerden [1950] 299-300. All that has been identified of I.NAM.GIŠ.ḪUR.AN.KI.A has now been published in Livingstone [1986] pp. 17-49; see also Al-Rawi-George [1991/92].

The "first" Tablet of I.NAM.GIŠ.ḪUR.AN.KI.A, preserved on BM 47860, which was copied by Bēl-nādin-apli in -487, is concerned with the volume of storehouses and the temple, Ekur, of Enlil at Nippur. The second section, K 2164+2195+3510, and two fragments of the third, K 2670 and K 170+Rm 520, were all copied by the well-known scribe, Nabû-zuqup-kēnu, who was active during the reigns of Sargon and Sennacherib, in the decades immediately before and after -700. It is the second part that is of interest to us.

The obverse begins: "The Moon on day 1 3;45," where 3;45 is the number of UŠ that the moon is visible on the first day of an equinoctial month according to table A of Tablet 14 of *Enūma Anu Enlil*, a table which, like the Ekur of section 1 of I.NAM.GIŠ.ḪUR.AN.KI.A, is associated with

Nippur. There follow various mathematical speculations on the Moon on various other days of the month. But the reverse presents exactly the same data as are found in MUL.APIN II ii 44 - iii 12 except that, whereas MUL.APIN indicated that the periods of time in UŠ between sunset and either moonset or moonrise is 4 times the minas of water that measure the length of the night, I.NAM.GIŠ.ḪUR.AN.KI.A includes the statement within each line that x (minas) times 4 = y (UŠ).

### *5. Ziqpu Star Texts*

Two lists of *ziqpu* stars, as we have seen, are preserved in MUL.APIN I iv 1-30. The later *ziqpu* star lists have a different structure. The important tablet AO 6478, copied at Uruk in ca. -200, was first investigated in Kugler [1913] pp. 88-106; a full edition and translation of this tablet can be found in Thureau-Dangin [1913], with an astronomical commentary in Kugler [1914]. A further discussion of this text is offered in Schaumberger [1935] 353-354, and a final version with the parallel *ziqpu* star lists of K 9794, VAT 16437 (copied in the Seleucid period; from Babylon), and VAT 16436 (from Uruk) in Schaumberger [1952]. More recently Horowitz [1994] has added to the editions of such texts an edition of the variant BM 38369+38694, written in Neo-Babylonian script and found at Babylon. Clearly the existence of a copy of the original version from Assurbanipal's library, K 9794, guarantees that the original was already written by the early seventh century B.C.; Schaumberger [1952] 224 notes the suggestive fact that the right ascension of the first star in the list, ŠU.PA (= $\alpha$ Boötis), was 182;48° in -700, exactly 180° in -933; the date of the *ziqpu* star list most probably lies in the first three centuries of the last millennium B.C.

The accompanying table is based on AO 6478, but adopting the readings and star identifications of Schaumberger. Note that AO 6478 shares with List Va of MUL.APIN, which also begins with ŠU.PA, the *ziqpu* stars numbered I, III, VII, IX, XIV, XVI, XVIII, XX, XXI, and XXV, on the assumption that the identifications are correct; while on the same assumption it shares with List Vb, which begins with the Shoulder of the Panther (= $\gamma$ Cygni), the *ziqpu* stars numbered VIII (identified with $\beta$ Cygni by Schaumberger), IX, X, XI (all named after the same anatomical parts of the Panther), XIII and XIV (in reverse order), XVIII/XIX, XXI - XXIV, XXV, I, and VII. Clearly the several lists are closely connected, but the same names do not always designate the same stars.

| No. | Weight in mina | UŠ | Star from which | *bēru* | Star to which |
|---|---|---|---|---|---|
| I | 1 ½ | 9 | Yoke | 16,200 | Rear Harness[44] |
| II | 2 | 12 | Rear Harness | 21,600 | Circle |
| III | 2 ½ | 15 | Circle | 27,000 | Star from the Doublets |
| IV | 5/6 | 5 | Star from the Doublets | 9,000 | Star from the Triplets |
| V | 1 2/3 | 10 | Star from the Triplets | 18,000 | Single Star |
| VI | 1 2/3 | 10 | Single Star | 18,000 | Lady of Life |
| VII | 3 1/3 | 20 | Lady of Life | 36,000 | Shoulder of Panther |
| VIII | 1 2/3 | 10 | Shoulder of Panther | 18,000 | Bright Star of its Breast |
| IX | 3 1/3 | 20 | Bright Star of its Breast | 36,000 | Knee |
| X | 3 1/3 | 20 | Knee | 36,000 | Heel |
| XI | 1 2/3 | 10 | Heel | 18,000 | 4 Stars of the Stag |
| XII | 2 ½ | 15 | 4 Stars of the Stag | 27,000 | Dusky Stars |
| XIII | 2 ½ | 15 | Dusky Stars | 27,000 | Bright Star of Old Man |
| XIV | 1 2/3 | 10 | Bright Star of Old Man | 18,000 | *naṣrapu* |
| XV | 2 ½ | 15 | *naṣrapu* | 27,000 | Crook |
| XVI | 1 2/3 | 10 | Crook | 18,000 | Hand of Crook |
| XVII | 5 | 30 | Hand of Crook | 54,000 | Twins |
| XVIII | 5/6 | 5 | Twins | 9,000 | Rear Twin |
| XIX | 3 1/3 | 20 | Rear Twin | 36,000 | Crab |
| XX | 3 1/3 | 20 | Crab | 36,000 | 2 Stars from Head of Lion |
| XXI | 1 2/3 | 10 | 2 Stars from Head of Lion | 18,000 | 4 Stars from his Breast |
| XXII | 3 1/3 | 20 | 4 Stars from his Breast | 36,000 | 2 Stars from his Thigh |
| XXIII | 1 2/3 | 10 | 2 Stars from his Thigh | 18,000 | Single Star from his Tail |
| XXIV | 1 2/3 | 10 | Single Star from his Tail | 18,000 | Frond |
| XXV | 4 1/6 | 25 | Frond | 45,000 | Harness |
| XXVI | 1 1/3 | 8 | Harness | 14,400 | Yoke |
| Sum: | 60 2/3 | 364 | | 655,200 | |

The weights (of water) regulate a water-clock in which 1 mina of water measures 0;1 days in contrast to the older tradition which used a water-clock in which 1 mina measured 0;10 days. A sixtieth of a day is, of course, a ghaṭikā or nāḍikā in Indian astronomy; this unit of time is first encountered in India in the *Jyotiṣavedāṅga* of Lagadha, which we have already seen to have been strongly influenced by Mesopotamian astronomy. The weights are expressed in terms of 5/6, (1 1/3, 1 ½), 1 2/3, (2), 2 ½, 3 1/3, 4 1/6, and 5 minas = 5/6 × 1, 5/6 × 2, 5/6 × 3, 5/6 × 4, 5/6 × 5, and 5/6 × 6, with 1 1/3,

[44] This translates *nattullu*, sometimes written ŠUDUN.ANŠE, which is part of a harness according to CAD N s.v.

1 ½, and 2 excepted. This is because 60 ghaṭikās = 360 UŠ so that 1 ghaṭikā = 6 UŠ, and 5 UŠ = 5/6 ghaṭikās. The introduction into this simple scheme of 1 1/3 (= 8 UŠ), 1 ½ (= 9 UŠ), and 2 (= 12 UŠ) means that instead of the nychthemeron equaling 60 ghaṭikās it appears to equal 60 2/3 ghaṭikās; 1 1/3 + 1 ½ + 2 = 4 5/6, which is 5/6 × 5 + 2/3. This shows that the primary data were the UŠ, usually taken to be multiples of 5 but in three cases—for numbers I, II, and XXVI, at the beginning and end of the table—taken to be other than 5. Note that the first three entries for UŠ—9, 12, 15—form an arithmetical progression. This has led to a nychthemeron measured by 364° and 60 2/3 ($=\frac{364}{6}$) ghaṭikās. We use the UŠ to determine the right ascensions of the stars as follows.

| No. | UŠ | RA | Star | RA | δ |
|---|---|---|---|---|---|
| I | | 180° | α Boötis | 182;40° | +33;22° |
| II | 9 | 189° | ε Boötis | 191;12° | +40;27° |
| III | 12 | 201° | α Cor. Bor. | 205;6° | +38;17° |
| IV | 15 | 216° | β Herculis | 218;58° | +30;27° |
| V | 5 | 221° | δ Herculis | 231;29° | +31;10° |
| VI | 10 | 231° | μ Herculis | 240;38° | +31;58° (ϑ Herculis 246;22°,+40;28°) |
| VII | 10 | 241° | α Lyrae | 256;42° | +39;21° |
| VIII | 20 | 261° | β Cygni | 265;38° | +25;36° (γ Cygni 281;31°, +34;17°) |
| IX | 10 | 271° | α Cygni | 287;32° | +38;1° |
| X | 20 | 291° | α Lacertae | 311;36° | +38;5° |
| XI | 20 | 311° | λ Androm. | 324;7° | +32;32° |
| XII | 10 | 321° | α Cassiop. | 337;28° | +41;44° |
| XIII | 15 | 336° | h Persei | 355;38° | +42;45° |
| XIV | 15 | 351° | α Persei | 9;25° | +36;57° (η Persei 1;40°, +42;5°) |
| XV | 10 | 1° | b,c Persei | 18;54° | +36;38° |
| XVI | 15 | 16° | α Aurigae | 33;25° | +37;39° |
| XVII | 10 | 26° | ϑ Aurigae | 45;53° | +31;32° |
| XVIII | 30 | 56° | α Gemin. | 69;53° | +32;21° |
| XIX | 5 | 61° | β Gemin. | 74;10° | +29;27° |
| XX | 20 | 81° | ε Cancri | 89;55° | +24;40° |
| XXI | 20 | 101° | ε Leonis | 105;29° | +32;34° |
| | | | μ Leonis | 106;51° | +35;8° |
| XXII | 10 | 111° | η Leonis | 112;54° | +26;49° |
| | | | ζ Leonis | 113;46° | +33;47° |
| | | | α Leonis | 114;25° | +22;13° |

| | | | | | |
|---|---|---|---|---|---|
| XXIII | 20 | 131° | δ Leonis | 130° | +33;20° |
| | | | ϑ Leonis | 131;5° | +28;19° |
| XXIV | 10 | 141° | β Leonis | 141;3° | +28;31° |
| XXV | 10 | 151° | γ Com. Ber. | 150;49° | +42;52° |
| XXVI | 25 | 176° | η Boötis | 175;47° | +33;29° |
| I | 8 | 184° | α Boötis | 182;40° | +33;22° |

For number V one would expect ζ Herculis (RA = 225;19°, δ = +39;33°), and the entries for V to XVI should be lowered by one. The list is seriously off for XVIII and XIX, but becomes correct again for the rest. The corrected form of the table, with just right ascensions, is:

| No. | UŠ | RA | Star | RA | Δ RA |
|---|---|---|---|---|---|
| I | | 180° | α Boötis | 182;40° | |
| II | 9 | 189° | ε Boötis | 191;12° | 8;32° |
| III | 12 | 201° | α Cor. Bor. | 205;6° | 13;54° |
| IV | 15 | 216° | β Herculis | 218;58° | 13;52° |
| | 5 } 15 | 221° | <ζ Herculis> | 225;19° | 6;21° } 12;31° |
| V | 10 } | 231° | δ Herculis | 231;29° | 6;10° } |
| VI | 10 | 241° | μ Herculis | 240;38° | 9;9° |
| VII | 20 | 261° | α Lyrae | 256;42° | 16;4° |
| VIII | 10 | 271° | β Cygni | 265;38° | 9;56° |
| IX | 20 | 291° | α Cygni | 287;32° | 21;54° |
| X | 20 | 311° | α Lacertae | 311;36° | 24;4° |
| XI | 10 | 321° | λ Androm. | 324;7° | 12;31° |
| XII | 15 | 336° | α Cassiop. | 337;28° | 13;21° |
| XIII | 15 | 351° | h Persei | 355;38° | 18;10° |
| XIV | 10 | 1° | η Persei | 1;40° | 6;2° |
| XV | 15 | 16° | b,c Persei | 18;54° | 17;14° |
| XVI | 10 | 26° | α Aurigae | 33;25° | 14;31° |
| XVII | } 55 | | ϑ Aurigae | 45;53° | (12;28°) } 36;28° } 56;30° |
| XVIII | 30 } | 56° | α Gemin. | 69;53° | 24° } |
| XIX | 5 } | 61° | β Gemin. | 74;10° | 4;17° } 20;2° } |
| XX | 20 } | 81° | ε Cancri | 89;55° | 15;45° } } |
| XXI | 20 | 101° | ε Leonis | 105;29° | 15;34° |
| XXII | 10 | 111° | α Leonis | 114;25° | 8;56° |
| XXIII | 20 | 131° | ϑ Leonis | 131;5° | 16;40° |
| XXIV | 10 | 141° | β Leonis | 141;3° | 9;58° |
| XXV | 10 | 151° | γ Com. Ber. | 150;49° | 9;46° |
| XXVI | 25 | 176° | η Boötis | 175;47° | 24;58° |
| I | 8 | 184° | α Boötis | 182;40° | 6;53° |

When we compare this with our analysis of List Vb of MUL.APIN, we note deviations for numbers IX (γ Cygni), XV (α Persei), and VII (ϑ Herculis). Certainty in these identifications is, of course, unattainable.

The *bēru* or distances between the stars are computed on the assumption that 1 UŠ = 1800 = 30,0 *bēru*; in time measures, of course, 30 UŠ = 1 *bēru*. The total circumference of the circle of *ziqpu* stars should then be 6,0 × 30,0 = 3,0,0,0 = 648,000 *bēru*; the difference between the text's number and this, 7,200, equals 1800 × 4. It was reasonably suggested in Kugler [1914] that the *bēru* are computed on the assumption that the diameter of the Moon (or Sun) equals 1000 *bēru*; then the diameters measure 360 UŠ : 648 *bēru* = 0;33,20°.

VAT 16436, published in Schaumberger [1952] 226-227, lists the *ziqpu* stars, beginning with "the Bright Star from its Breast" (which cannot be the original beginning since there is nothing for "its" to refer to), with the number of stars in each group and a number of cubits with, occasionally, an associated deity; numbers XVIII and XIX (α and β Geminorum) are combined into one. The cubits add up to 150,400; we don't know what they signify.

| Star | God | Number | Cubits | Stars |
|---|---|---|---|---|
| IX | | 1 | 5,760 | α Cygni |
| X | | 3 | 6,780 | α + β Lacertae + |
| XI | | 1 | 8,200 | λ Andromedae |
| XII | | 4 | 7,200 | α, β, γ, δ Cassiopeiae |
| XIII | | 5 | 8,200 | h + χ Persei + |
| XIV | | 1 | 7,200 | η Persei |
| XV | | 6 | 8,200 | b + c Persei + |
| XVI | $^{d}$BE | 1 | 14,400 | α Aurigae |
| XVII | $^{d}$DAM.KI.NUN.NA | 2 | 10,800 | ϑ + υ Aurigae |
| XVIII, XIX | Nabû and Nergal | 2 | 10,600 | α + β Geminorum |
| XX | | 10 | 14,600 | ε Cancri + (Praesepe) |
| XXI | | 2 | 14,600 | ε + μ Leonis |
| XXII | | 4 | 7,200 | α, γ, ζ, η Leonis |
| XXIII | | 2 | 21,600 | δ + ϑ Leonis |
| XXIV | | 1 | 7,200 | β Leonis |
| XXV | | 6 | 7,200 | γ + 13,14,16,17,18 Comae Ber. |
| II (for XXVI) | Enlil | 2 | 14,600 | η + υ Boötis |
| I | | 1 | 7,200 | α Boötis |
| XXVI (for II) | | 3 | 13,400 | ξ, ο, π Boötis |
| III | | 9 | 7,200 | α, β, γ, δ, ε, ϑ, ι, π, ρ Coronae Bor. |
| IV 2 ½ | | 3 | 18,000 | β + γ Herculis + |

| | | | |
|---|---|---|---|
| V | 2 | 13,260 | α + δ Herculis |
| VI 1 ½ | 2 | 10,600 | μ (or ϑ) Herculis + |
| VII | 3 | 3,600 | α, ε, ζ Lyrae |
| VIII | 2 | 10,800 | β + φ Cygni |

BM 38369+38694, published in Horowitz [1994] 92-93, gives an abbreviated list of *ziqpu* stars, with a total of 12 *bēru* (= 360 UŠ) in the "circle of the *ziqpu*".

| Stars from which | Stars to which | *bēru* |
|---|---|---|
| <ŠU.PA> | <She-goat> | <2 ½> |
| She-goat | Panther | <1> |
| Bright Star of Panther | Its Knee | 2/3 |
| Its Knee | Its Heel | ⌜2/3⌝ |
| Its Heel | Horn of the Stag | <½> |
| Horn of the Stag | Bright Star of Old Man | <5/6> |
| Old Man | Crook | <5/6> |
| Crook | Great Twins | <1> |
| Between the Great Twins | | 5 UŠ (= 1/6) |
| Great Twins | Crab | <2/3> |
| <Crab> | <Head of Lion> | <2/3> |
| Head of Lion | Breast of Lion | 10 UŠ (= 1/3) |
| Breast of Lion | 2 Stars of its Thigh | <2/3> |
| 2 Stars of its Thigh | Cluster of its Tail | ½ |
| Cluster of its Tail | ŠU.PA | 1 |

The remainder of the text refers to MUL.APIN: line 22 to list Va (I iv 4-6); lines 23-24 to I iv 2-3; 25-28 quote I iii 49-50; and 29 refers to the beginning of list Vb (I iv 10).

The purpose of the *ziqpu* star lists must have been to determine time intervals at night by observing their crossings of the meridian; see also II A 6. This explains why column I gives the weights of water in the water-clock that measure the intervals between (the culminations of) the *ziqpu* stars. This interpretation is confirmed by a letter from the time of Sargon II (Lanfranchi-Parpola [1990] 249:12'-15') in which of a storm at night it is said: "it started at (the culmination of) the Circle and subsided at (the culmination of) the Triplets." From *ziqpu* star III to V is an interval of 3 1/3 ghaṭikās or 1½ hours. Cf. also Parpola [1993] pp. 113-114 (Letter 149), wherein the time of a lunar eclipse in the morning of 14 Simānu is indicated by the culmination of the Shoulder of the Panther. The use of *ziqpu* stars to indicate the times of lunar eclipses and other phenomena in the Diaries will be described in detail below.

For texts relating the oblique ascensions of the zodiacal signs to *ziqpu* intervals see II C 1.1d below.

### *6. The GU Text*

This text, preserved on the tablet BM 78161, was inscribed in the period between the seventh and the fifth centuries B.C., and was probably found in the area of Babylon or Sippar, though its precise provenance has not been recorded. It was published with a translation and interpretation by Pingree-Walker [1988]; a different interpretation was offered in Koch [1992]. We shall describe the text and the two interpretations in sequence.

The text, inscribed on both sides of the tablet which is complete at top and bottom, is divided into twenty sections by the Sumerian word GU, meaning “string”, which concludes each section. All but the first and last sections contain (a) either the name of a *ziqpu* star or a time interval; (b) one or more star names; and (c) the word GU. The first section contains (b) and (c), the last (a) and (b), so that the tablet begins and ends in the middle of a list of strings given in a longer text. Furthermore, the *ziqpu* stars mentioned are numbers XVIII to XXVI and I to VII of the list in AO 6478 given above (in three sections time intervals are given). Since these *ziqpu* stars are listed in the same order as in AO 6478, it is clear that the “strings”, at least at the top, are in the order of meridian transits. It is also probable that strings headed by *ziqpu* stars VIII to XVII (and perhaps other strings headed by time intervals) were also included in the original; *ziqpu* star XVII would have been (a) in our text’s first string. The most likely number of strings is thirty, so that we are missing a third of the original.

Noting the usual relation of the first element of each “string” to the *ziqpu* star list and the use of time-intervals in place of missing *ziqpu* stars (the objective seems to have been to have differences in right ascension of about 10°), Pingree and Walker concluded that the stars in the “strings” were supposed to cross the local meridian simultaneously. This would mean that the stars in each “string” have essentially the same right ascension, but within each “string” the declinations should be successively lower. This scheme works quite nicely for the first nine “strings”, but thereafter there is a dislocation of the *ziqpu* stars and their “strings”. We believe this to be due to the inept use by the compiler of the GU text of two sources: the *ziqpu* star list, and lists of “strings” of stars that simultaneously culminate. This solution, revised in the light of the new identifications of *ziqpu* stars, Koch’s criticisms, and other considerations, is presented in the following table, with the right ascensions and declinations computed for -700; the letters A to T refer to the twenty “strings” of the text, while the Roman numerals I to XXI refer to the reconstructed “strings”.

| "Strings" | Star name | Identification | RA | δ |
|---|---|---|---|---|
| I (47°) | [Hand of Crook (*ziqpu* XVII) | ϑAurigae | 45;53° | +31;32°] |
| | A1. True Shepherd of Anu | β Orionis | 46;52° | -15;4° |
| | A2. Rooster | μ Leporis | 48;27° | -22;56° |
| II (56°) | B1. Feet and hands of the front | ϑ Gem. (hand) | 59;8° | +31;38° |
| | Great Twin | μ Gem. (foot) | 55;43° | +18;48° |
| | B2. Right hand and rear heel of | ξ Orionis | 56;53° | +8;29° |
| | True Shepherd of Anu | κ Orionis | 55;28° | -14;31° |
| | B3. Middle of Rooster | α Leporis | 53;55° | -23;20° |

There is no *ziqpu* star whose RA is 56°; between "strings" II and III is *ziqpu* XVIII:

| | | | | |
|---|---|---|---|---|
| | [Front star of Great Twins | α Geminorum | 69;53° | +32;21°] |
| III (74°) | C1. Rear star of Great Twins (*ziqpu* XIX) | β Geminorum | 74;10° | +29;27° |
| | C2. Little Twins | λ Geminorum | 70;37° | +16;32° |
| | C3. Arrow | α Canis Maioris | 71;48° | -17;34° |
| IV (86°) | D1. Crab (*ziqpu* XX) | ε Cancri | 89;55° | +24;40° |
| | D2. Snake | β Cancri | 86;49° | +13;10° |
| | D3. Bow | η Canis Maioris | 84;31° | -27;17° |
| | D4. Left foot of constellation of Arrow | π Puppis | 85;38° | -35;10° |
| V (95°) | E1. 5 UŠ after Crab | | ca. 95° | |
| | E2. Jupiter which stands behind Crab in front of Lion | | | |
| | E3. Constellation of Arrow | ρ Puppis | 93;14° | -19;48° |
| VI (104°) | F1. Two stars of the head of | μ Leonis | 106;50° | +35;8° |
| | Lion (*ziqpu* XXI) | ε Leonis | 105;30° | +32;34° |
| | F2. Middle of Snake | ϑ Hydrae | 102;37° | +9;54° |
| | F3. Elbow of Arrow | α Pyxis | 104;1° | -26;19° |
| | F4. Hands of Eridu | γ Velorum | 101;35° | -41;43° |
| VII (113°) | G1. Four stars of the chest of | ζ Leonis | 113;46° | +33;47° |
| | Lion (*ziqpu* XXII) | γ Leonis | 115;17° | +30;25° |
| | | η Leonis | 112;54° | +26;49° |
| | | α Leonis | 114;26° | +22;12° |
| | G2. Right front foot of Lion | π Leonis | 112;59° | +17;55° |
| | G3. Middle of Snake | κ Hydrae | 112;35° | -4;59° |

| | | | | |
|---|---|---|---|---|
| | G4. Eridu | λ Velorum | 112;33° | -34;54° |
| VIII (123°) | H1. ½ *bēru* after the four of its chest (1/3 *bēru* better) | | ca. 128° (ca. 123°) | |
| | H2. Foot in the middle of Lion | ρ Leonis | 121;7° | +20;40° |
| | H3. Hand of Ninmaḫ | φ Velorum | 126;20° | -43;29° |
| IX (132°) | I1. Two stars of the rump of Lion (*ziqpu* XXIII) | δ Leonis<br>ϑ Leonis | 130;1°<br>131;4° | +33;20°<br>+28;20° |
| | I2. Tail of Raven | α Crateris | 132;26° | -5;33° |
| | I3. Bite of Harrow | μ Velorum | 134;34° | -36;45° |

(At this point the compiler did not find a set of simultaneously culminating stars in his second source so that the *ziqpu* star of "string" XI is the first star of "string" K)

| | | | | |
|---|---|---|---|---|
| X (141°) | J1. Single star in the tail of Lion (*ziqpu* XXIV) | β Leonis | 141;2° | +28;31° |
| XI (150°) | J2. Middle of Raven | γ Corvi | 150;11° | -3;2° |
| | J3. Hand of Ḫabaṣirānu | δ Centauri | 150;57° | -36;13° |
| | K1. Erua (*ziqpu* XXV) | γ Comae Beren. | 150;8° | +42;51° |
| XII (167°) | K2. Barley-stalk | α Virginis | 166;46° | +3;42° |
| | K3. Šullat and Ḫaniš | ε Centauri | 168;41° | -38;41° |
| | K4. Mercury which stands with Barley-stalk in front of Raven | | ca. 166° | |
| | L1. ½ danna after Erua | | ca. 165° | |

(The next "string" should have had a RA of 176°; the GU text has only the *ziqpu* star, at the beginning of "string" M. Note that the *ziqpu* star of "string" XIII has been displaced to the beginning of "string" N, that of "string" XIV to the beginning of "string" O.)

| | | | | |
|---|---|---|---|---|
| (176°) | M1. Harness (*ziqpu* XXVI) | η Boötis | 175;47° | +24;58° |
| XIII (185°) | L2. Sitting Gods | μ Virginis | 186;19° | +8;1° |
| | L3. Front pan of Scales | α Librae | 187;16° | -2;32° |
| | L4. Numušda | κ Centauri | 185;47° | -28;37° |
| | L5. Saturn which stands in front of Scales | | | |
| | N1. ŠU.PA (*ziqpu* I) | α Boötis | 182;50° | +33;36° |
| XIV (192°) | M2. Middle of Scales | β Librae | 194;24° | +3;17° |

| | | | | |
|---|---|---|---|---|
| | M3. [...] horn of Scorpion | γ Scorpii | 189;14° | -12;3° |
| | M4. Eye of Mad Dog | γ Lupi | 193;31° | -28;44° |
| | O1. Second star of Harness (*ziqpu* II) | ε Boötis | 191;12° | +40;27° |
| XV (202°) | N2. Middle of Scales | γ Librae | 197;59° | -2;41° |
| | N3. Stars of head of Scorpion | δ Scorpii | 202;34° | -11;23° |
| | N4. [Midd]le of Mad Dog | η Lupi | 199;26° | -26;52° |
| | P1. Circle (*ziqpu* III) | α Coronae Bor. | 205;6° | +38;17° |
| XVI (210°) | O2. Star of chest of Scorpion | α Scorpii | 208;30° | -16;25° |
| | O3. Upraised tail of Scorpion | $\mu^1$ Scorpii | 210;43° | -28;47° |
| | O4. Rear foot of Mad Dog | $\zeta^2$ Scorpii | 210;7° | -33;5° |

(There is no *ziqpu* star whose RA falls between 205° and 218°.)

| | | | | |
|---|---|---|---|---|
| XVII (220°) | P2. Eye of Zababa | η Ophiuchi | 220;22° | -8° |
| | P3. Šarur (and) Šargaz | λ Scorpii | 220;10° | -30° |
| | | υ Scorpii | 219;28° | -30;2° |
| | Q1. From the side (*ziqpu* IV) | β Herculis | 218;58° | +30;26° |
| XVIII (233°) | Q2. Middle of Zababa | ν Ophiuchi | 233;25° | -5;1° |
| | Q3. Left hand of Pabilsag which is on the bow | δ Sagittarii | 233;8° | -25;43° |
| | Q4. Bark | ε Sagittarii | 232;26° | -30;17° |
| | R1. Triplets (*ziqpu* V) | δ Herculis | 231;28° | +31;10° |
| XIX (241°) | R2. Shin of Zababa | η Serpentis | 240;54° | +0;11° |
| | R3. [Right hand of Pabi]lsag which is on the arrow | φ Sagittarii | 239;44° | -24;31° |
| | S1. Single star of the knee of She-goat (*ziqpu* VI) | μ Herculis | 240;36° | +31;58° |
| XX (250°) | S2. Foot of Zababa | λ Aquilae | 250;51° | -4;31° |
| | S3. Pabilsag above Bark | ι Sagittarii | 250;36° | -43;5° |

(There is no *ziqpu* star with a RA of 250°; the compiler of the GU text simply gives the next serial *ziqpu* star.)

| | | | | |
|---|---|---|---|---|
| (256°) | T1. Crook of She-goat (*ziqpu* VII) | α Lyrae | 256;44° | +39;20° |
| XXI (265°) | T2. Bright star of Eagle | α Aquilae | 264;43° | +6;1° |

("String" XXI may have continued:

| | | | |
|---|---|---|---|
| [Goat-Fish | β Capricorni | 266;22° | -18;50°] |
| [Shoulder of Panther (*ziqpu* VIII) | β Cygni | 265;38° | +25;36°]) |

Koch [1992] criticizes this interpretation because the width of the "strings" is sometimes quite large, because it must be assumed that the compiler was inept in putting together two sources, and because it leads to the identification of some star-names with different stars than he expects; see further Koch [1993]. Koch proposes instead that the "strings" connect stars that lie on the same altitude circles, and he turns to observational reports in the *Diaries* to substantiate his claim that the Babylonians recorded the distances of planets from fixed stars measured (according to his interpretation) on altitude circles. Koch computes the time of night (on 28/29 January and 28/29 May of -650), the azimuth, and the altitude of the stars that he had identified as having been referred to in the GU text; we record his results in the following table (asterisks indicate that the identifications according to the two interpretations are the same).

| "String" | Stars | Azimuth | Altitude | Time at Babylon 28 Jan. -650 |
|---|---|---|---|---|
| A | γ Orionis | 355;19° | +57;19° | 18;55$^h$ |
| | ι Leporis | 355;39° | +38;49° | |
| B | *μ Geminorum | 14;47° | +76;7° | 19;59$^h$ |
| | *ϑ Geminorum | 14;12° | +89;18° | |
| | ν Orionis | 13;25° | +67;35° | |
| | λ Eridani | 17;33° | +39;59° | |
| | 64 Eridani | 17;50° | +35;37° | |
| | 60 Eridani | 18;4° | +31;10° | |
| C | *β Geminorum | 0;14° | +87;3° | 20;58$^h$ |
| | β Canis Minoris | 358;8° | +66;44° | |
| | *α Canis Maioris | 3;36° | +40;47° | |
| D | *ε Cancri | 358;50° | +82;10° | 22$^h$ |
| | 1 Hydrae | 355;40° | +58;35° | |
| | 30 Monocerotis | 355;20° | +58;28° | |
| | 2 Hydrae | 354;56° | +58;26° | |
| | ρ Puppis | 356;7° | +37;34° | |

| | | | | |
|---|---|---|---|---|
| E | 5 uš behind the Crab | | | 22;19$^{h}$ |
| | Jupiter | | | |
| | 11 Canis Maioris | 29;2° | +37;7° | |
| | Pulk$_{55}$ 1121m/ | 27;58° | +36;55° | |
| | Paris 8170 | 28;55° | +36;44° | |
| F | *μ Leonis | 199;21° | +87;16° | 23;4$^{h}$ |
| | ζ Hydrae | 23;22° | +68;21° | |
| | π Puppis | 17;40° | +19;41° | |
| | ν Puppis | 19;40° | +10;13° | |
| G | *ζ Leonis | 203;22° | +88;43° | 23;34$^{h}$ |
| | ο Leonis | 23;47° | +74;56° | |
| | ϑ Hydrae | 26;46° | +65;19° | |
| | α Carinae | 18;46° | -0;3° | |
| | | | | 29 Jan -650 |
| H | Half a *bēru* behind the four of its chest | | | 0;33$^{h}$ |
| | *ρ Leonis | 29;21° | +76;29° | |
| | 196G. Puppis *Q* | 22;57° | +9;29° | |
| I | *δ Leonis | 149;35° | +89;8° | 0;42$^{h}$ |
| | Pictoris 140 | 328;5° | +38;58° | |
| | ε Centauri | 330;53° | +10;20° | |
| J | *β Leonis | 335;18° | +85;28° | 1;17$^{h}$ |
| | β Corvi | 337;20° | +45;59° | |
| | Centauri *M* | 334;46° | +14;44° | |
| K | *γ Comae Berenices | 159;38° | +79;2° | 2;22$^{h}$ |
| | *α Virginis | 337;35° | +59;4° | |
| | β Centauri | 348;7° | +9;59° | |
| | α Centauri | 342;8° | +6;4° | |
| | Mercury | | | |
| L | Half a *bēru* behind Erua | | | 3;22$^{h}$ |
| | υ Virginis | 333;16° | +67;9° | |
| | *α Librae | 332;39° | +51;22° | |
| | ω Lupi | 338;8° | +23;10° | |
| | Saturn | | | |

| | | | | |
|---|---|---|---|---|
| M | υ Boötis | 323;37° | +87;5° | 3;28$^h$ |
| | γ Librae | 320;25° | +46;55° | |
| | λ Librae | 320;54° | +40;3° | |
| | σ Scorpii | 319;50° | +31;41° | |
| | Scorpii *H* | 325° | +22;35° | |
| N | *α Boötis | 159;2° | +87;52° | 4;15$^h$ |
| | *γ Librae | 336;24° | +52;3° | |
| | λ Librae | 334;58° | +45;19° | |
| | *δ Scorpii | 334;49° | +42;23° | |
| | *η Lupi | 344;1° | +28;30° | |
| O | π$^1$ Boötis | 336;14° | +86;59° | 4;27$^h$ |
| | *α Scorpii | 333;50° | +36;40° | |
| | *μ$^1$ Scorpii | 337;4° | +24;27° | |
| | *ζ$^2$ Scorpii (correcting Koch's ξ$^2$) | 339;9° | +20;42° | |
| P | *α Coronae Borealis | 168;47° | +84;17° | 5;45$^h$ |
| | υ Ophiuchi | 349;45° | +58;13° | |
| | *λ Scorpii | 346;43° | +25;57° | |
| | *υ Scorpii | 347;21° | +26;3° | |
| | | | | 28 May -650 |
| Q | *β Herculis | 357;24° | +87;44° | 22;42$^h$ |
| | η Ophiuchi | 357;32° | +49;11° | |
| | ϑ Ophiuchi | 357;16° | +39;37° | |
| | ϑ Arae | 357;16° | +13;4° | |
| R | α Herculis | 1;35° | +78;22° | 23;22$^h$ |
| | ξ Serpentis | 2;30° | +48;6° | |
| | Sagittarii *X* | 2;56° | +35;26° | |
| | | | | 29 May -650 |
| S | *μ Herculis | 11;46° | +89;50° | 0;10$^h$ |
| | μ Sagittarii | 8;17° | +40;6° | |
| | ε Telescopi | 12;1° | +15;26° | |
| T | η Lyrae | 181;34° | +85;12° | 1;48$^h$ |
| | *α Aquilae | 1;28° | +63;13° | |

We find Koch's solution extremely doubtful for two main reasons. The first is that he uses a large number of extremely obscure stars to fill out his "strings". The second, and more important, is that under his hypothesis the text has no utility: it records phenomena seen at different times on different nights, and the phenomena are of no particular interest anyway. If the text means what Koch supposes it to mean, there is, so far as we can see, absolutely no reason to have written it and no reason to have kept it; without dates in the sidereal year (which could not be expressed in the Babylonian calendar) and times of night it would be impossible for anyone to observe a second time these azimuth circles except by chance. Moreover, the choice of azimuth circles makes the presence in the text of several planets and many *ziqpu* stars, and especially the several time-intervals, impossible to explain satisfactorily.

Our solution, on the other hand, provides a purpose (to measure what we now would call right ascensional differences of planets and the Moon, and thence to measure time-differences at night) that perfectly fits in with the presence of *ziqpu* stars and time-intervals. Its main handicap is that it requires the assumptions that the text was composed by combining two independent sources: a list of *ziqpu* stars with their right ascensional differences, such as we have in the tablets discussed in II A 5, and lists of simultaneously culminating stars, of which we have no examples; and that the compiler did a very poor job. The resulting lists of stars forming "strings", we believe, is quite persuasive that this interpretation provides the correct understanding of the text since in almost all cases known stars line up in the right place; though we admit that some of our choices may be incorrect, the alternative choices should not disturb the overall scheme of the text. That scheme is reflected in III 5 of Hipparchus' Τῶν Ἀράτου καὶ Εὐδόξου Φαινομένων ἐξήγησις, where various stars are named as lying on each of 24 hour circles (i.e., meridian circles). Also connected with this scheme is the use of polar longitudes and co-declinations as coordinates for stellar positions by Hipparchus (see Neugebauer [1975] 283 and 340) and the use of polar longitudes and latitudes in Indian astronomy (see Pingree [1978] 565, 578, 610, and 625). For the use of *ziqpu* stars to measure right ascensional differences in Babylon see II A 5 and II C 1.1d.

Finally, we can test the extent to which some Babylonian constellations taken over by the Greeks retained their original shapes by comparing the data in the GU text with those in Ptolemy's star catalogue (*Almagest* VII-VIII).

| Description in GU text | Identification | Description by Ptolemy |
|---|---|---|
| Hand of Front Great Twin | ϑ Geminorum | Star in the left forearm of the advance twin |
| Feet of Front Great Twin | μ Geminorum | To the rear of this (η Gem on the forward foot of the advance twin) on the same foot |
| Rear star of Great Twins | β Geminorum | Reddish star on head of rear twin |
| True Shepherd of Anu | β Orionis | Bright star in left foot |
| Right hand of True Shepherd of Anu | ξ Orionis | Rear, double star on the southern side (in right hand) |
| Rear heel of True Shepherd of Anu | κ Orionis | Star under right, rear knee |
| Two stars of head of Lion | μ Leonis<br>ε Leonis | Northernmost and southernmost of two stars in head |
| Four stars in chest of Lion | ζ Leonis<br>γ Leonis<br>η Leonis | } Northernmost, middle, and southernmost of 3 stars in neck |
| | α Leonis | Star on the heart |
| Right front foot of Lion | π Leonis | Star on left [front] knee |
| Foot in middle of Lion | ρ Leonis | Star on left armpit |
| Two stars of rump of Lion | δ Leonis | Rearmost of two stars on rump |
| | ϑ Leonis | Southernmost of two stars in buttocks |
| Single star of tail of Lion | β Leonis | Star on end of tail |
| Front pan of Scales | α Librae | Bright star (at tip of southern claw) |
| Middle of Scales | β Librae | Bright star (at tip of northern claw) |
| Horn of Scorpion | γ Scorpii | |
| Stars of head of Scorpion | δ Scorpii | Middle of 3 bright stars in forehead |
| Star of chest of Scorpion | α Scorpii | Middle of 3 bright stars in body |
| Upraised tail of Scorpion | $\mu^1$ Scorpii | In second [tail-]joint |
| Eye of Mad Dog | γ Lupi | Rearmost of 2 stars over shoulder-blade |
| Middle of Mad Dog | η Lupi | Southernmost of 2 stars in neck |
| Rear foot of Mad Dog | $\zeta^2$ Scorpii | Northern star in third [tail-]joint of Scorpius |
| Left hand of Pabilsag which is on the bow | δ Sagittarii | Star in [bow-]grip held by left hand |
| Bark | ε Sagittarii | Star in southern portion of bow |
| Right hand of Pabilsag which is on the arrow | φ Sagittarii | One in advance of this (star on left shoulder), over the arrow |
| Pabilsag above Bark | ι Sagittarii | Star on right hind lower leg |

We feel that the many coincidences serve to validate our interpretation.

A tablet from Uruk, W 22281a, published by Hunger [1976a] pp. 99-100 (no. 95), seems to be related to the GU text; it lists constellations that cross the meridian just before sunrise on the 15th of every month (it breaks off in Tešrītu); it remarks that the same constellations will cross the meridian after sunset six months later. The choice of the 15th day associates the text with those in which the colures occur on I 15, IV 15, VII 15, and X 15; it is not intended to be precise. Nor are the lists of constellations intended to have the accuracy of the "strings" in the GU text, though the use of right ascensional differences in months VII and VIII points to the beginnings of such a tradition. Because of this laxity we do not believe it to be profitable arbitrarily to identify which stars are meant. In the following summary we simply state the Sun's right ascension when its longitude is 0°, 30°, 60°, 90°, 120°, 150°, and 180°. Note that the constellations are named in the order of decreasing latitude, sometimes beginning with a *ziqpu* star or its surrogate.

| Month | RA of Sun | Constellations |
|---|---|---|
| I | 0° | Middle of Goat-Fish |
| II | 27;50° | Aquarius |
| III | 57;44° | Foot of Panther |
| | | Field |
| | | Swallow |
| IV | 90° | Horn of Lulim |
| | | Anunītu |
| | | Middle of the Fish |
| V | 122;16° | *naṣrapu* |
| | | Shoulder of Old Man |
| | | Bright star in front of Enmešarra |
| | | Middle of Hired Man(?) |
| VI | 152;10° | 15° east of Hand of Crook |
| | | Twins |
| | | True Shepherd of Anu |
| VII | 180° | 5° east of Crab |
| | | ——————— |
| | | Arrow |

Month VI corresponds roughly to string B of the GU text, month VII to string D.

The obverse of BM 34790 (LBAT 1502), of which a transliteration is published in Donbaz-Koch [1995] pp. 75-76, lists in the left-hand column nine pairs (first and last broken) of AN.TA.GUB.BA and AN.TA.ŠUR.RA constellations and in the right-hand column 17 *ziqpu* stars—i.e, one for each of the stars above the eastern or western horizon just before sunrise or just after sunset; see the end of II A 2.

| AN.TA.GUB.BA | AN.TA.ŠUR.RA | *ziqpu* | No. |
|---|---|---|---|
| | | 2 from his Tail(?) | XXIII |
| Rooster | | Single Star from his Tail | XXIV |
| | ———— | Frond | XXV |
| NUN.KI | | 15° after Frond | |
| | Great One | 1° after Rear Harness | XXVI |
| Nin-maḫ | | Yoke | I |
| | ———— | Front Harness | II |
| $^{d}$PA and $^{d}$LUGAL | | Circle | III |
| | Anunītu(?) and Fish | From the Doublets | IV |
| Šarur and Šargaz | | Triplets | V |
| | ———— | Single Star | VI |
| Pabilsag | | Lady of Life | VII |
| | ? | Shoulder of Panther | VIII |
| [Goat-F]ish(?) | | Bright star of its Breast | IX |
| | King | 10° after bright star of its Breast | |
| ———— | | 4° before Twin(?) | |
| | ———— | 10° after .... | |

See II A 5.

### *7. The DAL.BA.AN.NA Text*

A substantial fragment of this text (K 6490, a part of exemplar A; K 2079, a part of exemplar B; K 2070+2084, a part of exemplar C; and Sm 1171, exemplar E) was published in transliteration in Weidner [1915] pp. 112-115, with an interpretation on pp. 115-118. The complete text, as far as it is preserved, was published in transliteration from eight exemplars in Walker [1995], most of which were found in Kuyunjik, though one is a Babylonian copy. Since all exemplars except the last (which is unplaced) contain the comments *ḫe-pí* (break) or *ḫe-pí eš-šú* (new break) in the same place when they overlap, it is clear that they all go back to a single incomplete copy of an already broken original. An attempt to interpret this text was made in Koch [1995]. Though admirably ingenious, we believe that Koch's solution is too complex in structure to correspond to real astronomical activity and too arbitrary in its definitions of technical terms and in its identifications of star-names to be, in any case, convincing.

The text contains various phrases of which it is difficult to understand the precise significance, though they clearly in some way describe the situations of the stars. The most common is: 2 DAL.BA.AN.NA DAL.BA.AN.NA (2 intervals, interval). It occurs in B, C, E, F, G, H, L, M, N, S, V, X, Y, AA(?), a, e, h, j, k, n, o, t, y, and aa. It may well mean, as Weidner and Koch assume, that the interval (in right ascension ?) between two stars is twice the interval between one of the two and a third, but that is hard to demonstrate since there remain many problems with the identifications of the individual stars. The fact that three stars in the appropriate constellations can be found that fit the description, with the larger interval either the first or the second in a set of three stars, proves that there are many stars in the sky, but not necessarily that the ones selected by the modern scholar are those intended by the ancient astronomer. A less common phrase employing the word for interval is: 2 DAL.BA.AN.NA MIN (2 intervals, ditto). It occurs in Q, R, U, b, and, perhaps, c. Koch's explanation that in these cases the star-names refer to three pairs of neighboring stars (what he calls "double stars") and that the two intervals between these pairs are equal is difficult to sustain; the star names of Q and R are missing, only one of those in U, b, and c can be interpreted as "double" (in b, on p. 61, Koch has inadvertently omitted the second star, MUL SAG GÍR.TAB); and one star name in c is missing. Perhaps related to 2 DAL.BA.AN.NA MIN is the word *mál-ma-liš* (equally), found directly after it in b and c, but also occurring in M and N after 2 DAL.BA.AN.NA DAL.BA.AN.NA and after a short break in P. Perhaps, then, 2 DAL.BA.AN.NA MIN means 2 DAL.BA.AN.NA DAL.BA.AN.NA.

Other phrases refer to the shape of the configuration of stars. GU SI.SÁ DÚB.BA (a taut, aligned(?) string) occurs in B, G, J, N, U, h, n, o, s, t, and aa, while GU TU.LU (a slack string) is found in E, F, H, and k. One might expect the first to refer to a strict alignment in which a taut string passes through all the named stars, the second to a less strict alignment. Koch's identifications do not satisfy this expectation. He takes the words SI.SÁ DÚB.BA to mean, "taut, bent", and perceives no difference between a GU SI.SÁ DÚB.BA and a GU TU.LU despite the obvious contrast between SI.SÁ and TU.LU. But he notes that a star (or the middle of something) is called *maḫ-ru* (first) in E, F, G, H, L, e, h, n, o, t, and y; and probably in V, X, Y, a, j, and k; all of these except L, V, X, Y, e, j, and y also are said to be a string, but L, X, Y, e, and j have breaks, usually filled in by Koch, and V and y read *ḫe-pí* (break) in the crucial location. There remain a number of sections that have strings but no *maḫ-ru* star: B, J, N, U, s, and aa. B, U, and aa are broken at the end, but J, N, and s appear to be complete. The match is not perfect, but Koch claims that the position of the "bent line" is anchored at the *maḫ-ru* star on the meridian in the cases of G, h, n, o, and t. But, if we examine the right ascensions of the stars as determined by Koch, we find the *maḫ-ru* star in E has the least, that in F the most, that in G the most, that in H the least; in L the middle (MURUB$_4$-*tum*) is *maḫ-ru*; the *maḫ-ru* star in a has the most right ascension; in e the middle of the two bases, whatever that is, is *maḫ-ru*; the *maḫ-ru* star in t is in the middle, and that in y has the most right ascension. Given Koch's identifications, we see no clear relationship between the *maḫ-ru* stars and the order in which the stars in a "string" cross the meridian; nor do the "strings" "bend" except at the middle when there are three or four stars. We do not understand Koch's argument for the meaning of *maḫ-ru*.

Other geometrical and non-geometrical configurations are also found in the list. The shape of D, f, i, and u is said to be a SAG.DÙ or triangle; one way to understand this is in contrast to the taut or slack string. The star-names of D and f are too broken to be helpful; the identification of those of i and u are not without doubt. In q is written: SAG.DÙ AŠ [x] x GÍD.DA (triangle, long ....); Koch suggests reading SAG.DÙ S[AG]?-⸢BI⸣ GÍD.DA (triangle, long width). This section is, in any case, problematical. It names three stars of Zababa and a fourth star, *lubuštu* (clothing), but then states: "3 stars Triangle, long ...". The three stars of Zababa would form a triangle, but not one with a long ... . Koch identifies the Star of Clothing, which also occurs in c, with α Herculis, but, under the circumstances, this seems quite arbitrary.

The SAG.BI GÍD.DA (long width) that Koch wishes to restore in q is found in v, where four stars are named: MUL LI$_9$.SI$_4$ (α Scorpii), MUL GÌR GÙB GU$_4$.ALIM (left foot of the Bison), MUL ZI.BA.AN.NA IGI-*ú* (front star of Scales), and MUL SUḪUŠ AB.SÍN (basis of Furrow). The text (according to Koch's translation) proceeds: "4 stars; the whole for observation; long width; Lisi and the basis of Furrow are the swelling, the front star of Scales (and) left foot of Bison the dividing line". The long width would seem to be between the basis of Furrow and α Scorpii, which is the "swelling" (but *dikšu* is a mathematical term referring to the difference between the radii of two concentric circles); and the line between the other two stars must somehow divide this, but we do not know which star is the left foot of the Bison. The description remains obscure.

In that description occurs the strange phrase: "the whole for observation." The second part of this, *a-na ṣu-ub-bi-i* ("for observation"), follows a break in S and d where ŠÁR ("the whole") may have been written. We are not at all convinced that ŠÁR in this text means "the whole". The sign ŠÁR can have several readings, and it is uncertain which one might fit here. *ana ṣubbî* may indeed mean "for observing" although we don't see what that should mean here. The significance is unclear, as is that of DÙ ŠÚ [x] ("the whole sets") in g, and also that of MAŠ ŠÁR ("half of the whole") in K and r. DÙ is indeed a frequent logogram for *kal-* "all (of)", but it is usually followed by a pronominal suffix or a noun in the genitive, e.g., *kalušu* "all of him". So

DÙ-*šú* here may be "all of it/him"; but then the more important word is lost in the following break. "The whole sets" is too uncertain to base star identifications on it, not to speak of the grammatical problem that in Akkadian *kal-* "all (of)" needs a referent which would be missing in Koch's translation. In passing, we remark that according to Koch's interpretation two different words are used here for "the whole", which does not inspire confidence. MAŠ ŠÁR, according to the structure of the other sections, should be a geometrical shape. MAŠ may indeed mean "half", but how "half of the whole" should carry the meaning which Koch finds here is unclear to us.

The term SAG.KI.TA (lower width) is used in mathematical texts to refer to the lower side of a trapezoid; it is to this figure that SAG.KI.TA GÍD.DA (long lower width) should refer in d, after the phrase: "[the whole(?) is for ob]serving." Unfortunately, this section names only three stars, which cannot make a trapezoid, and the names of only two are preserved.

Less specific are the terms GÁN ÁB (cow pasture) in p and w and GÁN GIM Ú (field like a forearm) in P, S, and AA. Section p has three stars; it breaks after GÁN Á[B]. Section w has four stars; it ends MUL GÌR ÙZ *ḫe-pí*, where one expects *maḫ-ru*. If that was the word now replaced by *ḫe-pí*, this is one of the cases where a *maḫ-ru* star is not associated with a "string". In any case, the shape of the "cow pasture" cannot be determined. Similarly, the shape of the "field like a forearm" eludes us. The "forearm" is the well-known KÙŠ = *ammatu* "cubit"; but it is not at all sure that the sign (Ú = KÙŠ) is to be understood here in this way. It could just as well be a phonetic complement to GIM whose reading is equally uncertain; while "like" is a possible meaning, it is not the only one; another one is "to build, make". To express in Akkadian that a field has the shape of a forearm (whatever that may mean) would require a relative clause and a verb for "to be like", neither of which is to be found here. The end of P, which has three stars, is: GÁN GIM ⌜Ú⌝ [....] *mál-ma-liš*; *mál-ma-liš* usually follows 2 DAL.BA.AN.NA DAL.BA.AN.NA or 2 DAL.BA.AN.NA MIN. The end of S, which also had three stars (the name of one is replaced by *ḫe-pí*), is: GÁN GIM Ú ⌜2 DAL⌝.B[A.AN.N]A DAL.BA.AN.NA MUL *ḫe-pí a-na ṣu-ub-bi-i ḫe-pí*; the MUL suggests the name of a *maḫ-ru* star, replaced along with *maḫ-ru* ŠÁR by the first *ḫe-pí*, but the second *ḫe-pí* replaces something unique. And AA, which names two stars followed by a break in which a third star may have been named, has: GÁN GIM Ú 2 DAL.BA.AN.NA followed by a break. This evidence indicates that the phrases "field like a forearm" and "2 intervals, interval" or "2 intervals, ditto" are compatible. If Koch's identifications of the stars are correct, the forearm is constructed of two stars of approximately identical declination and a third at a considerably higher or lower declination; the figure would be almost vertical when its midpoint lay on the meridian.

Section m originally named four stars, of which the name of the fourth is missing; it ends: [x x] x NA BI KUN BI x [....]. NA BI KUN BI means: "its position(?) its tail." Koch restores: [GU TU?]-[LU?] NA BI KUN BI [KUR?]. We see no justification for calling this a "slack string", nor do we understand the meaning of "its position [reaches] its tail." The only part of this translation that has some probability is KUN.BI "its tail", because KUN is a relatively rare sign. If Koch's restoration were correct, the damaged signs would have to fit LU and KUR respectively. This, according to Walker's transliteration, does not seem to be the case. NA may be just the end of a group of signs which is partly broken. The whole group of signs cannot, in our opinion, be used to restore the missing star name and to draw conclusions from it.

Section I, after naming three stars (the name of the second is missing), writes: 3 MUL$^{meš}$ ták si (written DAG SI by Koch). This should mean something like: "3 stars (whose) location is thriving" if SI = SI.SÁ; Koch takes it to mean: "3 stars: ordered course." We do not understand the significance of either translation. Moreover, DAG does not serve as a logogram for *rapādu* "to run around"; only the compound ŠU-DAG is translated by *rapādu* in lexical texts. In any case, *rapādu* is planless running, the opposite of "ordered course". And DAG can mean "dwelling place" (Akk. *šubtu*). What it is here remains uncertain.

Finally, there is the problem posed by MUL dil-bat in A, B, E, F, H, J, M, T, and k. In an astronomical context one expects this phrase to signify the planet Venus, though it may be thought that a planet is inappropriate for mention in this text. It is remarkable that each time it occurs it is followed by the name of a constellation, the "Old Man", "Anunītu", or the "Bull of Heaven". Koch's explanation is that MUL dil-bat means "the first star", taking dil-bat as a writing of *maḫ-ru*. It is not clear on what basis he does this, but dil-bat is a Sumerian equivalent of Akkadian *nebû* (or *nabû*) "bright, brilliant". And there is a "bright star of the Old Man" (= $\alpha$ or $\eta$ Persei ?) and a "bright star of the Bull of Heaven" (= $\alpha$ Tauri ?) listed in CAD N/2 148. Koch is simply wrong here.

Koch's table II, which shows the azimuths and altitudes of the stars that he has identified at different hours of different nights (3 January, 4 January, 15 May, 16 May, 3 September, and 4 September) of -650 at Nineveh, does not seem to us plausibly to reflect the structure of this text. No Mesopotamian astronomer would be able to reconstruct the observations upon which the text is based, so that the text would lose its meaning immediately after it was composed.

We present in the following table the names of the Dalbanna text, and note beside each star name Koch's identification. Those identifications which we believe to be supported from other sources are marked with an asterisk, those contradicted by a previous identification with an obelisk.

| | | |
|---|---|---|
| A. | Right hand of True [Shepherd of Anu] | ν Orionis |
| | [Bre]ast of Old Man | *η Persei |
| | Bright star of Bull of Hea[ven] | γ Tauri |
| B. | Bright star of Bull [of Heaven] | γ Tauri |
| | Crown of True Shepherd of Anu | λ' Orionis |
| | Star [....] | ϑ Geminorum |
| | 3 stars. Str[aight], taut string. 2 intervals, interv[al]. | |
| C. | Crown of Panther | α Cephei |
| | Panther | *α Cygni |
| | MUL GÚ.MUR ... [...] | o' Cygni |
| | [... 2 in]tervals, interval [....]. | |
| D. | Left. New [bre]ak. | |
| | Bright star of Anunī[tu]. | γ Pegasi |
| | [MUL .... ri]ght of Field. | α Pegasi |
| | 3 stars. Trian[gle ....] | |
| E. | Deleter | †α Cassiopeiae |
| | Bright star of A[nunītu] | γ Pegasi |
| | ... star from right of Fie[ld] | α Pegasi |
| | 3 stars. Slack string. [2 inter]vals, interval. Field fir[st]. | |
| F. | Pelvis of Pig | †α Lacertae |
| | Righ[t] foot [....] | ι Cygni |
| | Bright star of Anunī[tu] | γ Pegasi |
| | 3 stars. S[lac]k string. [2 inter]vals, interval. Bright star of Anunītu firs[t]. | |
| G. | Breast of Pig | o Andromedae |
| | Righ[t] foot [....] | ι Cygni |
| | [De]leter | †α Cassiopeiae |
| | 3 stars. [...] str[ing. 2 int]ervals, i[nterval]. Deleter f[irst]. | |
| H. | Right foot of Panther | ι Cygni |
| | Rai[nbow] | β Cassiopeiae |
| | Bright [sta]r of Old Man | ϑ Persei |
| | 3 stars. Slack string. [2 inter]vals, int[erval.] Right foot of Panther first. | |
| I. | Right foot of Panther | ι Cygni |
| | [....] | β Lacertae |
| | PA SI of Stag | δ Cassiopeiae |
| | 3 stars. ták-si. | |

| | | |
|---|---|---|
| J. | Toe of right foot of Panther | π Pegasi |
| | [Bright] s[tar of] Anunītu | γ Pegasi |
| | Head of Hired Man (*or* of Zababa) | α Arietis |
| | 3 stars. Straight, taut string. | |
| K. | [MU]L KA x x x | ε Cassiopeiae |
| | [....] | γ Cassiopeiae |
| | Fin of Anunītu | υ Piscium |
| | 3 stars. .... | |
| L. | [....] | γ Pegasi |
| | [PA S]I of Stag | δ Cassiopeiae |
| | Hip of Stag | γ Cassiopeiae |
| | Deleter | †α Cassiopeiae |
| | 4 sta[rs ..... 2 interva]ls, interval. Middle first. | |
| M. | Bright star of Bull of [Heaven] | γ Tauri |
| | [.....] | η, 27 Tauri |
| | Beginning of Bull of Heaven | ζ Tauri |
| | [.....] 2 intervals, [inter]val. Equally. | |
| N. | Front (star) of the Little Twins | β, γ Canis Minoris |
| | MU[L ....] | ζ Hyadum |
| | [MUL] x of the Lion above | μ Leonis |
| | 3 stars. Straight, taut string. [2 intervals, in]terval. Equally. | |
| O. | Back (star) of the Little Twins | α Canis Minoris |
| | [MUL ...] of the Lion below (fem.!) | ε Leonis |
| | Crab | δ Cancri |
| | New break | ϑ Leonis |
| | [... s]tars. Break. | |
| | Front (star) of the Little Twins, [Bea]rd of Li[on], Crab, and Head of Snake, Bird(?). Break. | |
| P. | King | *α Leonis |
| | Breast of Sna[ke$^{?}$] | 30 Monocerotis, 1,2 Hyadum |
| | [....] | omitted |
| | [x]-*ḫi-it* of Arrow | 18 Monocerotis |
| | 3 stars. Field .... [....] Equally. | |
| Q. | [.....] | |
| | [L]ion | |

| | | |
|---|---|---|
| | Break | |
| | 3 sta[rs ....] 2 intervals, ditto. | |
| R. | Twin[s ....] | |
| | 3 [stars .....] 2 intervals, ditto. | |
| S. | Back (star) of the Great Twins | χ Geminorum |
| | Back (star) of the Great Twins (!) | *β Geminorum |
| | Break. | α Canis Maioris |
| | 3 stars. Field .... 2 int[erva]ls, interval. Star break for observing break. | |
| T. | Deleter | †α Cassiopeiae |
| | Bright star of Anunītu | γ Pegasi |
| | Break | 72 Pegasi or δ Cassiopeiae |
| U. | Crook | ζ, η Aurigae |
| | Right foot of Old Man | ρ Persei |
| | Left foot of [x x] Twin | *μ Geminorum |
| | [....] Straight, taut string. 2 inter[va]ls, ditto [x x]. | |
| V. | MUL PA 4-šú | 321G, 328G, 340G, υ Centauri |
| | Wa[go]n | η Ursae Maioris |
| | King x [......] | †α Centauri |
| | [x] stars. Break. 2 interva[ls, interva]l. S[tar .....] | |
| W. | Right foot [......] x | †γ Lupi |
| | Šullat x [.....] | 321G, 328G, 340G, υ Centauri |
| | Shou[lder .....] | ζ Ophiuchi |
| X. | Left foot of Ḫabaṣirānu | †κ Centauri |
| | Šullat [.....] | †321G, 328G, 340G, υ Centauri |
| | [.....] | †α Centauri |
| | 3 stars. [...... 2 inter]vals, interval. L[eft] foot [....] | |
| Y | Left foot of Mad Dog | *$\zeta^2$ Scorpii |
| | L[eft] foot [.....] | σ Ophiuchi |
| | [.....] | κ Ophiuchi |
| | 3 stars. [..... 2] intervals, interval. Star x [.....] | |
| Z. | Right foot of the Bison | κ Ophiuchi |
| | Left foot [.....] | *$\zeta^2$ Scorpii |
| | [.....] | χ Lupi |

| | | |
|---|---|---|
| AA. | Left foot of Bison | σ Ophiuchi |
| | Left foot of Mad Dog | *$\zeta^2$ Scorpii |
| | [.....] | ϑ Lupi |
| | [... stars]. Field .... 2 intervals, [.....] | |
| BB. | Right hand of Mad Dog | χ Lupi |
| | [.....] | |

Reverse

| | | |
|---|---|---|
| a. | Front (star) of [Scal]es | *α Librae |
| | Furrow | α Virginis |
| | Tail of Rav[en] | Pi 140 |
| | [3 stars .... 2 inte]rvals, interval. [Front (star) of] Sc[ales ....] | |
| b. | Ba[ck (star) of ....] | †β Librae |
| | Head of Scorpion | *δ Scorpii |
| | Base of Upright of Sc[orpion] | *$\mu^1$, $\mu^2$ Scorpii |
| | [3 star]s. [.....] 2 intervals, [ditto$^?$]. Equally. | |
| c. | [.....] | η Scorpii |
| | Lisi | *α Scorpii |
| | Šarur and Šargaz | *λ, υ Scorpii |
| | 3 stars [..... 2] intervals, [ditto $^?$]. Equally. | |
| d. | [.....] | γ Librae |
| | Front (star) of [S]cales | *α Librae |
| | Right horn of Scorpion | †σ Scorpii |
| | [3 stars .... for o]bservation. Lower [wid]th long. | |
| e. | Star x [x x] | 7 Andromedae |
| | [.....] | β Cassiopeiae |
| | Head of Horse and Swallow | β, ε Pegasi |
| | 4 stars [...... 2 in]tervals, interval. Head of Swal[low ...] x interval. Middle (and$^?$) base both first. | |
| f. | Base of Upright of Scorpion | *$\mu^1$ Scorpii |
| | Star x [x x] | γ Ophiuchi |
| | [Star x] of Bison | κ Ophiuchi |
| | 3 stars. Triangle. | |
| g. | Šullat | †β Centauri(!) |
| | Ḫaniš | †α Centauri |

| | | |
|---|---|---|
| | Right foot of B[ison] | κ Ophiuchi |
| | [3 s]tars. All of it ? [...] | |
| h. | Head of Scorpion | *δ Scorpii |
| | Left hand of Bison | Serpentis d |
| | Šullat | †α Centauri |
| | [3 stars]. Straight, ta[ut] string. 2 intervals, interval. [Šul]lat first. | |
| i. | Crown of Zababa | δ Ophiuchi |
| | Shoulder of Zababa | ζ Ophiuchi |
| | Head [....] | β Scorpii |
| | [3] stars. Tri[angle]. | |
| j. | Sting of Scorpion | ϑ Scorpii |
| | Head of Bison | †η Serpentis |
| | Šullat | †β Centauri(!) |
| | 3 [stars .....] 2 intervals, interval. Sti[ng of Scorpion ....] | |
| k. | Right foot of Zababa | †ξ Ophiuchi |
| | Left foot of Zababa | †μ Sagittarii |
| | Star x x x [......] | η Serpentis |
| | [.....] Slack string. 2 intervals, interval. Star x [.....] | |
| l. | Right foot of Zababa | †ξ Ophiuchi |
| | Left foot of Zababa | †μ Sagittarii |
| | Crow[n ....] | δ Ophiuchi |
| m. | Left hand of She-goat | ϑ Lyrae |
| | Foot of She-goat | δ Herculis |
| | Crown of Panther | α Cephei |
| | Star [.....] | β Librae |
| | 4 star[s x x] x its ... its tail x [....] | |
| n. | Left hand of Mad Dog | ϑ Lupi |
| | Center of Mad Dog | δ Lupi |
| | Ri[ght] foot [of Bison] | κ Ophiuchi |
| | 3 stars. Straight, taut string. 2 intervals, int[erval]. Right foot of Bison fir[st]. | |
| o. | Shoulder of Zababa | ζ Ophiuchi |
| | Star of Clothing | α Herculis |
| | Left foot of Zababa | †μ Sagittarii |
| | 3 stars. Straight, ta[ut] string. 2 intervals, interval. Shoulder fir[st]. | |

| | | |
|---|---|---|
| p. | Left shoulder of Zababa | †ν Ophiuchi |
| | Star of Clothing | α Herculis |
| | Right foot of Zababa | †ξ Ophiuchi |
| | 3 star[s]. Cow pasture. | |
| q. | Crown of Zababa | δ Ophiuchi |
| | Right star of Zababa | †ξ Ophiuchi |
| | Left foot of Zababa | †μ Sagittarii |
| | Star of Clothing | α Herculis |
| | 3 stars. Triangle AŠ [x] x long. | |
| r. | Eagle | ε Aquilae |
| | Goat-Fish | β Capricorni |
| | Sting of Pabilsag | α Coronae Austrinae |
| | 3 stars. Half[?] ... | |
| s. | Head of Bison | †η Serpentis |
| | Stars of Middle of Bison | β, γ Ophiuchi |
| | Lef[t] foot of [Bi]son | σ Ophiuchi(!) |
| | Midd[le x x ] x | η Telescopi |
| | 4 stars. Straight, taut string. | |
| t. | Right hand of Bison | ζ Serpentis |
| | Šarur | υ Scorpii (! mistake for *λ) |
| | Šargaz | *υ Scorpii |
| | Sting of Pabilsag | α Coronae Austrinae |
| | 3 stars. Straight, taut string. 2 intervals, interval. Right hand of Bison first. | |
| u. | Left hand of Bison | σ Ophiuchi(!) |
| | Lisi | *α Scorpii |
| | Front (star) of Scales | *α Librae |
| | 3 stars. Triangle. | |
| v. | Lisi | *α Scorpii |
| | Left foot of Bison | σ Ophiuchi(!) |
| | Front (star) of Scales | *α Librae |
| | Base of Furrow | α Virginis |
| | 4 stars. ... for observation. Its width long. Lisi and Base of Furrow the "swelling", the front (star) of Scales (and) Left foot of Bison the dividing line. | |
| w. | Shoulder of Panther | †κ Cygni |
| | Crown of Panther | α Cephei |

| | | |
|---|---|---|
| | Left hand. New break | Serpentis d |
| | Zababa | ζ Ophiuchi |
| | 4 stars. Cow pasture. Foot of She-goat break. | |
| x. | Basis of left foot of Panther | μ Cygni |
| | Bright star of Anunītu | γ Pegasi |
| | Hand break | ϑ Lyrae |
| y. | She-goat | †δ Herculis |
| | Right foot of Zababa | †ξ Ophiuchi |
| | Eagle | ε Aquilae |
| | 3 stars. Break. 2 intervals, interval. Eagle first. | |
| z. | Mouth of Dog | ϕ Herculis |
| | She-goat | †δ Herculis |
| | Head of Wildcat | ι Telescopi |
| | Eagle | ε Aquilae |
| | [x x] | |
| aa. | Great One | †α Lyrae(!) |
| | Tongue of Dog | τ Herculis |
| | SAR $U_5$ x [.....] | γ Serpentis |
| | 3 stars. Straight, taut string. 2 interval[s, .... Star] x x [.....] | |

Despite the many breaks, both old and "new" and recent, in this text and its several tablets, Koch has filled almost everything in by applying one of his five types (with nineteen sub-types) of "intervals" to each section. Since the existence of such a complex system as the five types depends entirely on Koch's conjecture based on star identifications, since almost none of the star-names in this text occur elsewhere, and since in 48 cases not enough of the star-name is left to be able to know what it was called, much less what it signified, Koch's solution is highly arbitrary, and seems to us to make implausible assumptions concerning the nature of Mesopotamian star-lore. We do not believe that we know enough yet about either the technical descriptions (the intervals, the triangles, the taut and slack strings, the cow pastures, etc.) or the names of individual stars to venture an interpretation. We would, however, suggest that the "strings" ought to be comparable to the alignments quoted from Hipparchus and provided by himself in VII 1 of Ptolemy's *Almagest*; but actual parallels cannot be adduced because of the insecurity of the identifications of the stars being referred to.

## 8. *Time-keeping Texts*

### 8.1. *The Ivory Prism*

This small fragment (BM 123340) from the middle of a four-sided prism contains two columns (one blank on face A) on each side. It was found by Loftus at Nineveh, and so was made before the destruction of that city in -611. Originally published as "rules of a game" in Lenormant [1873] no. 86, and as an "augural staff" in Sayce [1887], the explanation of two of its faces was first discovered by Fotheringham [1932], then repeated in Langdon [1935] pp. 55-64 and van der Waerden [1950] 295-297 and [1954] 25-27, and rejected by Neugebauer [1947] 40, who believed that the prism contains information about shadow-lengths. It was argued by Smith [1969] that Fotheringham's explanation of faces C and D is correct, and that faces A and B are consistent with it.

Fotheringham's solution for faces C and D was that they record for each month in the Babylonian "ideal" calendar the times measured in *bēru* and UŠ when, by day and by night, each of twelve seasonal hours shall have lapsed. Therefore, each representative nychthemeron in each ideal month contains 24 hours; each seasonal hour is a twelfth part of the length of daylight or of the length of nighttime on each of those nychthemera. The equinoxes are located in Nisannu and Tešrītu, and the ratio of the longest to the shortest length of daylight is 2 : 1. This is the arrangement of, for instance, MUL.APIN. The numerical entries of faces C and D are restored in the following; numbers that are underlined are present on the fragment of the prism.

Column C
Nisannu day, Nisannu night
Tešrītu night, Tešrītu day
6 *bēru* (= 180 UŠ)

| | | |
|---|---|---|
| 1 | | 15 UŠ |
| 2 | 1 *bēru* | |
| 3 | 1 *bēru* | 15 UŠ |
| 4 | 2 *bēru* | |
| 5 | 2 *bēru* | 15 UŠ |
| 6 | 3 *bēru* | |
| 7 | 3 *bēru* | 15 UŠ |
| 8 | 4 *bēru* | |
| 9 | 4 *bēru* | 15 UŠ |
| 10 | 5 *bēru* | |
| 11 | 5 *bēru* | 15 UŠ |
| 12 | <u>6 *bēru*</u> | |

Ajjaru, Ulūlu day
Araḫsamnu, Addaru night
6 *bēru* 20 UŠ (= 200 UŠ)

| | | |
|---|---|---|
| 1 | | 16;40 UŠ |
| 2 | 1 *bēru* | 3;20 UŠ |
| 3 | 1 *bēru* | 20 UŠ |
| 4 | 2 *bēru* | 6;40 UŠ |
| 5 | 2 *bēru* | 23;20 UŠ |
| 6 | 3 *bēru* | 10 UŠ |
| 7 | 3 *bēru* | 26;40 UŠ |
| 8 | 4 *bēru* | 13;20 UŠ |
| 9 | 5 *bēru* | |
| 10 | 5 *bēru* | 16;40 UŠ |
| 11 | 6 *bēru* | 3;20 UŠ |
| 12 | 6 *bēru* | 20 UŠ |

Ajjaru, Ulūlu night
Araḫsamnu, Addaru day
5 *bēru* 10 UŠ (= 160 UŠ)

| | |
|---|---|
| | 13;20 UŠ |
| | 26;40 UŠ |
| 1 *bēru* | 10 UŠ |
| 1 *bēru* | 23;20 UŠ |
| 2 *bēru* | 6;40 UŠ |
| 2 *bēru* | 20 UŠ |
| 3 *bēru* | 3;20 UŠ |
| 3 *bēru* | 16;40 UŠ |
| 4 *bēru* | |
| 4 *bēru* | 13;20 UŠ |
| 4 *bēru* | 26;40 UŠ |
| 5 *bēru* | 10 UŠ |

Column D

Simānu, Abu day
Kislīmu, Šabāṭu night
7 *bēru* 10 UŠ (= 220 UŠ)

| | | |
|---|---|---|
| 1 | | 18;20 UŠ |
| 2 | 1 *bēru* | 6;40 UŠ |
| 3 | 1 *bēru* | 25 UŠ |
| 4 | 2 *bēru* | 13;20 UŠ |
| 5 | 3 *bēru* | 1;40 UŠ |
| 6 | 3 *bēru* | 20 UŠ |
| 7 | 4 *bēru* | 8;20 UŠ |
| 8 | 4 *bēru* | 26;40 UŠ |
| 9 | 5 *bēru* | 15 UŠ |
| 10 | 6 *bēru* | 3;20 UŠ |
| 11 | 6 *bēru* | 21;40 UŠ |
| 12 | 7 *bēru* | 10 UŠ |

Simānu, Abu night
Kislīmu, Šabāṭu day
4 *bēru* 20 UŠ (= 140 UŠ)

| | |
|---|---|
| | 11;40 UŠ |
| | 23;20 UŠ |
| 1 *bēru* | 5 UŠ |
| 1 *bēru* | 16;40 UŠ |
| 1 *bēru* | 28;20 UŠ |
| 2 *bēru* | 10 UŠ |
| 2 *bēru* | 21;40 UŠ |
| 3 *bēru* | 3;20 UŠ |
| 3 *bēru* | 15 UŠ |
| 3 *bēru* | 26;40 UŠ |
| 4 *bēru* | 8;20 UŠ |
| 4 *bēru* | 20 UŠ |

| | Du'ūzu day<br>Ṭebētu night<br>8 *bēru* (= 240 UŠ) | | Du'ūzu night<br>Ṭebētu day<br>4 *bēru* (= 120 UŠ) | |
|---|---|---|---|---|
| 1 | | 20 UŠ | | 10 UŠ |
| 2 | 1 *bēru* | 10 UŠ | | 20 UŠ |
| 3 | 2 *bēru* | | 1 *bēru* | |
| 4 | 2 *bēru* | 20 UŠ | 1 *bēru* | 10 UŠ |
| 5 | 3 *bēru* | 10 UŠ | 1 *bēru* | 20 UŠ |
| 6 | 4 *bēru* | | 2 *bēru* | |
| 7 | 4 *bēru* | 20 UŠ | 2 *bēru* | 10 UŠ |
| 8 | 5 *bēru* | 10 UŠ | 2 *bēru* | 20 UŠ |
| 9 | 6 *bēru* | | 3 *bēru* | |
| 10 | 6 *bēru* | 20 UŠ | 3 *bēru* | 10 UŠ |
| 11 | 7 *bēru* | 10 UŠ | 3 *bēru* | 20 UŠ |
| 12 | 8 *bēru* | | 4 *bēru* | |

In the three columns on faces A and B the boxes in which the numbers are entered are sufficiently large to contain 2½ lines of the entries in the columns on faces C and D. One can see all or part of five boxes in the single column on face A. Since there is a constant difference of 0;10 between entries in each column, one can easily reconstruct:

| A | B | |
|---|---|---|
| [13;]⌜50⌝ | [1]⌜3⌝ | ⌜12;20⌝ |
| 13;40 | 12;50 | 12;10 |
| 13;30 | 12;40 | 12 |
| 13;2[0] | [1]2;30 | 11;50 |
| [13;10] | | |

Smith noticed that the first column of B follows the column of A, and the second column of B follows the first column of B, to form a single function diminishing at the rate of 0;10 per step from 13;50 to 11;50. Considering that 12 UŠ is $\frac{1}{15}$ of an equinoctial night of 180 UŠ or 6 *bēru* and that 14 UŠ is $\frac{1}{15}$ of a night of 210 UŠ or 7 *bēru*, exactly in the middle between the night of the winter solstice (8 *bēru*) and that of the spring equinox, he concluded that this table represents the lengths of a fifteenth of a night. Smith further assumed that the solstice and equinox occurred on the first days of Ṭebētu and Nisannu, so that the night of 7 *bēru* would occur on 15 or Full Moon of Šabāṭu; instead, of course, the Babylonians took the solstice and the equinox to occur on 15 Ṭebētu and 15 Nisannu, so that the night of 7 *bēru* occurred on 1 Addaru, and Smith's speculations on pp. 77ff. are meaningless.

The question remains as to whether his interpretation of faces A and B as fifteenths of nights is credible. There would be twelve nights selected, equally spaced, during the 45 nychthemera between 1 Addaru and 15 Nisannu. We do not see how this could work because each interval is 3 ¾ nychthemera. The nightly difference is 30 UŠ / 45d = 0;40 UŠ/d. Moreover, Smith's theory requires an unequal number of entries in each column, an asymmetrical arrangement, and an inexplicable continuation beyond the entry, 12, for the vernal equinox. For all these reasons we believe that it cannot be accepted.

What we would note is that there is a pattern in the numbers; those in column 1 of B are 0;50 less than those in the column in A; and those in column 2 of B are 0;40 less than those in column 1 of B. We do not know the significance of this.

### *8.2. The Report on Seasonal Hours*

The tablet containing this text (K 2077+3771+11044 + BM 54619) written in Neo-Babylonian script by [...]-Gula in the eponymy of Bēl-šadûa (-649), was found, like the Ivory Prism, at Nineveh. It was published in Reiner-Pingree [1974/77]. The report provides the length of a seasonal hour ($\frac{1}{12}$ of daylight) on the 15th and the 1st of every month in the ideal calendar beginning with 15 Duʾūzu, the summer solstice. The ratio of the longest to the shortest length of daylight is again 2 : 1, so that the table forms a linear zigzag function in which:

| (seasonal hour) | (daylight) |
|---|---|
| M = 20 UŠ | M = 240 UŠ = 8 *bēru* |
| m = 10 UŠ | m = 120 UŠ = 4 *bēru* |
| μ = 15 UŠ | μ = 180 UŠ = 6 *bēru* |
| d = 0;50 UŠ | d = 10 UŠ |
| P = 24 | P = 24 |

The text, addressed to the king, claims that this table is written on no tablet and unknown to people, and the author (?) declares that he will show it to the Hittite *ša rēš šarri* official, presumably in Syria. Since the division of the nychthemeron into 24 hours and the computation of seasonal hours is first attested in Egypt on the cenotaph of Seti I (-1302 to -1289) (see Neugebauer-Parker [1960-1969] vol. 1, pp. 116-121), these two texts from Nineveh probably represent a Mesopotamian adaptation of the Egyptian practice.

An additional piece that joins this tablet, BM 54619, includes on its reverse a table of fifteenths of a night for every fifteenth night beginning with that of the winter solstice. It forms a linear zigzag function whose parameters are:

M = 16 UŠ (16 × 15 = 240 UŠ)
m = 8 UŠ (8 × 15 = 120 UŠ)
μ = 12 UŠ (12 × 15 = 180 UŠ)
d = 0;40 UŠ
P = 24.

## B. First Millennium: Observations and Predictions

### *1. Sargonid Period*

#### *1.1. Observations and Predictions in Letters and Reports*

The Letters and Reports concerning celestial omens sent from scholars located throughout Mesopotamia to the Assyrian king at Nineveh have been described in I. Here we wish only to glean from the records of the seventh century B.C. the references to observations and the evidence they provide for contemporary understanding of the predictability of celestial phenomena. These are found in the Letters published by Parpola [1993] and in the Reports published by Hunger [1992]. Concerning the observers and their distribution over Mesopotamia see Oppenheim [1969]. Many of the Reports are redated by De Meis-Hunger [1998]. Their proposed dates are noted after the excerpts from the relevant Reports.

The observation of the Sun and the Moon in opposition to each other in the middle of the month was frequently made; see, e.g., the five reports of Nabūʾa of Assur (Hunger [1992] pp. 82-83 (Reports 134-138)) and the unusual observation of the opposition on the twelfth day (Hunger [1992] p. 52 (Report 88)). Often, in place of observation we find predictions, both of the day of opposition and of the day of the New Moon.

1. Hunger [1992] pp. 34-35 (Report 57): The Moon will be seen with the Sun on the 14th day. From Nabû-aḫḫe-eriba.

2. Hunger [1992] pp. 35-36 (Report 60): The Moon will be seen with the Sun in Ṭebētu on the 14[th] day. The Moon will complete the day in Šabāṭu; on the 14th day it will be seen with the Sun. [The Moon] will reject the day [in] Addaru; [on the xth day] it will be seen with the Sun. [The Moon] will complete the day in Nisannu. [From Na]bû-aḫḫe-eriba.

3. Hunger [1992] p. 39 (Report 67): [The Moon] will be seen on the 15th [day with] the Sun; it will omit [an eclipse], it will not make it. From [Nabû-aḫḫe]-eriba.

4. Hunger [1992] p. 136 (Report 244): The Moon will be seen in Ajjaru on the 16th day together with the Sun. --- From Nergal-eṭir. 8 April -670.

5. Hunger [1992] pp. 140-141 (Report 252): In intercalary Addaru on the 14th day the Moon will be seen with the Sun. From Nergal-eṭir.

6. Hunger [1992] pp. 142-143 (Report 255): In Abu on the 16th day the Moon will be seen together with the Sun. --- From Nergal-eṭir.

7. Hunger [1992] pp. 150-151 (Report 271): On the 1st day I wrote to the king as follows: On the 14th day the Moon will be seen together with the Sun. (Now), on the 14th day the Moon and the Sun were seen together: reliable speech. --- From Nergal-eṭir.

8. Hunger [1992] pp. 162-163 (Report 294): If the Moon reaches the Sun and follows it closely, and one horn me[ets] the other, ---. On the 14th day the Moon and the Sun will be seen with each other. --- At the beginning of the year an eclipse [was seen] ...; on the 14th day (the Moon) wa[s seen] with the Sun. --- From Nabû-iqiša of Borsippa.

9. Hunger [1992] pp. 232-233 (Report 410): Earlier, I wrote to the king my lord as follows: "On the 14th day one god will be seen with the other." If the Moon reaches the Sun and [follows it closely], ---. [On the 14th day one god] is seen [with the other]: --- [From] Rašil.

10. Hunger [1992] p. 270 (Report 489): The Moon and the Sun will be seen together on the 14th day. --- From Rimutu.

11. Hunger [1992] p. 274 (Report 499): If the Moon does not wait for the Sun but sets, ---. On the 15th day it is seen with the Sun. In Nisannu (the Moon) will complete the day; on the 14th day the Moon will be seen with the Sun. --- From Šumaya.

12. Hunger [1992] p. 282 (Report 506): [The 30th day] will complete [the length of the month. On the 14th day] (the Moon) will be seen [with the Sun].

13. Hunger [1992] p. 286 (Report 516): [the Moon will com]plete the day in Addaru; on the 14th day it will be see[n together] with (the Sun). [The Moon] will reject the day in Nisannu; [the Moon] will reject the day in Ajjaru; [the Moon] will be close to rejecting in Simānu.

14. Hunger [1992] pp. 286-287 (Report 517): If the Moon reaches the Sun and follows it closely, and one horn meets the other, ---. On the 14th day one god will be seen with the other.

15. Hunger [1992] pp. 297-298 (Report 545): The Moon will be seen [...] on the ⌜15th⌝ day.

We do not know on what bases these predictions were made. The wordings of 2 ("complete the day" means that the month is 30 days long, "reject the day" indicates that it is 29 days long), 7, 8, 9, 11, 12, 13, and 14 indicate that the criteria for predicting the day of the conjunction included the length of the previous month and the distance of the Moon from the Sun at last visibility; neither is sufficient for a correct solution of the problem.

But this concern with the horns of the Moon at last visibility recalls the use of a protruding right horn to predict a lunar eclipse in Tablet 20 of *Enūma Anu Enlil.* In only one case (Hunger [1992] p. 162 (Report 293)) does the observer, Nabû-iqiša of Borsippa, reveal how he knew that the Moon and the Sun were seen together on the 14th: The king my lord must not say as follows: "(There were) clouds; how did you see (anything)?" This night, when I saw (the Moon's) coming out, it came out when little of the day (was left), it reached the region where it will be seen (in opposition with the Sun). This is a sign that is to be observed. In the morning, if the day is cloudless, the king will see; for one *bēru* of daytime (the Moon) will stand there with the Sun.

Naturally, on many of these observations of oppositions, an eclipse of the Moon was seen in the following night; sometimes it is clear that it was expected, but often it was not. For the eclipse-cycles available in the seventh century B.C. see II C 1.1.

1. Parpola [1993] p. 106 (Letter 128): To the 'farmer', my lord: your servant Nabû-šumu-iddina, the foreman of the collegium of ten (scribes) of Nineveh. --- On the 14th day we were watching the Moon; the Moon was eclipsed.

2. Parpola [1993] p. 108 (Letter 134): To the king, my lord: your servant Babu-šumu-iddina of Kalḫu. --- [The Moon was e]clipsed [on the xth day, during] the evening watch; [...] culminated, [the ... wind] was blowing.

3. Parpola [1993] p. 109 (Letter 137): To the king, my lord: your servant, the foreman of the collegium of ten (scribes) of Arbela. --- An eclipse (of the Moon) occurred [on the 1]4th [day of ...], during the morning watch.

4. Parpola [1993] p. 113 (Letter 147): Concerning the watch for the eclipse about which the king, our lord, wrote: we kept watch; the clouds were dense. On the 14th day, during the morning watch, the clouds dispersed, and we were able to see. The eclipse took place.

5. Parpola [1993] pp. 113-114 (Letter 149): A lunar eclipse took place on the 14th of Simānu, [during] the morning watch. It started in the south and cleared up in the south. Its right side was eclipsed. It was eclipsed in the area of the Scorpion. The Shoulder of the Panther was culminating. An eclipse of two fingers took place.

6. Parpola [1993] p. 129 (Letter 168): To the king of the lands, my lord: your servant Zakir. --- The Moon made an eclipse on the 15th day of Ṭebētu, in the middle watch. It began in the east and shifted to the west.

7. Parpola [1993] p. 282 (Letter 347): To the king, my lord: your servant Mar-Issar. --- Concerning the lunar eclipse about which the king, my lord, wrote to me: it was observed in the cities of Akkad, Borsippa, and Nippur. What we saw in Akkad corresponded to the other [observation]s.

8. Parpola [1993] pp. 286-287 (Letter 351): To the king, my lo[rd]: your servant Mar-Issar. --- The substitute king, who on the 14th day sat on the

throne in [Ninev]eh and spent the night of the 15th ⸢day⸣ in the palace o[f the kin]g, and on account of whom the eclipse took place, --- This eclipse which occurred in the month Ṭebētu ...

9. Parpola [1993] p. 308 (Letter 371): To the king of the lands, my lord: your servant Kudurru. --- After the king, my lord, had gone to Egypt and the eclipse took place in Duʾūzu, ...

10. Parpola [1993] p. 309 (Letter 372): The eclipse of the Moon which [took place] in Du[ʾūzu(?)] ...

In numbers 2, 5, and 6 some of the ominous features of the eclipse are reported; in the Reports one frequently finds only the relevant omens that fit the features of the observed eclipse. Such examples are: Hunger [1992] pp. 5-7 (Report 4); pp. 62-63 (Report 103); p. 117 (Report 208); pp. 168-169 (Report 300); pp. 192-193 (Report 336); and pp. 293-294 (Report 535). But sometimes Reports contain more straightforward accounts of the observations, but still embedded in interpretation.

1. Hunger [1992] p. 154 (Report 279): The night of the 13th day there was no observation; let the son of the king sleep. The n[ight] of the 14th day there was no observation; let the son of the king sleep. The ni[ght of] the 15th day there was no observation; let the son of the king sleep. The night of the 16th day a great observation: the Moon made an [ecli]pse; this eclipse of the Moon ...

2. Hunger [1992] pp. 178-179 (Report 316): I could not let the king, my lord, hear the words about the eclipse from my mouth. Now in compensation I send a written report to the king, my lord --- Simānu means ... the 14th day means ... The beginning, where (the eclipse) began, we do not know. (The Moon) pulled the amount of the eclipse to the south and west ... it became clear from the east and north ... it covered all of (the Moon) ... Because of the evening watch ... which covered all of it ... In the eclipse [of the Moon] Jupiter stood there ...

3. Hunger [1992] pp. 262-263 (Report 469): The eclipse of the Moon which took place in Araḫsamnu began [in the east] ...

For many of the oppositions the occurrence or non-occurrence of a lunar eclipse was predicted; there was a prediction that a lunar eclipse would *not* occur in no. 3 of the list of oppositions of the Sun and the Moon. Other cases are:

1. Hunger [1992] p. 52 (Report 87): [Concerning] the observation of the Moon [... ...] ... it will let [an eclipse] pass by; it will [not] make it. [From Ba]lasī.

2. Hunger [1992] p. 181 (Report 321): The eclipse will pass by, (the Moon) will not make it. Should the king say: "What signs did you see?" [... the god]s did not see each other [...] for the night [......]. [The eclipse] will pass by (and) [the Moon] will be seen [together with] the Sun. From Munnabitu.

In these cases the non-occurrence of an eclipse seems to be connected with the observation of the Moon and the Sun together on the 14th. The predicted eclipse is presumably to occur in the following month. That this was a criterion is confirmed by two Reports:

1. Hunger [1992] pp. 180-181 (Report 320): If the Moon becomes late at an inappropriate time and is not seen --- It is seen on the 16th day. --- one god is not seen with the other on the 14th or 15th day. If the Moon in Ulūlu is seen with the Sun neither on the 14th nor on the 15th day: --- Within one month the Moon and the Sun will make an eclipse; i.e. each month on the 14th day one god will not be seen with the other. If on the 16th day the Moon and the Sun are seen together: --- The [ki]ng must not become negligent about these observations of the Mo[on]; let the king perform either a *namburbi* or [so]me ritual which is pertinent to it. From Munnabitu.

2. Hunger [1992] p. 251 (Report 447): [In Aj]jaru (the Moon) will complete the day; [on the 14]th day the Moon and the Sun will be seen together. The 13th [day], the night of the 14th day, is the [da]y of the watch (to be held), and there will be no eclipse. I guarantee it seven times, an eclipse will not take place. I am writing a definitive word to the king. From Ṭab-ṣilli-Marduk, nephew of Bel-naṣir.

A prediction that may have resulted from such a procedure is recorded in Report 250 (Hunger [1992] p. 139): In Addaru on the 14th day the Moon will make an eclipse. --- From Nergal-eṭir. That other curious means of prediction were employed, however, is seen from Report 388 (Hunger [1992] pp. 222-224): On the 14th day the Moon will make an eclipse. --- (Already) when Venus became visible, I said to the king my lord: "An eclipse will take place." From Rašil the older, servant of [the king]. Cf. the entrance of Venus into the Moon's *šurinnu* in Tablet 20 of *Enūma Anu Enlil.* Allusions to other curious criteria for predicting eclipses are given in Hunger [1992] p. 219 (Report 382): The Moon disappeared on the 24th day. If the Sun on the day of the Moon's disappearance is surrounded by a halo, there will be an eclipse of Elam. In Kislīmu there will be a watch for an eclipse. The halo which surrounded the Sun and the Moon which disappeared appeared for the eclipse watch of Kislīmu.

Apparently watches were regularly kept for predicted eclipses; the king was informed in almost all cases by Letter. For lunar eclipses see the following:

1. Parpola [1993] p. 20 (Letter 26): To the 'farmer', my lord: your servant Issar-šumu-ereš. --- Concerning the wa[tc]h (for the lunar eclipse) about which the ['farmer'], my [lord], wrot[e to m]e: [...]. "If it should occ[ur], what is the word about it?" The 14th day (signifies) the Eastland, the month Simānu (signifies) the Westland, and the relevant "decision" (pertains) to Ur. And if it occurs, the (interpretation concerning the) region it afflicts and the wind blowing will be quoted as well.

2. Parpola [1993] p. 35 (Letter 46): To the king, my lord: your servant Balasī. --- Concerning the watch of the Moon about which the king, my lord, wrote to me, (the eclipse) will pass by, it will not occur.

3. Parpola [1993] p. 53 (Letter 71): Concerning the watch for a lunar eclipse about which the king, my lord, wrote to me, its watch will be (kept) [on the deci]ded [night]; (but) [whether] its [wat]ch should be during sunset, [we have not been able to de]cide.

4. Parpola [1993] p. 58 (Letter 78): [Concerning the watch for a lunar eclipse] about which the king, [my lo]rd, wrote to me, its watch will be (kept) tonight, in the morning watch. The eclipse will occur during the morning watch.

5. Parpola [1993] p. 107 (Letter 132): To the king, my lord: your servant Nergal-šumu-iddina. --- Concerning the watch about which the king, my lord, wrote to me, the Moon let the eclipse pass by, [it did not occ]ur.

6. Parpola [1993] p. 107 (Letter 133): To the king, my lord: your servant Nergal-šumu-iddina. --- The Moon [skipped the eclipse on the 14]th day. On the 15[th day ... we]nt.

7. Parpola [1993] p. 108 (Letter 135): To the king, my lord: your servant Babu-šumu-iddina. --- Concerning the watch about which the king, my lord, wrote to me, neither the Moon nor the eclipse were seen.

8. Parpola [1993] p. 120 (Letter 159): [To the king], my lord: your servant Bel-naṣir. [In Tešrītu(?)] there was no watch for the Moon. [On] the 28th of [Tešrītu(?)] the Sun made an eclipse; in Araḫsamnu the Moon [let] the eclipse [pass] by. Now then, in Nisannu, ...

As is clear from numbers 3 and 4 in the above list, the observers felt that they could predict the watch of the night in which the lunar eclipse would occur. Another instance is in Report 487 (Hunger [1992] p. 269): I wrote [to the k]ing, my lord, [as follows: "The Moon] will make an eclipse." [Now] it will not pass by, it will occur. ---: Ajjaru means Elam, the 14th day means Elam, the morning watch means [Elam]. Month, [day, watch all of them] refer to E[lam ...] --- From Nadinu. The interpretation here depends on omens such as *Enūma Anu Enlil* 21 II 1 (Rochberg-Halton [1988a] p. 235); cf. 16 II 11 (Rochberg-Halton [1988a] p. 87). Here again cf. the predictions of lunar eclipses on the 14th day of each month in Tablet 20, though that text does not attempt to predict the watch.

At least one Report indicates that sometimes the watch of the Moon was kept many nights during the month. It seems to be a form of Diary without numerical or planetary information.

Hunger [1992] pp. 116-117 (Report 207): [On the 29th(?) day] a possible(?) solar eclipse. [In month ... on the 1(?)]st [day] the Moon became visible. [The night] of the 11th day [was cloudy(?); in the] morning [watch] the Moon came out. [In the daytime of] the 11th day there was much [...]; the Moon set. [The ni]ght of the 12th day [was cloudy(?); in] the morning [watch] the Moon came out. [The daytime of the 1]2th day was cloudy; the setting of the Moon was not visible. The night of the 13th day [was cloudy(?); in] the morning [watch] the Moon came out. [The daytime of the 1]3th day was cloudy; (the Moon) was not visible; [... an eclipse(?) of] the Moon passed by. The night of the 14th day [was cloudy(?); in the mor]ning [watch] the Moon came out; the disk did not wane. [--- On the] ⌜15⌝th day the setting of [the Moon] was not visible; solstice(?).

There are a few observations of solar eclipses reported.

1. Parpola [1993] p. 107 (Letter 131): [To the king], my lord: [your servant] Nergal-šumu-iddina. --- The Sun was [eclipsed on the 29(?)]th [day, in the m]iddle of the day.

2. Parpola [1993] p. 113 (Letter 148): A certain Akkullanu has written: "The Sun made an eclipse of two fingers at sunrise."

3. Hunger [1992] p. 5 (Report 3): [Mannu-k]i-Harran [wrote to me] today: "The Sun was [eclips]ed on the 29th; [what day do you have] today? We have [the ...th]." 15 April -656.

4. Hunger [1992] pp. 63-64 (Report 104): On the 28th day, at 2½ *bēr[u* of the day ......] in the West [......] it also cover[ed ......] 2 fingers toward [......] it made [an eclipse], the east wind(?) [......] the north wind ble[w ...]. --- From [Akkullanu]. 15 April -656.

For another solar eclipse described by the relevant omens see Report 384 (Hunger [1992] pp. 220-221).

Some criteria, presumably non-mathematical, were also used to predict solar eclipses, as the following examples demonstrate.

1. Parpola [1993] pp. 34-35 (Letter 45): To the king, my lord: your servant Balasī. --- Concerning the watch of the Sun about which the king, my lord, wrote to me, it is (indeed) the month for a watch of the Sun. We will keep the watch twice, on the 28th of Araḫsamnu and the 28th of Kislīmu. Thus we will keep the watch of the Sun for 2 months. Concerning the solar eclipse about which the king spoke, the eclipse did not occur. I shall look again on the 29th and write (to the king).

2. Parpola [1993] p. 35 (Letter 46): To the king, my lord: your servant Balasī. ---. Concerning the watch of the Sun about which the king, my lord, wrote to me, does the king, my lord, not know that it is being closely observed? The day of tomor[row] is the only (day left); once the watch is over, (this eclipse), too, will have passed by, it will not occur.

3. Parpola [1993] pp. 130-131 (Letter 170): Concerning the solar eclipse, about which the king wrote to me: "Will it or will it not take place? Send definite word!" An eclipse of the Sun, like one of the Moon, never escapes me; should it not be clear to me and should I have failed (to observe it), I would not find out about it. Now since it is the month to watch the Sun and the king is in the open country, for that reason I wrote to the king: "The king should pay attention."

4. Parpola [1993] p. 300 (Letter 363): On the 27th day the moon stood there; on the 28th, 29th, and 30th we kept watch for the eclipse of the Sun, (but) he let it pass by and was not eclipsed. On the 1st the Moon was seen (again); the (first) day of Du'ūzu is (thus) fixed.

5. Hunger [1992] p. 28 (Report 47): On the 29th [of Araḫsamnu we observed] the Sun: [the S]un set; it let [the eclipse] pass by. [The Moon] will reject [the day] in Kislīmu. [From] Nabû-ahhe-eriba.

There are seven types of observations and predictions with respect to the planets: their being with a particular constellation; their heliacal rising or setting; their retrogression; their conjunctions with other planets or the Moon; their being in the halo of the Moon; their being visible during an eclipse; and their motion over a period of time. We shall present these for each planet, without repetitions when two or more planets are involved. Types I and II are related to periodic phenomena, but are not sharply enough defined. Most of the phenomena observed are of no use whatsoever in establishing a mathematical description of planetary motion.

*Saturn*

I. Being with a particular constellation (some of doubtful meaning):

1. Hunger [1992], p. 24 (Report 39): Saturn stands in the Crab.

2. Hunger [1992], p. 104 (Report 180): If the Lion is black ---. The black star is [Saturn].

II. Heliacal rising or setting

1. Hunger [1992], pp. 184-185 (Report 324): Satu[rn] became visible [in]side of the Lion. If the Lion is dark ---. If a planet rises in Abu ---. From Ašaredu the elder. 4/5 August -674.

III. Retrogression. No examples.

IV. Conjunctions with other planets or the Moon.

1. Parpola [1993] pp. 35-36 (Letter 47): To the king, our lord: your servants Balasī and Nabû-ahhe-eriba. ---. Concerning the planets [Satur]n and [Mars] ---. There is (still a distance of) about 5 fi[ngers] left; it (the conjunction?) is not y[et] certain. At this time we keep observing and shall write to the king, our lord. It (Mars) moves about a finger a day.

2. Hunger [1992], pp. 28-29 (Report 48): Mars passed below Saturn. --- [From Nabû-ahhe]-eriba. 6 or 10 June -668.

3. Hunger [1992], p. 55 (Report 95): Tonight Saturn approached the Moon. 3 April -667(?).

4. Hunger [1992], pp. 61-62 (Report 102): As to Mars which [moved(?)] towards [Saturn], it will absolutely not come close [......]. ---. From Akkullanu. 15 March -668

5. Hunger [1992], p. 77 (Report 125): [Mar]s comes close to Saturn. [From] Bulluṭu.

6. Hunger [1992], p. 98 (Report 166): Saturn entered the Moon. From Aššur-šarrani. See De Meis-Meeus [1991].

7. Hunger [1992], p. 164 (Report 297): Saturn ---. If a Sun disk stands above the Moon --- If the Sun [stands] in the manz[azu of the Moon] ---. Saturn [stands together with(?) the Moon]. ---. From Nabû-iqiša of Bor[sippa].

8. Hunger [1992], p. 186 (Report 327): [Ma]rs comes close to Saturn ---. From Ašaredu the elder, servant of the king. 15 March -668.

9. Hunger [1992], pp. 274-275 (Report 500): Saturn did not approach Venus for (less than) 1 cubit. 23 April or 7 May -677.

V. In the halo of the Moon.

1. Parpola [1993], pp. 94-95 (Letter 113): To the king of the lands, my lord: your servant Bel-uš[ezib]. ---. Saturn stands in the halo of the Moon. ---. The Crab stands in the halo of the Moon.

2. Hunger [1992], pp. 24-25 (Report 40): Saturn stands in the halo of the Moon. ---. Because the Moon was seen on the 13th day, because of that Saturn stood in the halo of the Moon. --- [If Sat]urn in front of the King [......]. From Nabû-ah[he-eriba].

3. Hunger [1992] pp. 25-26 (Report 41): Saturn stands in the halo of the Moon. If the Moon is surrounded by a halo, and Mars stands in it, ---. If the Moon is surrounded by a halo, and the King stands in it, ---. [The King] stands in the halo of the Moon. [From Nabû]-ahhe-eriba.

4. Hunger [1992] pp. 29-30 (Report 49): [If] the Moon is surrounded by a halo, and Mars [stands in it], ---. Saturn stands in the halo of the Moon. ---. Mars will reach Saturn. ---. From [Nabû]-ahhe-[eriba]. 15 March -668(?).

5. Hunger [1992] p. 49 (Report 82): If Mars keeps going around a planet, ---. Mars remained four fingers distant from Saturn, it did not come close. It did not reach it. --- If the Moon is surrounded by a halo, and Mars stands in it, ---. If the Moon is surrounded by a halo, and the Field stands in it, ---. The Field (means) the Furrow. From Balasī. 15 March -668.

6. Hunger [1992] pp. 67-68 (Report 110): On the 14th day --- Saturn stood inside the halo of the Moon. --- Saturn stands with the Moon on the 14th day ---. [Saturn] stands in the halo of the Moon. --- [Saturn] stands [with] the Moon. [From Akkull]anu.

7. Hunger [1992] p. 92 (Report 154): If the Sun stands in the halo of the Moon, ---. If the Moon is surrounded by a halo, and Ninurta stands in it, ---. From Nabû-mušeṣi.

8. Hunger [1992] pp. 99-100 (Report 168): If the Moon is surrounded by a halo, and the Sun stands in the halo of the Moon, ---. If the Moon is surrounded by a halo, and Mars stands in it, ---. If the Moon is surrounded by a halo, and two stars stand in the halo of the Moon, ---. If Mars and a

planet confront each other and stand there, ---. From Š[um]aya.[45] 15 March -668.

9. Hunger [1992] p. 100 (Report 169): [If] the Moon is surrounded by a halo, and Mars [stands] in it, ---. If the Sun stands in the halo of the Moon, ---. If two stars st[and] in the halo of the Moon, ---. From Bamaya. 15 March -668.

10. Hunger [1992] p. 105 (Report 181): If the Moon is surrounded by a halo, and Sa[turn] stands in it, ---. If ditto the King stands in it, ---. [If ditt]o Mars stands in it, ---. From Urad-Ea.

11. Hunger [1992] pp. 124-125 (Report 228): If the Moon is surrounded by a halo, and the Sun [stands in it(?) ---].

12. Hunger [1992] p. 169 (Report 301): [If the Moon] is surrounded by a halo, and the Cr[ab] stands in it, ---. If the Sun stands in the halo of the Moon, ---. [If] the Moon is surrounded by a halo, and the King stands in it, ---. From Zakir.

13. Hunger [1992] pp. 169-170 (Report 302): If the Moon is surrounded by a halo, and the True Shepherd of Anu stands in it, ---. [If the Sun] stands [in] the halo of the Moon, ---. [If the Moon] is surrounded [by a halo], and the King stands in it, ---. From Zakir.

14. Hunger [1992] p. 179 (Report 317): If the Moon is surrounded by a halo, [and the Crab] stands [in it], ---. If the Sun [stands] in the halo [of the Moon], ---. If the Moon is surrounded by a halo, and [two stars stand in the halo] with the [Moon], ---. If the Moon is surrounded by a halo, [and Mars] stands in it, ---. From Munn[abitu].

15. Hunger [1992] pp. 219-220 (Report 383): If the Moon is surrounded by a halo, and [two stars stand in the halo with the Moon], ---. If the Sun [stands] in the halo [of the Moon], ---. The Moon was surrounded by a halo, [and Saturn stood in it]. If the Moon is surrounded by a halo, and [Mars stands in it], ---. If the Moon (is surrounded) by a halo, and the Yoke stands in it, ---. The Yoke means Mars. ---. From Rašil [the elder], servant of the king. 15 March -668.

16. Hunger [1992] p. 230 (Report 403): If the Moon is surrounded by a halo, and [the Sun] stands in it, ---. From Rašil, son of Nu[rzanu].

17. Hunger [1992] p. 238 (Report 416): [If the Moon] is surrounded by a halo, and [Ma]rs stands in it, ---. [If] the Moon is surrounded by a halo, and the Sun stands in it, ---. [Satu]rn stands [in the halo of] the Moon. [If the Moon] is surrounded by a halo, and [two] stars s[tand] in the halo with the Moon, ---. [From] Nabû-iqbi [of] Cutha. 15 March -668.

18. Hunger [1992] p. 271 (Report 491): If the Moon is surrounded by a halo, and Mars stands in it, ---. [If ditto, and the Sun] stands [in it], ---. [Ma]rs reaches Saturn. ---. Mars --- is bright and carries radiance ---. Saturn

45 Name proposed by Koch-Westenholz [1995] p. 182 n. 1

--- is faint, and its radiance is fallen. --- From Šapiku of Borsippa. 15 March -668.

19. Hunger [1992] p. 278 (Report 501): [If the Sun] stands in the halo of the Moon, ---. Kislīmu, 14th day, year 1, Sargon king of Babylon (= 27 November -708).

VI. Being visible during an eclipse.

1. Parpola [1993] pp. 41-42 (Letter 57): To the king, my lord: your servant Balasī. ---. Jupiter, Venus, and [Sa]turn were present [during the (lunar) eclipse].

VII. Motion over a period of time.

1. Hunger [1992] pp. 221-222 (Report 386): [Saturn] stands in the Scales [in front of the Scor]pion; it will move [from] the Scales [into] the Scorpion; ---. --- [until it] comes out from the Sc[orpion]. ---. [From Rašil] the elder, [servant of] the king.

*Jupiter*

I. Being with a particular constellation.

1. Hunger [1992] pp. 29-30 (Report 49): Jupiter entered the Bull of Heaven. ---. From [Nabû]-ahhe-[eriba]. 15 March -668.

2. Hunger [1992] p. 270 (Report 489): If Jupiter has awesome radiance, ---. The King stands either to the right or to the left of Jupiter. Now it stands to the left of Jupiter for 3 fingers. From Rimutu.

II. Heliacal rising or setting.

1. Parpola [1993] p. 50 (Letter 67): To the king, [my lord]: your servant Nabû-[ahhe-eriba]. --- the observ[ation] of J[upiter(?)] and Me[rcury(?)] which, in the same [day], came forth in succession, ---. They [are at a distance] and will keep away from each other.

2. Parpola [1993] pp. 76-78 (Letter 100): [To] the king, my lord: [your servant] Akkullanu. --- Jupiter retained its position (KI.GUB); it was present for 15 more days (after the solar eclipse of 15 April -656):

3. Hunger [1992] pp. 5-7 (Report 4): In Simānu on the 5th day Jupiter stood in [...] where the Sun shines forth; [it was] bright, and its features were red; its rising [was as perfect as the rising of the Sun] ---. If Jupiter [car]ries radiance, ---. [If Jupiter] becomes steady in the morning, ---. From [Issar-šumu-ereš]. Cf. nos. 6 and 8.

4. Hunger [1992] pp. 32-33 (Report 24): If Jupiter [becomes visible(?)] in Abu, ---. If Jupiter [rises] in the path of [Enlil], ---.

5. Hunger [1992] p. 50 (Report 84): [If J]upiter [becomes visible] in Duʾūzu, ---. [If] Jupiter [rises] in the path [of Enlil], ---. If Jupiter [passes] behind [the Twins], ---.

6. Hunger [1992] p. 73 (Report 115): If Jupi[ter in Simānu] approaches and stands where the S[un shines] forth, (if) it is bri[ght and its features] are red, its rising is as perfect as [the rising of the Sun], ---. If Jupiter becomes

steady in the morning, ---. If Jupiter carries radiance, ---. If Jupiter is bright, ---. If Jupiter becomes visible in the path of Anu, ---. From Bulluṭu. Cf. nos. 3 and 8.

7. Hunger [1992] p. 99 (Report 167): Jupiter [becomes visible in the East and stands] in the sky all year. Now until Duʾūzu [---] additional days in [...]. ---. From Aššur-šumu-idd[ina]. Cf. no. 11.

8. Hunger [1992] pp. 100-101 (Report 170): [If Jupiter] becomes steady in the morning, ---. [If Jupiter in] Simānu approaches and stands where the Sun shines forth, (if) it is [br]ight and its features are red, its [ri]sing is as perfect as the rising of the Sun, ---. If Jupiter becomes visible in the path of Anu, ---. [From) Bamaya. Cf. nos. 3 and 6.

9. Hunger [1992] p. 108 (Report 184): [If Jupiter] becomes visible in Tešrītu, ---. [If Jupiter] becomes steady in the morning, ---.

10. Hunger [1992] p. 118 (Report 211): [If] Jupit[er rises] in the path of [Enlil], ---. If Jupiter at its appearance [is red, ---] ---. At its appearance the north wind [blew]. If Jupiter is br[ight], ---.

11. Hunger [1992] p. 138 (Report 248): Jupiter [becomes visible] in the [East and] stands in the sky for a year.[46] [---. L]ast year (on) this day J[upiter] did not [...] its year. Now, until the 10th day of Kislīmu, it has s[tood for ...] additional [days]. ---. From Nergal-eṭir. Cf. no. 7.

12. Hunger [1992] p. 142 (Report 254): [If Jup]iter becomes steady in the morning, ---. [If Jupi]ter carries radiance, ---. [If J]upiter is bright, ---. If Jupiter becomes visible in Ajjāru, ---. [From] Nergal-eṭir. Cf. no. 15. 29 April -669.

13. Hunger [1992] p. 160 (Report 289): If Jupiter becomes visible in Ulūlu, ---. If the stars of the Lion ---. [Ju]piter [stands in(?) the L]ion. 28 August -667.

14. Hunger [1992] p. 184 (Report 323): [If J]upiter passes to the West, ---. It is seen in front of the Crab. If Jupiter stands in the path of Enlil, and its light --- carries radiance, ---. At the beginning of your kingship Jupiter was seen in his manzāzu. ---. From Ašaredu, son of Damq[a].

15. Hunger [1992] p. 185 (Report 326): [If J]upiter becomes visible in Ajjāru, ---. [If J]upiter becomes visible in the p[ath of Anu, ---] ---. [If J]upiter is red at its appearance, ---. [If] Marduk (Jupiter? Mercury?) reaches the Stars, ---. [From A]šaredu the elder, [servant] of the king. Cf. no. 12. 3 May -669.

16. Hunger [1992] p. 187 (Report 329): Jupiter is passing to the West, ---. [Ju]piter stands in the sky for additional days. [From] Ašaredu the elder, servant of the king.

---

46 Cf. MUL.APIN II i 60.

17. Hunger [1992] p. 204 (Report 356): [If in Araḫsamnu) Jupiter becomes visible, ---. [If Jupiter] becomes visible in the *mešḫu* of Pabilsag, ---. [From] Aplaya of Borsippa. Cf. no. 18. 27 November -674.

18. Hunger [1992] p. 210 (Report 369): If Jupiter becomes visible in Araḫsamnu, ---. If Jupiter stands in the *mešḫu* of Pabilsag, ---. If the same star comes close to IN.DUB.AN.NA, ---. IN.DUB.AN.NA is the *mešḫu* of Pabilsag. From Nabû-šuma-iškun. Cf. no. 17. 27 November -674.

19. Hunger [1992] pp. 256-257 (Report 456): Jupiter stood there one month over its expected time (*adannu*). (If) Jupiter passes to the West, ---. Jupiter stood there one month over its expected time. From Bel-le'i, descendant of Egibi, exorcist. 30 October -674.

To be connected with this is Report 339 (Hunger [1992] p. 194): [Ju]piter remained ste[ady in] the sky for a month of da[ys].

III. Retrogression.

1. Hunger [1992] pp. 278-280 (Report 502): If Jupiter reaches and passes the King and gets ahead of it, (and if) afterwards the King, which Jupiter had passed and got ahead of, reaches and passes Jupiter, moving to its setting, ---. 26 November -678(?).

IV. Conjunctions with other planets and the Moon.

1. Parpola [1993] p. 64 (Letter 84): [To the king, my lord: your] servant [Akkullanu]. ---. Jupiter stood behind the Moon. ---. At this time, I am waiting for [the occultation] ---. The area behind the Moon should be shown to a eunuch who has a sharp eye; there is less than a span left to close. He should sta[nd] in shadow to observe.

2. Hunger [1992] p. 60 (Report 100): If Jupiter stands inside the Moon, ---. If Jupiter enters the Moon, ---. If Jupiter comes out to the back of the Moon, ---. 27 April -675(?).

3. Hunger [1992] p. 119 (Report 212): [If Venus] reaches Jupiter and [follows, variant]: approaches and stands, ---. [If Jupiter] reaches [Venus] and passes her, ---. [If Venus] comes close to [Jupite]r, ---. 8 April -670(?).

4. Hunger [1992] p. 120 (Report 214): If Jupiter [passes] to the right [of Venus, ---] ---. If Venus [reaches] Jupiter and follows, variant: [approaches and stands], ---. 8 April -670(?).

5. Hunger [1992] p. 136 (Report 244): If Jupiter goes with Venus, ---. From Nergal-eṭir. 8 April -670(?).

6. Hunger [1992] p. 160 (Report 288): Mars comes close to Jupiter. ---. From Nabû-iqiša of Borsippa.

7. Hunger [1992] p. 248 (Report 438): If Jupiter ent[ers] the Moon, ---. If the Moon covers Jupiter, ---. From Ṭabiya. 27 April -675(?).

8. Hunger [1992] p. 251 (Report 448): If Jupiter passes to the right of Venus, ---. If Jupiter ... to the right of Venus [and] stands, ---. From Ṭab-ṣilli-[Marduk], son of Bel-upahhir.

9. Hunger [1992] pp. 262-263 (Report 469): After it (a lunar eclipse in Araḫsamnu) Jupiter ent[ered] the Moon three times. 27 April -675(?).

V. In the halo of the Moon.

1. Hunger [1992] p. 8 (Report 6): If the Moon is surrounded by a halo, and Jupiter stands in it, ---. If the Moon is surrounded by a halo, and the Crab stands in it, ---. From Issar-šumu-ereš.

2. Hunger [1992] p. 42 (Report 71): The night of the 2nd day, Jupiter stood in the halo of the Moon. ---. From Nabû-ahhe-eriba.

3. Hunger [1992] pp. 54-55 (Report 93): If the Moon is surrounded by a halo, and Jupiter [stands] in [it], ---. From Balasī.

4. Hunger [1992] pp. 89-90 (Report 147): This night the Moon was surrounded by a halo, [and] Jupiter and the Scorpion [stood] in [it]. ---. From Nabû-mušeṣi.

5. Hunger [1992] p. 116 (Report 205): If the Moon [is surrounded] by a halo, and Jupiter [stands] in [it], ---. If the Moon is surrou[nded] by a halo, and the King [stands] in it, ---.

6. Hunger [1992] p. 120 (Report 215): If Jupiter [in Simānu stands] in the halo of the Moon, ---. A halo of Simānu ---.

7. Hunger [1992] p. 153 (Report 277): [If the Moon] is surrounded by a halo, and Jupiter stands in it, ---. From Nergal-eṭir.

8. Hunger [1992] p. 210 (Report 370): If the Moon is surrounded by a halo, and Jupiter stands in it, ---. If the Frond stands with [...], ---. From Nabû-šuma-iškun. 23 December -674(?).

9. Hunger [1992] pp. 227-228 (Report 398): [If the Moon is surrounded by a halo, and Jup]iter stands in it, ---. [If the Moon is surrounded by a halo, and Jupite]r stands in it, ---. [From Raš]il the elder, servant of the king.

VI. Being visible during an eclipse; see Saturn VI 1.

1. Parpola [1993] p. 57 (Letter 75): To the king, my lord: your servant Nabû-ahhe-eriba. ---. Jupiter and Venus were present during the (lunar) eclipse until it cleared.

2. Hunger [1992] pp. 168-169 (Report 300): [If in its eclipse] (i.e., the lunar eclipse in Pabilsag on 14 Simānu = 22 May -677) Jupiter stands there, ---. [......] Zakir. Cf. nos. 3 and 4.

3. Hunger [1992] pp. 178-179 (Report 316): In the eclipse [of the Moon] (on 14 Simānu) Jupiter stood there: ---. From Munnabitu. Cf. nos. 2 and 4.

4. Hunger [1992] pp. 293-294 (Report 535): [If in this eclipse] (in Pabilsag on 15(!) Simānu) Jupiter stands there, ---.

VII. Motion over a period of time.

1. Parpola [1993] pp. 298-299 (Letter 362): ... [did not ap]pear, and the day [---. Last year] it became visible on the 22nd of Ajjāru in the [Old Man]; it disappeared in Nisannu of the [present] year, on the 29th day. Jupiter [remains invisible] from 20 to 30 days; now it kept itself back from the sky

for 35 days. It appeared on the 6th of Simānu in the area of the True Shepherd of Anu, exceeding its term (edānu) by 5 days.

With this goes Letter 363 (Parpola [1993] p. 300): Concerning Jupiter about which I previously wrote to the king, my lord: "It has appeared in the path of Anu, in the area of the True Shepherd of Anu," it was low and indistinct in the haze, (but) they said: "It is in the path of Anu," ---. Now it has risen and become clear; it is standing under the Chariot in the path of Enlil.

2. Hunger [1992] p. 50 (Report 84): Jupiter [is visible(?) behind the Twins]. Now I wrote to the king, my lord, as soon as I observed [it]. Later, once (it) has moved on and approached the Crab, I shall copy another report and let the king, my lord, hear (it).

*Mars*

I. Being with a particular constellation.

1. Hunger [1992] p. 27 (Report 45): If the Wolf [reaches] the Lion, ---. The Wolf is [Mars]; ---. Mars [reaches the Lion]. ---. From Nabû-ahhe-eriba.

2. Hunger [1992] pp. 37-38 (Report 64): If the Strange One (Mars) [comes close to] the Twins, ---. From Nabû-ahhe-eriba.

3. Hunger [1992] p. 48 (Report 80): [If M]ar[s stands] in the Crab, ---. From Bal[asī].

4. Hunger [1992] pp. 48-49 (Report 81): [If M]ars en[ters the Lion and stands], ---. Mars stood in [the Lion. If Ma]rs [stands] in the East, ---. Mars stands [in] the Lion. [From] Bal[asī].

5. Hunger [1992] p. 121 (Report 216): Mars stan[ds] below the right foot of the Old Man. It has not ente[red] it, but stan[ds] in its area.

6. Hunger [1992] pp. 121-122 (Report 219): If the Plow [comes close] to the Scorp[ion], ---. Mars [came close] to [the Scorpion].

7. Hunger [1992] p. 218 (Report 380): Mars reached the Crab. [---] ---. [From] Rašil the elder, servant of the king. April -674(?).

8. Hunger [1992] pp. 228-229 (Report 400): [If Mars] comes close to the Old Man, ---. [From Rašil] the elder, [servant of the king].

9. Hunger [1992] p. 255 (Report 452): If the Cra[b comes close to the Plow], ---. If the Strange One (Mars) [comes close] to the Cr[ab], ---. If Mars [comes close] to [the Crab], ---. From Ahhešā of Uruk.

10. Hunger [1992] pp. 278-280 (Report 502): Mars stands in the Scorpion. 26 November -678(?).

11. Hunger [1992] p. 280 (Report 503): If the Strange One comes close to Enmešarra, ---.

II. Helical rising and setting.

1. Parpola [1993] pp. 76-78 (Letter 100): [To] the king, my lord: [your servant] Akkullanu. ---. [M]ars has appeared in the path of Enlil at the feet of

the Old Man; it is faint and of a whitish color. I saw it on the 26th of Ajjāru when it had (already) risen high.

2. Parpola [1993] p. 313 (Letter 381): In the evil of Mars which exceeded its expected time (*adannu*) and ap[peared] in the Hired Man, ---.

3. Hunger [1992] p. 8 (Report 7): Twice or thrice we watched for Mars today (but) we did not see (it); it has set. ---. From Issar-šumu-ereš.

4. Hunger [1992] pp. 72-73 (Report 114): [Ma]rs became visible in Duʾūzu ---. Mars rises scintillating and its radiance is yellow --- If Mars becomes faint, ---; if it becomes bright, ---. If Mars goes behind Jupiter, ---. From Bulluṭu. 25 June -669.

5. Hunger [1992] p. 88 (Report 143): [If] the Red [planet] (Mars) is bright, ---. [If Mars becomes visible in Duʾūzu], ---. [From Nab]û-mušeṣi.

6. Hunger [1992] p. 195 (Report 341): [If] in Duʾūzu Mars becomes visible, ---. [If] a planet stands in the North, ---. If the Strange One (Mars) comes close to the Twins, ---. From Ašaredu the younger. 25 June -669.

7. Hunger [1992] p. 239 (Report 419): If Mars becomes visible in Ulūlu, ---. If the Red planet [becomes visible], ---. Mars at its appearance carries radiance. From Nabû-iqbi.

III. Retrogression and stations.

1. Parpola [1993] pp. 80-81 (Letter 104): Mars has gone on into the Goat-Fish, halted, and is shining very brightly.

2. Hunger [1992] p. 31 (Report 52): Mars stands inside Pabilsag; ---. Mars became stationary in Pabilsag and stood there. Afterwards, in Simānu, it will turn and move forw[ard. From Nabû-ahhe-eri]ba. 15 April -666(?).

3. Hunger [1992] p. 32 (Report 53): Mars has turned around, started moving, and is going forward in the Scorpion.

IV. Conjunction with other planets or the Moon. See Saturn IV 1, 2, 4, 5, and 8, and Jupiter IV 6.

1. Parpola [1993] pp. 45-46 (Letter 63): To the king, [our] lo[rd]: your servant[s] Bal[asī] and Bamaya. ---. If the Stars flare up and go before Venus, ---. The Stars as a planet is Mars.

2. Hunger [1992] p. 30 (Report 50): [The Stars] will come close to the Moon. ---. [The Stars are] Mars. ---. [From Nabû]-ahhe-eriba.

3. Hunger [1992] p. 42 (Report 72): If the Stars come close to the front of the Moon and stand, ---. The Stars are ⌈Mars⌉. ---. From Nabû-ahhe-eriba.

4. Hunger [1992] pp. 173-174 (Report 311): If Mars comes close to the front of the Moon, ---. From Zakir.

5. Hunger [1992] p. 235 (Report 415): If Mars stands(?) in [...] to the left of Venus, ---. From Rašil [---].

6. Hunger [1992] p. 268 (Report 484): If Mars comes close to the front of the Moon and stands, ---. From Nabû-eriba(?).

V. In the halo of the Moon. See Saturn V 3, 4, 5, 8, 9, 10, 15, 17, and 18.

1. Hunger [1992] p. 213 (Report 376): If the Moon is surrounded by a halo, and Mars [stands] in it, ---. If a star stands in the halo of the Moon, ---. Mars is the star [...] If the Moon is surrounded by a halo, and the Stars stand in it, ---. The Stars are Mars. ---. From Nabû-šuma-iškun.

2. Hunger [1992] pp. 233-234 (Report 412): If the Moon is surrounded by a halo, and Mars stands in it, ---. If the Moon is surrounded by a halo, and a planet stands in it, ---. [If the Moon] is surrounded by a halo, and the Field, behind which are the Stars, stands in it, ---. From Rašil.

VI. Being visible during an eclipse. No example.

VII. Motion over a period of time.

1. Parpola [1993] pp. 131-132 (Letter 172): "Mars has become visible; why have you not written?" Mars was sighted in the month of Abu; now it has approached to within 2;30 spans of the Scales. As soon as it has come close to it, I shall write its interpretation to the king, my lord. What was sighted now is Mercury in the Goat-Fish.

2. Parpola [1993] pp. 165-166 (Letter 206): [Concerning] Mars [about which the king, my lord], wrote to me, --- it moves [towards] the Fu[rrow ...].

3. Parpola [1993] pp. 205-207 (Letter 262): To the king, my [lord]: your servant Ma[rduk-šakin-šumi]. ---. Con[cerning the observation(?)] of Mars [about which the king, my lord], wrote to me: ---. It will reach [the Furrow(?)].

4. Parpola [1993] pp. 285-286 (Letter 350): [Ma]rs, which previously was approaching [Pabilsa]g, [......] to en[ter ---].

5. Hunger [1992] pp. 33-34 (Report 55): [M]ars did not approach the Fish of the Sky [...] ...; it departed from it. 14 December -666.

6. Hunger [1992] p. 51 (Report 85): Mars has departed, it goes forward; it will become stationary [in] the Scorpion. [To]morrow I shall inform the king, my lord. [From] B[alas]ī.

7. Hunger [1992] pp. 60-61 (Report 101): Mars has reac[hed] the Crab and entered it. I kept wat[ch]; it did not become stationary, it did not stop; it tou[ched] the lower part (of the Crab) and goes on. (Its) going out (of the Crab) remains to be s[een]. When it will have gone out, ---. Cf. no. 9.

8. Hunger [1992] p. 174 (Report 312): When Mars had completed the region (*qaqqar*) of [...], it we[nt out(?)] (and) slowed down ⸢in⸣ its course (*tāluku*). Until 5 days [---].

9. Hunger [1992] p. 260 (Report 462): [Ma]rs has entered the Crab ---; it will not stand in it, it will not become stationary and it will not tarry; it will move out quickly. From Bel-naṣir. Cf. no. 7.

10. Hunger [1992] pp. 262-263 (Report 469): Mars [stood] 7 months in the Twins (!!).

*Venus*

I. With particular constellations.

1. Parpola [1993] p. 66 (Letter 88): Venus [is approaching the Crab]; ---. If the She-goat [approaches the Crab, ---] ---.

2. Hunger [1992] pp. 142-143 (Report 255): Venus stands in front of the True Shepherd of Anu.

3. Hunger [1992] p. 155 (Report 282): If at the beginning of the year the Stars stand on the left(?) of Venus, ---. From Nergal-eṭir.

4. Hunger [1992] p. 218 (Report 380): [Venu]s(?) stands with the foot of the Old Man. April -674(?).

5. Hunger [1992] pp. 259-260 (Report 461): Venus sta[nds] in the Stars. ---. Venus stands in the Bull of Heaven. ---. From Bel-leʾi, descendant of Egibi, exorcist.

6. Hunger [1992] pp. 278-280 (Report 502): Venus stands in Pabilsag. 26 November -678(?).

II. Heliacal rising and setting.

1. Parpola [1993] p. 23 (Letter 31): Venus has arisen [at] the time of its [expected time (*adannu*)(?)].

2. Parpola [1993] pp. 37-38 (Letter 51): To the ki[ng, my lord]: your servant [Balasī]. ---. Concerning [Venus] about which the king, my lord, [wrote to me: "I am] told that it has [become visible]", the man who wrote (thus) to the king, [my lord], is in ignorance. He does not k[now] the expected time (*adannu*) [---] the prowli[ng(?) of Venus]. ---. Venus is not y[et] vis[ible]. Tonight, as I am sending [this] message to the king, [my] lord, we [see] only Merc[ury]; we do not [see] Venus. At this time it should be situated under [the Hired Man(?)] opposite [Saturn(?)]. Cf. no. 3.

3. Parpola [1993] p. 54 (Letter 72): [He who] wrote to the king, my lord: "Venus is visible, it is visible [in the month Ad]dāru," is a vile man, an ignoramus, a cheat! [And he who] wrote to the king, my lord: "Venus is [...] rising in the H[ired Man" does] not [speak] the truth. Venus is [not] yet visible! Cf. no. 2.

4. Parpola [1993] p. 55 (Letter 73): To the king, my lord: your servant Nabû-ahhe-eriba. ---. Mercury (signifies) the crown prince, and it is vis[ib]le in the Hired [Man]. Venus [is] visible in [Bab]ylon, in the home of [his] father; and the Moon will complete the day in the month Nisannu.

5. Parpola [1993] pp. 300-302 (Letter 364): On the 10th day of Duʾūzu Venus was visi[ble] in the Lion.

6. Hunger [1992] pp. 4-5 (Report 2): Venus (and) Mercury are about to set. ---. From the Chief Scribe.

7. Hunger [1992] p. 7 (Report 5): Venus se[t] in the West. If Venus in [Ṭebētu] from the 1st to the 30th day disappears in the West, ---. If Venus moves in the path of Ea and sta[nds, ---] ---. From Issar-šumu-ereš. 27 December -668(?).

8. Hunger [1992] p. 16 (Report 27): [Venus] made her [posit]ion perfect [......] she became visible quickly, ---. [If Ven]us stays in her position (*manzāzu*) for long, ---. If the rising of [Venus] is seen early, ---. From Issar-šumu-ereš.

9. Hunger [1992] pp. 17-18 (Report 31): This rain and thunder concerns the expected time (*adannu*) of the sighting of Venus ---. From Issar-šumu-er[eš].

10. Hunger [1992] pp. 19-20 (Report 36): If there is an earthquake in Šabāṭu, ---. This (earthquake) [was predicted] by the (event) when Venus disappeared and [......] ---. From Issar-šumu-[ereš].

11. Hunger [1992] p. 34 (Report 56): If Venus disappears in the East in Nisannu from the 1st to the 30th day, ---. If Venus keeps, variant: kept, changing her po[sition] (KI.GUB), ---. [If Venus] descends darkly to the horizon and sets, ---. From Nabû-ahhe-[eriba].

12. Hunger [1992] p. 56 (Report 96): [If Venus disappears in the West in Abu from] the 1st to the 30th day, ---. [If Venus in Abu] descends darkly to the horizon and sets, ---. [From Bala]sī.

13. Hunger [1992] pp. 88-89 (Report 145): Venus set in the East. If Venus gets a ṣirḫu, ---. This means she does not complete her days, but sets. If Venus disappears in the East in Nisannu from the 1st to the 30th day, ---. From [Nabû]-mušeṣi. 5 April -665.

14. Hunger [1992] p. 92 (Report 156): [If V]enus disappears [in the West] in [Abu from the 1st day] until the 30th day, ---. [Fr]om Nabû-mušeṣi.

15. Hunger [1992] pp. 102-103 (Report 175): Venus became visible in the West in the path of Enlil. ---. If Venus becomes visible in Simānu, ---. If Venus becomes visible in the path of Enlil, ---. Until the 5th or 6th day she will reach the Crab. ---. From Šumaya.

16. Hunger [1992] p. 137 (Report 246): Venus disappeared in the West. If Venus in Abu descends darkly to the horizon and sets, ---. If Venus in Abu, from the 1st to the 30th day, disappears in the West, ---. During this month it will become visible in the East in the Lion. From Nergal-eṭir.

17. Hunger [1992] pp. 137-138 (Report 247): [If Venus] becomes visible in Simānu, ---. [If the rising of Venus] is seen [early, ---] ---. Venus comes close to the Crab. ---. [From] Nergal-eṭir. 20 June -673(?).

18. Hunger [1992] p. 138 (Report 248): Venus ---. Venus rises ---. Venus rises ---. From Nergal-eṭir.

19. Hunger [1992] pp. 193-194 (Report 338): If in Kislīmu from the 1st day to the 30th day Venus disappears in the East, ---. The lord of kings will say: "The month is not (yet) finished. Why did you write me good or bad (results)?" The scribal art is not heard about in the market place. Let the lord of kings summon me on a day which is convenient to him, and I will investigate and speak to the king, my lord. From Ašaredu.

20. Hunger [1992] p. 198 (Report 349): If Venus [becomes visible] in Simānu, ---. [If Venus becomes visible] in the path of Enl[il, ---] ---.

21. Hunger [1992] p. 204 (Report 357): [If Ve]nus becomes visible in Šabāṭu, ---. Venus stands inside Anunītu. If Venus becomes visible inside the Field, ---. From Aplaya of Borsippa. Cf. no. 24.

22. Hunger [1992] p. 254 (Report 451): If Venus disappears in the East in Nisannu from the 1st to the 30th day, ---. From Ahhe[šā of] Uruk.

23. Hunger [1992] p. 294 (Report 536): [If Venus] se[ts] in Nisannu, ---. Venus [stands] in the Stars. ---. [From ...]a, son of Bel-ušallim.

24. Hunger [1992] p. 295 (Report 538): [If Venus] becomes visible [in Šabā]ṭu, ---. [Venus stand]s in Anunītu. Cf. no. 21. 14 February -679(?).

25. Hunger [1992] p. 295 (Report 539): If Venus [becomes visible] in [Šabāṭu], ---. If Venus [...] in [Šabāṭu], ---.

26. Hunger [1992] p. 295 (Report 540): If Venus in month [......] sets early, ---.

27. Hunger [1992] p. 296 (Report 541): If Venus is red, ---. If the light of Venus at its rising is like a clo[ud(?) ...]. If Venus at its rising until the end of Abu [---].

III. Retrogressions. No examples.

IV. Conjunctions with other planets or the Moon. See Saturn IV 9; Jupiter IV 3, 4, 5, and 8; and Mars IV 1 and 5.

1. Hunger [1992] pp. 30-31 (Report 51): Venus becomes visible in Pabilsag. --- Mercury stands in the Wes[t behin]d(?) Venus. ---. If Venus [becomes visible ...] in Kislīmu, ---. From Nabû-ahhe-eri[ba].

2. Hunger [1992] p. 43 (Report 74): Mercury [and Venus are apart] from each other, they pa[ss]ed and went [on]. From Nabû-ahhe-eriba.

V. In the halo of the Moon. No examples.

VI. Being visible during an eclipse. See Saturn VI 1 and Jupiter VI 1.

VII. Motion over a period of time.

1. Parpola [1993] p. 176 (Letter 224): Venus will reach the Furrow.

2. Hunger [1992] pp. 33-34 (Report 55): If (Venus) came close to the Breast of the Scorpion, it would be copied like this; now so far she has not approached it. When she will have approached (it) she will not come close, she will pass (it) at some distance. The planets pass like this above the stars which are in their path. 14 December -666.

3. Hunger [1992] pp. 169-170 (Report 302): [If Venus(?) --- from] the 1st to the 30th day [---]. Venus [...] days [.....] did not reach [...] and set, [---] ---. From Zakir.

4. Hunger [1992] p. 230 (Report 403): Venus [......] a month and 10 days [---]. If Venus [disappears] in Nisannu [from the 1st] to the 30th day in [the East, ---]. If Venus at her rising ---. From Rašil, son of Nu[rzanu].

5. Hunger [1992] p. 295 (Report 537): [If] Venus [disappears (Ω)] in [Nisannu, ---] ---. [If] Venus [becomes visible (Γ)] inside [the Field, ---] ---. 13 March -672.

*Mercury*

I. Being with a particular constellation. See Mars VII 2.

1. Hunger [1992] p. 89 (Report 146): Mercury stood inside the Lion. ---. From Nabû-mušeṣi.

2. Hunger [1992] pp. 136-137 (Report 245): Mercury stands together with the King. ---. From Nergal-eṭir.

3. Hunger [1992] p. 144 (Report 258): Mercury stands in [...]. From Nergal-eṭir.

II. Heliacal rising or setting. See Jupiter II 1 and Venus II 2, 4, and 6.

1. Parpola [1993] p. 37 (Letter 50): [To the k]ing, [our] lord: your [servants] Balasī and [Nabû]-ahhe-eriba. ---. [Concer]ning Mercury [about which] the king, our lord, [wr]ote to us: "I have heard it [can be seen in B]abylon," [he who] wrote to the king, our [lord], may really have observed it, his eyes must have fallen on it. We ourselves have kept watch (but) we have not observed it. [On]e day it might be too early, [anoth]er day it might be flat (on the horizon). [To see it] our [e]yes sho[uld have f]allen on it.

2. Parpola [1993] p. 60 (Letter 81): Mercury has not yet appeared.

3. Parpola [1993] p. 115 (Letter 152): [Concerning Venus(?) about which the king wrote]: "What (month) do you h[ave wh]en it became vis[ible]?", [it appeared] in the month Addaru. And as to what was s[aid to the king: "It is visible] in the Hired [Man]", the Hired Man will app[ear] in the East [either] tomorrow or the day af[ter tomorrow]. It is not Mercury; (Mercury) itself is visible [in the West].

4. Hunger [1992] p. 30 (Report 50): [Mer]cury became visible on the 16th [of Si]mānu. ---. [From Nabû]-ahhe-eriba.

5. Hunger [1992] p. 42 (Report 73): If the Yoke is low and dark when it comes out, ---. [The Yoke] (means) Mercury; it is faint. ---. Mercury becomes visible in the Goat-Fish. From Nabû-ahhe-eriba.

6. Hunger [1992] p. 44 (Report 76): Mer[cury] is s[een ---]. From [Nabû-ahhe-eriba].

7. Hunger [1992] p. 50 (Report 83): Concerning Mercury, about which the king, my lord, wrote to me: yesterday Issar-šumu-ereš had an argument with Nabû-ahhe-eriba in the palace. Later, at night, they went and both made observations; they saw (it) and were satisfied. From B[a]lasī.

8. Hunger [1992] p. 72 (Report 113): Mercury became visible in the East. ---. From Bulluṭu.

9. Hunger [1992] pp. 92-93 (Report 157): Mercury became visible in the East. ---. From Nabû-mušeṣi.

10. Hunger [1992] p. 93 (Report 158): If a planet becomes visible in Du'ūzu, ---. If Ḫabaṣirānu flickers(?) when it comes out, ---. These (omens) are from Mercury. From Nabû-mušeṣi.

11. Hunger [1992] p. 141 (Report 253): Mercury stands inside the Swallow. ---. Mercury became visible in the East. ---. [From] Nergal-eṭir.

12. Hunger [1992] p. 145 (Report 259): [Merc]ury is seen at the appearance of the Moon. [From] Nergal-eṭir.

13. Hunger [1992] p. 155 (Report 281): [M]ercury became visible. ---. From Nergal-eṭir.

14. Hunger [1992] p. 185 (Report 325): Mercury became visible in the East in the region of the Furrow. ---. If in Ulūlu the Kidney becomes visible, ---. The Kidney is Mercury. From Ašaredu, son of Damqa, servant of the king. 4 September -674.

15. Hunger [1992] p. 218 (Report 381): M[erc]ury stands inside Pabilsag ---. Mercury [became visible] in the East; its expected time (*adannu*) will not pass; its [...] is made manifest. ---. [From Raš]il the elder, servant of the king.

16. Hunger [1992] p. 255 (Report 454): If a planet becomes visible in Ajjāru, ---. If the star of Marduk (Mercury) reaches the Stars, ---. [From] Bel-ahhe-eriba.

17. Hunger [1992] pp. 280-281 (Report 504): If a planet rises in Araḫsamnu, ---. Mercury stands in the Scorpion.

18. Hunger [1992] p. 281 (Report 505): Mercury [became vi]sible in the West in the Hired Man.

III. Retrogression. No examples.

IV. Conjunctions with other planets or the Moon. See Venus IV 1 and 2.

V. In the halo of the Moon. No examples.

VI. Being visible during an eclipse. No examples.

VII. Motion over a period of time.

1. Hunger [1992] pp. 54-55 (Report 93): The star of Marduk, Mercury, is going beyond its position (*manzāzu* = KI.GUB) and ascends. ---. From Bala[sī].

2. Hunger [1992] p. 152 (Report 274): [If] Mercury dis[appears] in the West, when it appears(?), ---; when it disappears, ---. From Nergal-eṭir.

3. Hunger [1992] pp. 268-269 (Report 486): [Me]rcury became visible in the West [wit]h the Stars. It will keep getting higher [...] into the Old Man. ---. From Nadinu. Cf. no. 4.

4. Hunger [1992] p. 280 (Report 503): (If) Mercury becomes visible in Nisannu ---. Mercury became visible in the Bull of Heaven, it reached(?) the Old Man.

As the system demanded, the Letters and Reports record only observations of ominous planetary phenomena. This material, then, demonstrates that observers throughout Mesopotamia were observing such phenomena during the seventh century B.C., and probably had been doing so for centuries. What they recorded, however, being undatable and imprecise with respect to time and longitude, was useless for the purpose of constructing mathematical models. Following the precedent of the δ section of the Venus Tablet of Ammiṣaduqa and the planetary section of MUL.APIN, these observers did utilize rough notions of the intervals between the Greek-letter phenomena. Thus, Jupiter II 7 and II 11 record an (expected) period of visibility (Γ→ Ω) of 1 year, and Jupiter VII 1 one of invisibility (Ω→Γ) of 30 days; the expected period is called an *adannu*, a concept also referred to in Mars II 2; Venus II 1, 2, 9, 13, 17, 19, and 26; and Mercury II 15. Jupiter II 7, 11, 16, and 19, and VII 1 also refer to "additional days" beyond the *adannu*. Finally, the month (Mars III 2 and Venus II 16 and 19) or the day (Venus II 15) or the constellation (Mars VII 6 and 9 and Venus II 16) on which the phenomenon will occur can be predicted, though we do not know with what accuracy. The *adannus* for Jupiter were given in MUL.APIN; perhaps those used for the other planets were derived from the same source. From a practical point of view, the observers seem to have had difficulty in identifying Mercury; see Mercury II 1 and 7.

### *1.2. Other Contemporary Observations*

Esarhaddon in an inscription found at Assur (Borger [1956] p. 2) states that in his first year Venus appeared in the West, in the Path of the stars of Ea, reached its <*ašar*> *niṣirti*, and disappeared, while Mars shone brightly in the Path of the stars of Ea. Since the edge of the Path of Ea cuts the ecliptic at about 210° and 330°, a first visibility of Venus in the West occurred on about 20 January -679 with a longitude of about 299°, it reached its *ašar niṣirti* in the Fish in February, and set in the West on about 14 October -679 with a longitude of about 200°. Mars was in the Path of Ea from about 18 December -680 till about 24 June -679; it was retrograde (and therefore bright) in the Path of Ea from about 4 September to 1 November -679. In another inscription, from Babylon (Borger [1956] p. 17), Esarhaddon reports that in his first year Jupiter approached the Sun in Simānu, had its first visibility, reached its *ašar niṣirti* in the month Pēt-bābi, and then had its first station. Jupiter set heliacally in the West on about 24 May -679 (the conjunction of Simānu had occurred on 30 April), rose heliacally on about 26 June -679, and reached its *ašar niṣirti* in the Crab in late September. The conjunction of the month here called Pēt-bābi occurred on 23 September. Jupiter's first station occurred on about 24 October. Thus, the statements fit the astronomical facts well, but are not presented with the details of position and date that would make them useful to an astronomer.

A step in this direction is provided by three of a group of four fragments of observational texts concerning Mercury published by Pingree-Reiner [1975]. The three are Rm 2, 361, K 6153, and Rm 2, 303, all from Nineveh; the fourth is a Neo-Babylonian tablet, BM 37467. The reports follow the pattern: in month *a* on day *b* Mercury became visible in the East/West, the Sun being *c* UŠ to the left/below; it remained in the East/West for *d* days, and then became invisible in the East/West on month *e* on day *f*; it remained invisible for *g* days. The Sun's being *c* UŠ to the left (below the Eastern horizon) or below (the western horizon) is a measure of the time between sunset and the first sighting of Mercury; it could have been measured with a water-clock or estimated from meridian-crossings of *ziqpu* stars or "strings". The use of the *ziqpu* stars is recommended in a tablet from Uruk, von Weiher [1998] 269 rev. 1-5.

However, since there are, at least in what survives of these fragments, no indications of regnal years that would provide the possibility of placing the observations in a chronological sequence, they are useless to those investigating the periodic motion of Mercury.

## 2. *The Diaries*

The Astronomical Diaries, or, more simply, Diaries are records of observations and computations made during each period of half a year (six or seven months). The oldest so far dated was inscribed in -651, but the series probably began in Babylon in the first year of Nabû-nāṣir, -746 (a date proposed for reasons given below), while the latest securely dated Diary is from -60. This means that the tradition of keeping the Diaries persisted through seven centuries—or even eight, if the Diaries continued to be kept till the end of cuneiform writing in the late first century A.D. During this time-span Babylonia was ruled by native Dynasties, Achaemenid Persians, Hellenized Macedonians, and Parthians, so that it is unlikely that the supporting institution was the state. There is some evidence, from the late second century B.C., that the observers for the Diaries were employed by the Temple of Marduk in Babylon, the Esagil; the relevant documents are discussed in van der Spek [1985] pp. 548-555.

The purpose of the compilation of the Diaries has been much debated. Two recent studies take opposite stands: Swerdlow [1998] argues that they were intimately connected with the Mesopotamian practice of reading celestial omens, while Slotsky [1997], following a suggestion by Pingree, interprets them as intended, as far as the celestial observations are concerned, for astronomical purposes. The arguments in favor of the latter hypothesis are as follows.

1. The omens appearing in *Enūma Anu Enlil*, the Reports, and the Letters include a majority that are in no sense periodic; they are phenomena caused

by distortions of light as it passes through the atmosphere or are horizon events that are either too trivial to be predicted (e.g., the planet changing its KI.GUB) or are not subject to mathematical analysis (e.g., an unnamed fixed star or planet being in the vicinity of a named planet when it rises or sets).

2. Of the Greek-Letter Phenomena only heliacal risings and settings occur as omens. The stations of the superior planets are of interest to the authors of the Reports and Letters, not, however, as omens; it was the retrogressions of the planets, especially from one constellation to another, that were ominous. The acronychal risings of the superior planets are not omens either. And ominous are the planets' relationships to the constellations, not their passings by of the individual Normal Stars.

3. The Diaries record periodic phenomena that were in fact regarded as predictable (they are full of predictions). All other celestial phenomena are omitted; the weather is reported because it affects observations. There are stray references to non-periodic events, like meteors, "falls of fire", and, from the Babylonian point of view, comets, and there are occasional citations of ominous events drawn from *Šumma izbu*, as there are to historical events.

4. The economic summaries and the records of the level of the river only indicate that the compilers of the Diaries regarded them as at least potentially periodic. There is no evidence that they believed or imagined that there was a connection between the occurrences of the Greek-Letter phenomena and the price of commodities.

5. The Diaries, then, should be viewed as an explicit expression of the Babylonians' belief that celestial motions are periodic, that they represent the order imparted to the heavens by Marduk according to *Enūma eliš*. We have seen that they had successfully constructed mathematical models to predict a number of phenomena beginning in the Old Babylonian period; the Diaries and the tradition they initiated, culminating in the mathematical astronomy of the ACT material, attest to the vigor of their belief in an ordered universe.

6. The Diaries treat periodic phenomena as predictable; this deprives them of their meaning as omens. For omens, celestial or otherwise, are sent to man as warnings by the gods. They must be seen, not computed, and they must occur randomly. The scribes of the Diaries certainly continued to believe in omens since they report some, but they cannot be shown to believe that the celestial and terrestrial phenonoma they primarily recorded were ominous. The reason for the inclusion of non-periodic phenomena such as historical events in the Diaries remains unclear to us.

The earliest reference to tablets containing Diaries is that made by Epping-Strassmaier [1890/1891] pp. 226-227 (of ZA 6) to Rm 845, a part of the Diary for -328; to BM 92688+92689, a part of the Diary for -273; and to Rm IV 397 (= BM 33837), part of the Diary for -232. They also provided copies of all three (*ibid.* 232, 234, and 236 respectively) and, in Epping-

Strassmaier [1892], transliterations and German translations of the last two. Kugler [1907] pp. 75-79 published BM 35195, which is what remains of the Diary for -378. The remains of the Diary for -567, preserved as VAT 4956, were used by Weidner [1912a] to prove that Babylonian "scientific" astronomy is older than the famous report of observations made in year 7 of Cambyses (-522) (for this text see II B 6). The Diary for -567 was published in transcription and German translation and with a commentary in Neugebauer-Weidner [1915].

For many years after 1915 no new Diaries were published, in large measure due to the inaccessibility of the tablets in the British Museum. They are described as a class and several tablets are identified by Sachs [1948] pp. 285-286; but at this time Sachs knew of only six or seven examples. A few years later Sachs [1952a] used such data as he could gather from some Diaries he knew (those for -424, -418, -384, and -322) to try to pin down the date of the introduction of the Sirius scheme and the Uruk solstice scheme: "less than a century after -380". Van der Waerden [1952/1953] pp. 220 and 224 claims that the Diary for -418 (VAT 4924, of which he publishes a copy of Weidner in Tafel XVIII) is the earliest text to refer to zodiacal signs rather than constellations. Sachs [1955] pp. xii-xxi lists 843 tablets from Diaries in the British Museum and elsewhere of which the earliest dated example is for -567, the latest, questionably, for -46; he also provides copies of most of them on pp. 44-149. Sachs [1974] provided a brief analysis of the distribution of the extant Diaries and of their contents. The Diaries for -163 and -86 are transliterated and translated in Stephenson-Walker [1985] pp. 21-40.

But it was only in the late eighties that the complete edition (transliterations, English translations, and photos) of all of the dated Diaries began with the publication of the first volume of Sachs-Hunger [1988-1996]; these three volumes will be supplemented in the future by a fourth containing the undated fragments. There are a total of about 1200 fragments of tablets of Diaries included in this edition.

An extended discussion of the first volume of Sachs-Hunger [1988-1996] was provided by Rochberg-Halton [1991]. Koch [1991/1992] argues that the dubious dates for BM 33478 (-440) and BM 40122 (-304) should be changed to -381 and -366 respectively, and that BM 47735 (-391) is correctly dated by Sachs-Hunger. In addition to the literature cited in Sachs-Hunger wherein specific Diaries are cited for the historical information they contain, the publication of the corpus has elicited the following: Geller [1990a], van der Spek [1993], and Gera-Horowitz [1997]; undoubtedly many more will be published in the future.

A Diary is a record of observed phenomena carefully chosen from the realms of the celestial, the atmospheric, and the terrestrial. Most of them come from Babylon, where the observers were, certainly at one time and probably throughout the duration of the project, employed by the Temple of Marduk, Esagil. However, one Diary comes from Uruk (-463), as may also that for -99, and in one Diary (-366) an observation that could not be made at Babylon was made in Borsippa. A Diary was dated during the Neo-Babylonian, Achaemenid, and Macedonian periods by the regnal year of the named king; the earliest Diary that mentions the year of the Seleucid Era is that for SE 9 (-302), though the anomalous Diary for -302/301 refers to year 4 (of Seleucus), which equals SE 10. But then the Diary for -299 refers to SE 12, and thereafter those in which the year number is preserved, even if they mention the name of the reigning king, use the Seleucid Era[47]. On one tablet from -140 (BM 34050) is the formula: year 107 (Arsacid Era = AE) which is year 171 (SE); this double dating continues through the series; see -137, -135, -132, -131, -119, -111, and -107. In Diaries later than -107 only the year in the Seleucid Era is attested; see -99, -97 (broken year number), -96, -95, -90, -87 (text C), -82, and -77. Exceptions, wherein double dates are given, are the Diaries for -88, -87 (text B; SE year restored), -79 (AE year number broken), and -75 (AE year number broken). The explanation for this variance in not obvious.

Since the clay that constitutes a tablet does not remain soft enough to receive impressions from the stylus for an extended period of time, the observations were originally noted on a small tablet for a few nights only or else, at times, on a wax tablet (see -384 X-XI). Thus we have tablets for just 1 night (-238; BM 55511), for 2+ nights (-170; BM 37284), for 3 nights (-201; BM 33808. -199; BM 33671. -184; BM 31581), for 3+ nights (-170; BM 32143), for 4 nights (-192; MNB 1879. -171; BM 77226), for 4+ nights (-157; BM 35979), for 5 nights (-167; BM 34000. -167; BM 41111. -142; BM 45632. -131; BM 35514), for 5+ nights (-175; BM 32572), for 6 nights (-201; BM 36591 [at end mentions events 20 days later]. -197; BM 36733. -195; BM 31804. -192; BM 31583), for 6+ nights (-198; BM 33992. -192; BM 41123. -171; BM 40119. -169; BM 31476), for 7 nights (-200; BM 36807), for 8 nights (-170; BM 47650), for 9 nights (-203; BM 34563. -195; BM 30830. -168; BM 35525), for 10 nights (-188; BM 40067. -171; BM 40574. -166; BM 35651), for 11 nights (-195; BM 34564. -190; BM 36857), for 12 nights (-170; BM 40057), for 12+ nights (-166; BM 45782), for 13 nights (-154; BM 40069), for 13 or 14 nights (-177; BM 41884+), for 14 nights (-183; BM 41037), for 14+ nights (-171; BM 31847), for 14 or 15

[47] See -291, -288, -286, -283, -281, -278, -277, -273, -270 (broken year number), -262, -261 (broken year number), -259, -255, -253, -247 (broken year number), -246, -245, -237, -234, -232, -230, -226 (broken year number), -222, -221, -214, -213 (broken year number), -212, -211 (broken year number), -209, -204, -203, -200, -198, -197, -195, -194, -193, -192, -191, -190, -189, -188, -187, -186, -185, -184, -183, -182, -180, -179 (broken year number), -178, -177 (broken year number), -175 (broken year number), -173, -171, -170, -168, -167, -166, -165, -164, -162, -161, -158, -157, -156 (broken year number), -155, -154, -152 (broken year number), -144, -142, -141 and -140 (king Arsaces named).

nights (-189; BM 34049), for 15 nights (-198; BM 32597. -194; BM 30739. -185; BM 32349+. -168; BM 33991. -155; BM 45627), for 16 nights (-190; BM 55539. -170; BM 34562. -155; BM 45768), for 17 nights (-193; BM 31539 [beginning broken]. -126; BM 34112), for 18 nights (-201; BM 41161. -180; BM 34702), and for 20 nights (-158; BM 45774). There is undoubtedly some significance in the fact that, aside from the tablet for 1 night in -238, all of these short Diaries are dated -203 or later.

The fact that BM 77226 (-171 $XII_2$ 11-14) and BM 40574 (-171 $XII_2$ 10-19) sometimes agree and sometimes disagree demonstrates that on some nights more than one observer was at work. This fact will also explain in part the several duplicates (-373; -366; -332; -291; -277; -253; -245; -197; -173; -171; -164; -163 [two different versions]; -161; -133; -118; -111; -108; -107; und -105), a few triplicates (-165; -132; and -119), and even some quadruplicates (-372; and -168). Clearly, cases like that of -163 are particularly significant for proving the existence of multiple observers.

The earliest Diaries that we have (-651 and -567) are for a whole year; so are some later ones (-382; -370; -286; -256; and -182). They and other Diaries for more than, say, three weeks were presumably compiled from short Diaries. A "standard" Diary from the fourth century B.C. on was for either 6 months[48] or, when there was an intercalary month $XII_2$, for VII-$XII_2$.[49] There are no extant Diaries for I-$VI_2$, but we do have one for II-$VI_2$ (-255) and another for $VI_2$-XII (-141). From the early third century on Diaries for I-VII also occur.[50]

From the early fourth century on the months were sometimes grouped in fours.[51] This practice was revived briefly in the late second century.[52] At times in the very late period there are groups of five months (cf. -329 IX-$XII_2$ noted above).[53] A more common type is the Diary for 1 month[54] or,

48 -381 I-VI; -373 VII-XII; -372 I-VI; -368 I-VI; -366 I-VI; -333 I-VI; -324 I-VI; -321 I-VI; -308 I-VI; -302 I-VI; -300 VII-XII; -293 I-VI; -291 I-VI; -289 I-VI; -288 VII-XII; -286 I-VI; -284 VII-XII; -273 I-VI and VI-XII; -270 VII-XII; -266 I-VI; -261 I-VI; -257 I-VI; -251 VII-XII; -246 I-VI; -245 I-VI; -232 VII-XII; -230 I-VI; -218 VII-XII; -210 I-VI; -209 I-VI; -202 I-VI and VII-XII; -199 I-VI; -197 VII-XII; -187 VII-XII; -186 VII-XII; -183 I-VI and VII-XII; -173 VII-XII; -170 I-VI and VII-XII; -165 I-VI; -164 VII-XII; -161 I-VI; -141 I-VI; -137 VII-XII; -134 VII-XII; -132 I-VI and VII-XII; -125 I-VI; -124 I-VI and VII-XII; -123 I-VI; -117 VII-XII; -111 I-VI; -107 I-VI and VII-XII; -90 VII-XII; -88 I-VI, and -77 I-VI and VII-XII.

49 -384; -381 (formerly -440); -375; -277; -266; -261; -253; -247; -234; -212; -193; -163; -112; and -108.

50 -277; -161; -158; -118; -105; -87; and -82.

51 -361 X-$XII_2$; -346 IX-XII; -338 I-IV; -332 V-VIII; -329 V-VIII (followed by IX-$XII_2$); -328 V-VIII; -322 IX-XII; -254 IX-XII; and -249 IX-XII.

52 -137 V-VIII and IX-XII; and -133 V-VIII and X-$XII_2$.

53 -156 I-V and VIII-XII; -144 V-IX; and -107 VIII-XII.

54 -418 VIII; -391 $XII_2$; -230 XII; -214 IX; -212 X; -193 II; -191 VIII; -183 II; -182 VIII; -181 IV; -180 I; -180 X; -179 I; -178 XI; -157 IV; and -104 VI.

from the second century, for 1 month and a few days of the next.[55] The final and rare type, from the middle of the third century on, is for 2 months.[56]

While full year Diaries may have served as archival copies in the early centuries, and 6 or 7 month Diaries almost certainly served that purpose from at least the early fourth century B.C., practical considerations led to the creation of Diaries having a variety of lengths. Less variety was allowed into the contents of the Diaries, though naturally some developments took place during the course of the six centuries for which we have evidence.

Particularly stable seem to have been the astronomical data recorded in the Diaries. This is remarkable in several respects: only phenomena which are periodic and therefore capable of being described mathematically are included, and almost all of the phenomena regarded as ominous in *Enūma Anu Enlil* (exceptional are halos around either luminary enclosing planets or constellations) and constantly observed, recorded, and interpreted for the court at Nineveh were assiduously ignored. These features lead us to conclude that the astronomical observations entered into the Diaries—and not only observations but, when they were impossible to make, rough estimates—were intended from the beginning to be the basis of a mathematical, predictive system. That beginning was most likely, as argued already by Sachs [1974] p. 44, the first year of Nabû-nāṣir, -746; the reasons are three.

1. The cuneiform "reports" of lunar eclipses in eighteen-year periods, which are most probably indices to and excerpts from the Diaries, begin in -746 (see Sachs [1955] LBAT 1413-1430 and 1436), except that LBAT 1422+1423+1424, 1425, and 1428 are based on the 'Saros' cycle.

2. Ptolemy in *Almagest* III 7 states that the beginning of the reign of Nabonassar is "the era beginning from which the ancient observations are, on the whole, preserved down to our own time."

3. The three lunar eclipses whose observations at Babylon are recorded by Ptolemy in *Almagest* IV 6 are dated from the first and second years of Mardokempadus (i.e., Marduk-apla-iddina), that is, -720 and -719. In V 14 he cites a lunar eclipse observed at Babylon in year 5 of Nabopolassar (Nabû-apla-uṣur) (-620); and elsewhere he refers to six other lunar eclipses observed at Babylon. In all likelihood these observation reports, presumably mostly assembled by Hipparchus, derive from the Diaries.

That someone in the middle of the eighth century B.C. conceived of such a scientific program and obtained support for it is truly astonishing; that it was designed so well is incredible; and that it was faithfully carried out for at least 700 years is miraculous.

---

55 -186 XI 1 - XII 4; -182 XI 1 - XII 7; -178 V 1 - VI 1+ and VI 1 - VII 3 and XI 1 - XII 10+; -164 II 1 - III 2+; -163 II 1 - III 12; and -162 V 1 - VI 6; cf. also -195 IX 16 - X 20.

56 -255 V-VI; -158 IV-V; -140 IX-X; -132 VI-VII; and -119 I-II.

*2.1. Planets*

The dates of the following "Greek-letter" phenomena are recorded: first visibility in the East (East and West for the inferior planets); last visibility in the West (East and West for the inferior planets); and first and second station and "opposition" (acronychal rising) for the superior planets only. Also, the planets' passings of the Normal Stars (see II B 2.3 below) and each other. Even if a "Greek-letter" phenomenon was not observed, its estimated time was recorded; thus the last appearance of Mercury in the East behind (the constellation) Pisces and of Saturn also behind (the constellation) Pisces on I 14 of -651 were not actually observed, though in that year the first station of Mars on I 17 near two stars in (the constellation) Scorpion was observed. On XII 19 Venus in (the constellation) Aries was observed 10 fingers behind Mars. In the Diary for -567 Jupiter's acronychal rising is recorded on I 11 or 12; on II 18 Venus passed by α Leonis. In -378 for the first time the time (11;30° = 0;46 h) between the first visibility in the East of a planet (Jupiter in Scorpius) and sunrise was recorded for VIII 16; this never became an inevitable practice. We do not find another example till -366 when, on III 10, there was 16° (= 0;48 h) between sunset and the setting of Mercury on the evening of its first visibility in the West in Leo, and, on IV 28, there was 22° (= 1;28 h) between Mars' first visibility near α Leonis and sunrise. Often, if the interval appears to the observer to be too long, he notes that the phenomenon should have been visible on a previous night or that it will be seen on a later night with the words: "(ideal) first (or last) appearance on the nth." For the superior planets we are sometimes given the time-degrees between Γ and sunrise or sunset and Ω; thus, in -378 VIII 16 from Γ of Jupiter to sunrise was 11;30°; on -332 VI 27 Γ of [Jupiter] to sunrise was 11°; in -324 III [18] Γ [of Saturn] to sunrise was 15°; and in -249 IX 27 Γ of Saturn [to sunrise] was 15°. A careful study of such observations might reveal the criteria that the Babylonians used to estimate the "ideal" dates of the phenomena.

Eventually it came to be the practice that the positions of the planets were recorded in a summary at the end of each month. The earliest attested development toward this is in the Diary for -463, where it is stated that Venus and Mercury were in Leo and Mars in Scorpius at the end of month V. However, in the Diary for -567 the positions of some planets are given at the beginning of the month. All the planets, in the order Jupiter, Venus, Mars, Saturn, and Mercury, are located with respect to the constellations (Jupiter was behind Cancer in the beginning of Leo) at the end of month X in the Diary for -453; the date of a Greek-letter phenomenon of Mercury is also recorded. In the Diary for -418 at the end of month I the positions of the planets are given in the same order as in the Diary for -453, with respect to zodiacal signs, it would appear (Jupiter and Venus were in the beginning of

Gemini). In this case all the planets were visible; in other months (II, III, XI, XII, and $XII_2$) the positions of planets which were not visible were not given. The position of Jupiter in month XII, in front of Cancer, seems still to refer to a constellation rather than a zodiacal sign. However, it is argued by van der Waerden [1952/1953] and others that this is the earliest cuneiform text to refer to zodiacal signs—i.e., arcs of 30° on the ecliptic named after twelve constellations. A similar ambiguity is found in the earliest dated Babylonian nativity, that for 12/13 January -409 (Rochberg [1998] pp. 51-55), wherein we are told that Σ of Venus occurred in front of Aquarius, and in the next nativity, that for 29 April -409 (Rochberg [1998] pp. 56-57), wherein we are informed that the Moon was in the lower part of the Pincer of the Scorpion (i.e., Libra). These phrases reflect the language of observations, which are made with respect to stars rather than to invisible boundaries of zodiacal signs. In the Diary for -384 at the end of month VII (where the order of the planets is Jupiter, Venus, Saturn, Mercury, and Mars) Saturn is located in a non-zodiacal constellation, the Chariot. However, the entries in the Diary for -381 seem all to refer to zodiacal signs; but in the Diary for -380 we find that Mars was in the Chariot in month XI. This is the last time that a non-zodiacal constellation is found in the Diaries in these monthly summaries except when an observation is being quoted. It is in a Diary for -384, month VII, that it was first recorded in one of these summaries that a planet had moved from one zodiacal sign to the next during the month. Only in the Diary for -277, month I, do we get a clear preference for the order Jupiter, Venus, Mercury, Saturn, and Mars; neither Jupiter nor Mercury was visible, and yet they are listed in their henceforth usually standard places (Mercury is omitted in the summary at the end of month VII in the Diary for -270; the order is Jupiter, Venus, Saturn, Mercury, Mars in month I in the Diary for -255; Mercury, being invisible, was placed at the end in month V in the Diary for -254; etc.). In the Diary for -212, month X, the day of Venus' entry into Sagittarius (the ninth) is mentioned; the days of the planets' entries into zodiacal signs are sometimes given thereafter. But Mars' entries into zodiacal signs are recorded for -318/7 in MNB 1856 (see II C 2.3), and the planets' entries into zodiacal signs are normally recorded in Almanacs (beginning from either -261/0 or from -219/8) and in some Normal Star Almanacs (beginning from -161/0), though it seems that these dates of entries are computed. However, the important study by Huber [1958] shows that the Babylonians could determine the beginnings of zodiacal signs in the sky from their distances from Normal Stars. The dates of entries into zodiacal signs in the Diaries are the results of observations.

*2.2. The Moon*

In the oldest Diary, that for -651, we are told only that on I 1 the Moon was bright and high, and that on XII 15 "one god (the Moon) was seen with the other (the Sun)." A seventh century "Diary" recording observations of the weather and of the Moon for a half-month without numerical information was edited as Report 207 in Hunger [1992] pp. 116-117; see II B 1. But in -567 we are given the time from sunset to moonset on III 1 (20° = 1;20 h) and are told that on III 15 the time from sunrise to moonset was 7;30° (= 0;30 h). The same two intervals are given for month XI and XII of -567 and for month VI of -463. The Diary for -391 gives the interval from sunset to moonset on VIII 1 (the entry for VIII 15 is lost) and then provides the interval (now broken) for moonrise to sunrise on VIII 27. In the broken Diary for -381 we find for the first time in our extant material the interval from moonset to sunrise on I 12 as well as the intervals from sunset to moonset on I 1 and VI 1 and the interval from sunset to moonrise on VI 16. It is only in the Diary for -372 that we apparently have the full "Lunar Six":

(1) Sunset to moonset, called na, on I 1; V 1; XI 1.
(2) Moonset to sunrise, called ŠÚ, on I 13.
(3) Sunrise to moonset, called na, on I 14; VI 15.
(4) Moonrise to sunset, called ME, on I 13 (sic); XI 14.
(5) Sunset to moonrise, called $GE_6$, on I 14 (sic).
(6) Moonrise to sunrise, called KUR, on IV 27; VI 27; VII 27; $XII_2$ 27.

The next Diary in which all six intervals are recorded in what we still have is that for -366:

(1) on II 1; III 1; V 1; VI 1; VII 1
(2) on I 13; III 13; V 12; VI 13(?)
(3) on I 14; II 15; III 14; VI 14
(4) on V 13
(5) on I 15; III 15
(6) on I 27; IV 27; V 27; VI 27

Thereafter, they are often attested, and must have normally been included in every Diary when visible. The records of first visibilities of the crescent Moon and of the "lagtimes" between sunset and moonset in the Diaries have been analyzed by Fatouhi-Stephenson-al-Dargazelli [1999].

In what remains of the Diary for -651 the main phenomenon involving the Moon that was of interest seems to have been its being surrounded by a halo. But the Moon's passing by the Normal Stars is sporadically recorded from -567 on (see I 8; III 5; III 8; III 9; III 10; III 16(?); X 18(?); XI 6; XII 2; and XII 7). The Moon and the planets are frequently said to be "above" , "below", "in front of" (to the West), or "behind" (to the East) a Normal Star, and the distances are given in cubits (1 cubit is supposed to equal 2;30°) and digits (1 digit is $\frac{1}{30}$ cubit or 0;5°). It is not at all clear how these directions and distances are to be interpreted, though a strong case for an ecliptic coordinate system has been made by Grasshoff [1999].

### 2.3. *The Normal Stars*

The Normal Stars are stars near the ecliptic—about 30 in number—whose "conjunctions" with the Moon and the planets are recorded in the Diaries and in the Normal Star Almanacs. The earliest attempt to identify them was made by Epping [1889a] pp. 117-133, who used two Normal Star Almanacs —BM 34033 (LBAT 1055) for -122 and BM 34032 (LBAT 1059) for -110. From these two sources he was able to identify 28 stars (actually 29 since he includes δ and β Scorpii as one number, XXIII). Epping's list is identical with Sachs-Hunger's except that Epping missed η and ϑ Cancri and has β instead of ϑ Leonis. The next list of Normal Stars was provided by Epping-Strassmaier [1892] pp. 224-225, who, by including data from the Diaries for -328, -273, and -232, increased the number to 33. The new stars are ϑ, ε, and γ Cancri, and ϑ Leonis. This list is repeated with notes by Kugler [1907] pp. 29-39.

The list of 31 Normal Stars in Sachs [1974] p. 46 omits ε Cancri and β Leonis, but provides co-ordinates (longitudes and latitudes) of each star for -600, -300, and 0. This list, with the addition of η Cancri and recomputed co-ordinates for -600, -300, and 0, is repeated in Sachs-Hunger [1988] pp. 17-19. We repeat here the translated names of the Normal Stars, their identifications, and their coordinates in -300, based on Sachs-Hunger.

| Name | Identification | λ | β |
|---|---|---|---|
| 1. Bright Star of Ribbon of Fishes | η Piscium | 354;52° | 5;14° |
| 2. Front Star of Head of Hired Man | β Arietis | 2;1° | 8;24° |
| 3. Rear Star of Head of Hired Man | α Arietis | 5;40° | 9;54° |
| 4. Bristle | η Tauri | 28;2° | 3;48° |
| 5. Jaw of Bull | α Tauri | 37;48° | -5;37° |
| 6. Northern rein of Chariot | β Tauri | 50;36° | 5;11° |
| 7. Southern rein or Chariot | ζ Tauri | 52;49° | 2;29° |
| 8. Front Star of Twins' Feet | η Geminorum | 61;31° | -1;10° |
| 9. Rear Star of Twins' Feet | μ Geminorum | 63;18° | -1;3° |
| 10. Twins' Star near Shepherd | γ Geminorum | 67;7° | -7;1° |
| 11. Front Twin Star | α Geminorum | 78;22° | 9;53° |
| 12. Rear Twin Star | β Geminorum | 81;38° | 6;30° |
| 13. Front Star of Crab to North | η Cancri | 93;27° | 1;21° |
| 14. Front Star of Crab to South | ϑ Cancri | 93;48° | -0;58° |
| 15. Rear Star of Crab to North | γ Cancri | 95;37° | 2;59° |
| 16. Rear Star of Crab to South | δ Cancri | 96;44° | -0;1° |
| 17. Head of Lion | ε Leonis | 108;43° | 9;32° |
| 18. King | α Leonis | 118;1° | 0;22° |

| | | | |
|---|---|---|---|
| 19. Small Star 4 cubits behind King | ρ Leonis | 124;26° | 0;1° |
| 20. Rump of Lion | ϑ Leonis | 131;25° | 9;39° |
| 21. Rear Foot of Lion | β Virginis | 144;42° | 0;39° |
| 22. Single Star in Front of Furrow | γ Virginis | 158;30° | 2;59° |
| 23. Bright Star of Furrow | α Virginis | 171;54° | -1;54° |
| 24. Southern Part of Scales | α Librae | 193;10° | 0;37° |
| 25. Northern Part of Scales | β Librae | 197;26° | 8;45° |
| 26. Middle Star of Head of Scorpion | δ Scorpii | 210;37° | -1;41° |
| 27. Upper Star of Head of Scorpion | β Scorpii | 211;13° | 1;18° |
| 28. Lisi | α Scorpii | 217;49° | -4;16° |
| 29. Bright Star on Tip of Pabilsag's Arrow | ϑ Ophiuchi | 229;26° | -1;31° |
| 30. Horn of the Goat-Fish | β Capricorni | 272;4° | 4;51° |
| 31. Front Star of Goat-Fish | γ Capricorni | 289;43° | -2;19° |
| 32. Rear Star of Goat-Fish | δ Capricorni | 291;28° | -2;11° |

We would just briefly note what has often been remarked on, that these stars are very unevenly distributed along the ecliptic. Twenty-two lie between 0° and 180°, only ten between 180° and 0°. Moreover, there is an enormous empty gap of 63° between star 32 and star 1. The reason(s) for such an extreme disequilibrium is/are difficult to comprehend. However, Roughton-Canzoneri [1992] have succeeded in identifying "Normal Stars" in Sagittarius called simply "the four stars in the East of Sagittarius" (which they identify with μ(13), 14, 15, and 16 Sagittarii) and "the four stars in the West of Sagittarius" (which they identify with $\nu^1$, $\nu^2$, $\xi^1$, and $\xi^2$ Sagittarii. They find these stars in Diaries, Goal-Year Texts, and Observational Texts, of which the earliest is dated -329. The statements concerning the cubits between the planets or the Moon and the "Normal Star" they show to be longitudinal differences, measured on the ecliptic without regard for the latitudes of either the planets or the Moon or the "Normal Stars", and they suggest that in most cases the planetary longitude was computed (planetary latitudes were not computed) and not observed. This result means that the Babylonians possessed catalogues of Normal Stars giving their longitudes. We have a fragment of such a catalogue, though the longitudes given in it are rounded off to integer numbers of degrees. BM 46083 was published by Sachs [1952]; the right column preserves the following information to which we have added the relevant information from the catalogue of Normal Stars recorded above.

| BM 46083 | | Normal Stars | |
|---|---|---|---|
| Loin(?) [of Lion] | 20(?) Leo(?) | 20. ϑ Leonis | 131;25° |
| Rear Foot of Lion | 30(?) (Leo? or?) 1 Virgo | 21. β Virginis | 144;42° |
| Root of Barley-Stalk(?) | 16 (Virgo) | 22. γ Virginis | 158;30° |
| Bright (star) of Barley Stalk? | 28(?) Virgo | 23. α Virginis | 171;54° |
| Southern Balance-Pan | 20 Libra | 24. α Librae | 193;10° |
| Northern Balance-Pan | 25 Libra | 25. β Librae | 197;26° |
| Upper (Star) | | 27. β Scorpii | 211;13° |

If we look at the difference in longitude in the two series, we see that the Babylonian longitudes are quite good, though their zero-point is naturally different from ours.

| BM 46083 | Normal Stars |
|---|---|
| 1. 10° or 11° | 13;17° |
| 2. 16° or 15° | 13;48° |
| 3. 12°(?) | 13;24° |
| 4. 22°(?) | 21;16° |
| 5. 5° | 4;16° |
| 6. (from Normal Star 20 to 25) 65° | 66;1° |

A study by Huber [1958] of some Almanacs and Normal Star Almanacs of -122/1 to -110/09 demonstrated the following beginnings of zodiacal signs to be in use at that time:

| | | |
|---|---|---|
| Taurus | 3°±0;45° before Normal Star 4 (η Tauri) | 0° = 24;17° – 25;47° |
| Gemini | 0;30° before or after Normal Star 7 (ζ Tauri) | 0° = 52;19° – 53;19° |
| Cancer | With Normal Star 12 (β Geminorum) | 81;38° |
| Leo | 4;48°±0;54° after Normal Star 17 (ε Leonis) | 112;37° – 114;25° |
| | 4;48°±0;54° before Normal Star 18 (α Leonis) | 112;19 – 114;7° |
| Virgo | 1;12°±0;54° before Normal Star 21 (β Virginis) | 142;36° – 144;24° |
| Libra | ca. 2° after Normal Star 23 (α Virginis) | ca. 173;54° |
| Scorpius | 5;35°±0;50° after Normal Star 25 (β Librae) | 191;1° – 192;41° |
| | 5;10°±0;47° after Normal Star 25 (β Librae) | 191;29° – 193;3° |
| | 5;29°±0;47° after Normal Star 25 (β Librae) | 191;10° – 192;44° |
| Sagittarius | 3°±0;45° after [Normal Star 30 (ϑ Ophiuchi)] | ca. 225;53° – 227;23° |
| Aquarius | 0;35° before or after Normal Star 32 (δ Capricorni) | 290;43° – 292;3° |

Comparing his results with BM 46083, one finds:

| | BM 46083 | Huber |
|---|---|---|
| β Virginis | 30° Leo/ 1° Virgo | 0;18° – 2;6° Virgo |
| α Virginis | 28° Virgo | ca. 28° Virgo |
| β Librae | 25° Libra | 23;35° – 25;37° Libra |

In other words, the Normal Star Catalogue agrees quite well with actual Babylonian practice. The mean difference between the Babylonian longitudes and longitudes counted from the Vernal Equinox in -100 was 4;28° ± 0;20°.

*2.4. The Sun*

The only aspects of solar motion mentioned in the Diaries are the dates of the occurrences of solstices and equinoxes and of the heliacal rising and setting and the acronychal rising of Sirius. In the fragments of the Diaries, of course, many records of solstices and equinoxes have been lost. The earliest record we have found is for a summer solstice in -567 III 9. This date may be based on a real observation, since the observer does not say (as he usually does for computed dates): "I did not watch," or it may be connected to the scheme found in BM 36731 (Neugebauer-Sachs [1967] pp. 183-190), which listed equinoxes, solstices, and heliacal risings and settings of Sirius for the years, apparently, from -615 to -587.

An autumnal equinox was recorded for -381 VI 27; the observer does not state: "I did not watch." If one extended the so-called abbreviated Uruk scheme (see Neugebauer [1948] and [1975] pp. 360-363) back to -381, which is year 7 of a nineteen-year cycle, and ignored the shift that Neugebauer [1948] and Sachs [1952a] hypothesized to exist at -198 (see Slotsky [1993]), the autumnal equinox was computed to fall on VI 27 (see Sachs [1952a]). However, this seems to be accidental; for the vernal equinox recorded for -375 $XII_2$ 7 ("I did not watch"), which was year 13 of a cycle, should have fallen on $XII_2$ 9 according to the scheme, and that recorded for -372 I 2 ("I did not watch"), which was year 16 of a cycle, should have fallen on $XII_2$ 12 according to the scheme. A perfect fit (or nearly so) begins in -330, which is year 1 of a cycle; in that year an autumnal equinox is recorded for VI 21, which fits the scheme; in -324 one on VI 27, which fits the scheme. The autumnal equinox recorded for -322 VI 18 should be VI 16, but the vernal equinox for -322 XII 25 and the autumnal equinox for -321 VI 30 fit. Nineteen years later, in -302, we find an autumnal equinox again "correctly" dated on VI 30. It is likely that the use of this scheme goes back to -368 since it is attested on the reverse of BM 36810+ published by Aaboe-

Sachs [1966] pp. 11-12 for at least years 2 to 12 of Artaxerxes III (-356 to -345). Unfortunately, the fragments of Diaries dated between -372 and -330 contain no reference to equinoxes or solstices.

| | SS | AE | WS | VE | Years | | | | |
|---|---|---|---|---|---|---|---|---|---|
| 1 | III 18 | VI 21 | IX 24 | XII 27 | -330 | -311 | -292 | -273 | -254 |
| 2 | III 29 | VII 2 | X 5 | $XII_2$ 8 | -329 | -310 | -291 | -272 | -253 |
| 3 | III 10 | VI 13 | IX 16 | XII 19 | -328 | -309 | -290 | -271 | -252 |
| 4 | III 21 | VI 24 | IX 27 | XII 30 | -327 | -308 | -289 | -270 | -251 |
| 5 | IV 2 | VII 5 | X 8 | $XII_2$ 11 | -326 | -307 | -288 | -269 | -250 |
| 6 | III 13 | VI 16 | IX 19 | XII 22 | -325 | -306 | -287 | -268 | -249 |
| 7 | III 24 | VI 27 | IX 30 | I 3 | -324 | -305 | -286 | -267 | -248 |
| 8 | IV 5 | VII 8 | X 11 | $XII_2$ 14 | -323 | -304 | -285 | -266 | -247 |
| 9 | III 16 | VI 19 | IX 22 | XII 25 | -322 | -303 | -284 | -265 | -246 |
| 10 | III 27 | VI 30 | X 3 | $XII_2$ 6 | -321 | -302 | -283 | -264 | -245 |
| 11 | III 8 | VI 11 | IX 14 | XII 17 | -320 | -301 | -282 | -263 | -244 |
| 12 | III 19 | VI 22 | IX 25 | XII 28 | -319 | -300 | -281 | -262 | -243 |
| 13 | III 30 | VII 3 | X 5 | $XII_2$ 9 | -318 | -299 | -280 | -261 | -242 |
| 14 | III 11 | VI 14 | IX 17 | XII 20 | -317 | -298 | -279 | -260 | -241 |
| 15 | III 22 | VI 25 | IX 28 | I 1 | -316 | -297 | -278 | -259 | -240 |
| 16 | IV 3 | VII 6 | X 9 | $XII_2$ 12 | -315 | -296 | -277 | -258 | -239 |
| 17 | III 14 | VI 17 | IX 20 | XII 23 | -314 | -295 | -276 | -257 | -238 |
| 18 | III 25 | VI 28 | X 1 | I 4 | -313 | -294 | -275 | -256 | -237 |
| 19 | IV 7 | $VI_2$ 9 | IX 12 | XII 15 | -312 | -293 | -274 | -255 | -236 |

Note that the table in Neugebauer [1975] p. 362 begins with -235 (SE 76).

Nor are there any records of the dates of the phases of Sirius during this period—more specifically, between -384 and -324. The earliest preserved mention of a phase of Sirius is its heliacal setting on -418 II 23. The later scheme for Sirius as established by Sachs [1952a] is ($\Gamma$ = heliacal rising, $\Omega$ = heliacal setting, and $\Theta$ = acronychal rising):

$\Gamma$ = SS + 21 tithis
$\Omega$ = SS – 44 tithis
$\Theta$ = SS + 191 tithis.

A tablet from Uruk, von Weiher [1998] no. 269 rev. 9-14 sets forth the following scheme:

$\Omega$ + 1 mo. 14 "d" = SS
SS + 20 "d" = $\Gamma$
$\Gamma$ + 2 mos. 13 "d" = AE
AE + 3 mos. 3 "d" = WS
WS + 3 mos. 3 "d" = VE
VE + 1 mo. [x "d"] = [$\Omega$]

This is identical to Sachs' scheme, except that Γ of Sirius occurs 20 instead of 21 tithis after the summer solstice. The distance from one colure to the next (except that we do not know the number for vernal equinox to summer solstice) is 93 tithis.

In -418 the Ω of Sirius was observed. In -384 a Θ of Sirius is recorded on X 1, which would, according to the scheme, date the SS on III 20 while the extension of the solstice scheme back to -384 places it at III 21. Moreover, there is a Γ of Sirius recorded for -324 IV 18 whereas the schematic SS occurred 24 instead of 21 tithis earlier; the Γ was clearly observed because the time-interval from Γ to sunrise (12° = 0;48 h) is given. Similarly, such an interval is given for the Γ recorded in -289 IV 13, which is 22 tithis after the schematic SS; the text correctly states that it should have been observed on IV 12. Other records are of a Θ in -322 IX 27 which would put the SS on III 16, its schematic date; an Ω in -293 II 23, which would put the SS at its schematic date, IV 7; and another Ω in -291 II 15, which would put the SS again at its schematic date, III 29. Thereby it is clear that when the Diaries in -330 began again to record solstices and equinoxes using schematic dates they also began to note the schematic dates of the phases of Sirius as well as, occasionally, the actual dates of observation. The scheme for a nineteen-year cycle is:

| | Γ | Θ | Ω |
|---|---|---|---|
| 1 | IV 9 | IX 29 | II 4 |
| 2 | IV 20 | X 10 | II 15 |
| 3 | IV 1 | IX 21 | I 26 |
| 4 | IV 12 | X 2 | II 7 |
| 5 | IV 23 | X 13 | II 18 |
| 6 | IV 4 | IX 24 | I 29 |
| 7 | IV 15 | X 5 | II 10 |
| 8 | IV 26 | X 16 | II 21 |
| 9 | IV 7 | IX 27 | II 2 |
| 10 | IV 18 | X 8 | II 13 |
| 11 | III 29 | IX 19 | I 24 |
| 12 | IV 10 | IX 30 | II 5 |
| 13 | IV 21 | X 11 | II 16 |
| 14 | IV 2 | IX 22 | I 27 |
| 15 | IV 13 | X 3 | II 8 |
| 16 | IV 24 | X 14 | II 19 |
| 17 | IV 5 | IX 25 | I 30 |
| 18 | IV 16 | X 6 | II 11 |
| 19 | IV 28 | X 18 | II 23 |

The other phenomena recorded as affecting the sun—its being surrounded by a halo, covered by clouds of various sorts, or in a "box"—are omens with no bearing on mathematical astronomy.

### *2.5. Eclipses*

The Diaries provide a full listing of both lunar and solar eclipses, observed and predicted. The very earliest extant Diary, that of -651, records a lunar eclipse (col. III 4'); this was presumably the eclipse of 2 July, when the Moon rose totally eclipsed at Babylon. The text is broken so that we are given no details (the restoration ⌜AN⌝-K[$U_{10}$] in col. II 1 is incorrect). The next surviving Diary, that for -567, records a prediction of a lunar eclipse on III 15; it is known to be a prediction because it states that it was omitted. The opposition in question occurred on 4 July at about 2 PM in Babylon. The prediction of an eclipse possibility was very good since the latitude of the Moon was less than 1° South. For a possible method of its prediction see van der Waerden [1940] p. 111. The next recorded lunar eclipse was in -370 VIII (the day-number is broken). This is the total eclipse of 11 November, which began at Babylon at about 7:30 PM and reached totality at about 8:40. The Diary gave the time of impact (broken) and the time of totality (21° = 1;24 h after sunset, which occurred at about 5 PM). The time from sunset to the beginning of totality was actually about 55° = 3;40 h. We cannot explain the discrepancy. But thenceforth, if not before, time intervals between sunset and the beginning of a lunar eclipse or between its end and sunrise and often the intervals between its phases are normally given, along with other factors whose interest derived from the fact that they were regarded as omens. A typical example from the fourth century B.C., though somewhat broken, is the lunar eclipse reported for -366 V 13 (30 August): "When 56° (= 3;44 h) were left to sunrise, lunar eclipse; on the south-east side [....] cubits behind the Rear basket of Aquarius it was eclipsed; the North wind blew. During its eclipse Mars(?) [....]." The eclipse began at 2:53 in the morning; somewhat less than three hours before sunrise. The longitude of the Moon was 331°; the "Rear basket of Aquarius", then, may be λ Piscium whose longitude was 324° and latitude +3;30°. During the eclipse the longitude of Mars was 128°, so that it was invisible; but the longitude of Saturn was 261° so that it was just setting. Unfortunately, the sign for Saturn does not look very much like that for Mars, so something not involving a planet was probably written here.

An example of a later report is that for the lunar eclipse of -123 V 14 (13 August): "[....] in 19° (= 1;16 h) of night it was completely covered; 24° (= 1;36 h) night maximal phase; [when it began to clear(?), in] 19° (= 1;16 h) of night it cleared from [.... to] West; 1;2° (= 4;8 h) onset, maximal phase, and clearing. Its eclipse [.... Mars,] Mercury, and Venus set; in the beginning of the maximal phase, Saturn came out; during clearing, Ju[piter came out ....] .... [....] opposite the Front basket of Aquarius it was eclipsed." The eclipse began at about 7:20 PM, less than half an hour after sunset; totality began at about 8:30 PM, an hour and a half after sunset; mid-eclipse was at about 9:15 PM, nearly two hours after sunset; and between the end of totality at about 10:4 PM and clearing at about 11:12 PM was 1:8 h. The duration of the eclipse was about 3;52 h. Some of these numbers are close to those in the Diary. The longitude of the Moon at opposition was 317°. The longitudes of Saturn, Jupiter, Mars, Venus, and Mercury were respectively 1°, 50°, 158°, 163°, and 159°; Saturn rose at about 9 PM, just after the beginning of totality; and Jupiter at about 11 PM, just before clearing; Mars and Mercury set at about 8 PM, Venus somewhat later, all of them shortly before the beginning of totality. The longitude of the Moon places the longitude of the Front basket of Aquarius also at about 317°; this suggests that the "Front basket of Aquarius" was one of the two fourth magnitude stars, $\varphi$ Aquarii (317;39° and -0;53°) and $\chi$ Aquarii (317;33° and -2;47°). The use of time-degrees and of the meridian transits of *ziqpu* stars to express the time of night at which the phases of lunar eclipses occur is discussed in II C 1.1d.

That the Diaries contained full-fledged reports of lunar eclipses from their beginnings in the first year of Nabû-nāṣir is strongly suggested by the existence of series of eclipse-reports beginning at the same date and by the eclipse-reports from the reigns of Marduk-apla-iddina and Nabopolassar cited by Ptolemy.

The earliest report of a solar eclipse in our extant Diary-fragments is dated -366 V 28; unfortunately, all details are missing. But for the solar eclipse in -357 XI (29 February -356) some details survive that demonstrate that the intervals were also recorded for these eclipses: "At 1,16° (= 5;4 h) before sunset, solar eclipse on the [....] side [....] and clearing; during its eclipse, the South wind which [....]." The eclipse began at about 12:45 PM and cleared at about 3:15 PM; the hour of the eclipse beginning was about 5 hours before sunset.

The excerpt from the Diary for -302/301 records a predicted solar eclipse on -302 VI 29 (25 September), "which was omitted." It was predicted to occur 1,18° (= 5;12 h) after sunset; the opposition actually occurred at about 1 AM at Babylon on 26 September.

As an example of a later solar eclipse report we cite that for -133 XI 27 (13 February -132): "[.... solar eclipse;] when it began [on the ....] and west side, in 20° (= 1;20 h) of day it made two thirds of the disk; when it began to clear, in 18° (= 1;12 h) of day [....] .... was visible(?); at 51° (= 3;24 h) (variant 50° = 3;20 h) before sunset; clouds were in the sky, the North wind blew." The eclipse began at about 1:50 PM and reached mid-eclipse at about 3:10 with a magnitude of 0;52; the time interval was 1;20 h as the text indicates, and the magnitude was slightly greater than 2/3. The beginning of the eclipse was about 3;55 hours (= 59°) before sunset, but from mid-eclipse to sunset was close to 2;35 hours (= 36°), double the interval reported in the Diary.

We do not here consider the weather phenomena (much of which is cited to explain the observers' failure to observe) and the terrestrial phenomena (e.g., prices, river levels, abnormal births, and historical events); though they have an important bearing on the question of the purpose of the Diaries, they are irrelevant to their astronomy. Concerning the prices and the variations in the level of the river see Slotsky [1997].

## *2.6. Excerpts from the Diaries*

A number of the tablets published in Sachs-Hunger [1988-1996] are excerpts rather than full Diaries. Thus VAT 4924 (-418) quotes planetary phenomena, Sirius' phases, "falls of fire", "historical" events, commodity prices, and summaries of planetary positions, from month I to month $XII_2$. At the end is stated: "excerpted for reading."

The earliest nativity omen or "horoscope" that we have, AO 17649 for 12/13 January -409, was published by Rochberg [1998] pp. 51-55. It excerpts from a surprisingly complete Diary not only the dates and "longitudes" (nearby constellations or zodiacal signs) of the Greek-letter phenomena of the planets that occurred for each before and after the nativity, but various extraneous information found in the Diary: "solstice on the 9th of Ṭebētu; <last visibility> on the 26th"; "dense clouds"; "an intercalary Addaru." We even find the characteristic formula for indicating an estimated date: "Duʾūzu the 30th, Saturn's first visibility in Cancer; high and faint; around the 26th (the ideal) first visibility."

Also apparently excerpted from a Diary is BM 38357, which contains reports of observed phenomena (eclipses, planetary phases, etc.) and historical events from years 14 to 19 [of Nabopolassar], -611/0 to -606/5.

BM 34616+ (-302/301) cites mainly lunar phenomena (including lunar and solar eclipses not visible at Babylon) from month I in 9 SE till month III in 10 SE, where the tablet breaks off. BM 55541 (-286) excerpts lunar and some planetary phenomena from month I till month VIII of 25 SE, where the tablet breaks off. And BM 140677, which is probably from Uruk, provides planetary phenomena, eclipses, and phases of Sirius from month VIII of 212 SE till month II of 214 SE.

## *2.7. Ptolemy's Use of the Diaries*

We have already noted several Babylonian eclipse observations that were utilized by Ptolemy in the *Almagest*. There are a number of other such observations preserved by the great Greek astronomer; we list them below. These observation reports, presumably derived from the Diaries, were compiled and arranged by Hipparchus according to Toomer [1988].

*2.7a. Lunar eclipses*

*Almagest* IV 6. Year 1 of Mardokempadus Thoth 29/30 (= 19/20 March -720), over an hour after moonrise; total eclipse. Mid-eclipse occurred 3 1/3 equinoctial hours before midnight at Alexandria.

Year 2 of Mardokempadus Thoth 18/19 (= 8/9 March -719), mid-eclipse at midnight at Babylon. Maximum obscuration 3 digits from the South. Mid-eclipse occurred 5/6 of an equinoctial hour before midnight at Alexandria. See also *Almagest* IV 9.

Year 2 of Mardokempadus Phamenoth 15/16 (= 1/2 September -719), beginning after moonrise. Maximum obscuration more than half the disk from the North. Beginning of eclipse occurred about 5 equinoctial hours before midnight, mid-eclipse 4 1/3 equinoctial hours before midnight at Alexandria.

*Almagest* V 14. Year 5 of Nabopolassar Athyr 27/28 (= 21/22 April -620), beginning at end of 11th hour at Babylon. Maximum obscuration was ¼ of diameter from the South. Eclipse began 5 seasonal hours after midnight and mid-eclipse occurred 6 seasonal hours, which correspond to 5 5/6 equinoctial hours at Babylon. Mid-eclipse occurred 5 hours after midnight at Alexandria.

*Almagest* IV 9. Year 20 of Darius Epiphi 28/29 (= 19/20 November -501) after 6 1/3 equinoctial hours of the night had passed. Maximum obscuration was ¼ of diameter from the South. Mid-eclipse occurred 2/5 of equinoctial hour before midnight at Babylon, 1¼ equinoctial hours before midnight at Alexandria.

Year 31 of Darius Tybi 3/4 (= 25/26 April -490) at middle of 6th hour of night at Babylon. Maximum obscuration was 2 digits from the South.

In all five cases it is clear that Hipparchus has taken from the Diaries rather vague Babylonian eclipse reports giving just the approximate time, in local seasonal hours, after sunset, or near midnight, or after midnight at Babylon, the maximum obscuration in digits or in a fractional part of the lunar disk, and the direction of the obscuration, converted the date in the Babylonian calendar into the corresponding date in the Egyptian calendar, and estimated the approximate times of impact and mid-eclipse in equinoctial hours at Alexandria. The records continue with an eclipse-triple which is treated in a different way, but is attributed directly to Hipparchus.

*Almagest* IV 9. Archonship of Phanostratos at Athens Poseideon, when half an hour of night remained. A small section of the disk was obscured from the summer rising-point (North-east); the Moon was still eclipsed when it set. The date corresponds to year 366 of Nabonassar Thoth 26/27 (= 22/23

December -382), 5½ seasonal hours after midnight at Babylon, 18½ equinoctial hours after noon at Alexandria.

Archonship of Phanostratos at Athens Skirophorion, which was Phamenoth 24/25, when first hour of night was well advanced. The Moon was eclipsed from the summer rising-point (North-east). The date corresponds to year 366 of Nabonassar Phamenoth 24/25 (= 18/19 June -381), about 5½ seasonal hours before midnight at Babylon, about 8¼ equinoctial hours after noon at Alexandria.

Archonship of Euandros at Athens, Poseidon I, which was Thoth 16/17, beginning after 4 hours of night; this was a total eclipse from the summer rising-point (North-east). The date corresponds to year 367 of Nabonassar Thoth 16/17 (= 12/13 December -381), about 2½ seasonal hours before midnight at Babylon, 10 1/6 equinoctial hours after noon at Alexandria.

None of these eclipses is mentioned in the surviving fragments of the Diaries for -382 and -381.

Clearly these reports have gone through a preliminary reduction by someone using the Athenian calendar (probably in the form proposed by Meton in which the vagaries of the civil calendar were replaced by the certainty of mathematically determined intercalations), but referring to magnitudes in an inadequate fashion and to directions in terms of the solar rising-points rather than the cardinal directions. Hipparchus is the one, given the statement of his role in the first eclipse-report, who converted the Athenian date into an Egyptian one with the years of Nabû-nāṣir. It may have been Ptolemy who converted the seasonal hours of the night at Babylon into equinoctial hours since noon at Alexandria.

*2.7b. Planets*

In *Almagest* IX 2 Ptolemy states: "most of the ancient [planetary] observations have been recorded in a way which is difficult to evaluate, and crude. For the more continuous series of observations concern stations and phases [i.e., first and last visibilities]. But detection of both of these particular phenomena is fraught with uncertainty: stations cannot be fixed at an exact moment, since the local motion of the planet for several days both before and after the actual station is too small to be observable; in the case of the phases, not only do the places [in which the planets are located] immediately become invisible together with the bodies which are undergoing their first or last visibility, but the times too can be in error, both because of atmospherical differences and because of differences in the [sharpness of] vision of the observers. In general, observations [of planets] with respect to one of the fixed stars, when taken over a comparatively great distance, involve difficult computations and an element of guesswork in the quantity measured, unless one carries them out in a manner which is thoroughly competent and knowledgeable. This is not only because the lines joining the observed stars do not always form right angles with the ecliptic, but may form an angle of any size (hence one may expect considerable error in determining the position in latitude and longitude, due to the varying inclination of the ecliptic [to the horizon frame of reference]); but also because the same interval [between star and planet] appears to the observer as greater near the horizon, and less near mid-heaven; hence, obviously, the interval in question can be measured as at one time greater, at another less than it is in reality. Hence it was, I think, that Hipparchus . . . did not even make a beginning in establishing theories for the five planets . . . . All that he did was to make a compilation of the planetary observations arranged in a more useful way."

*Almagest* IX 7. Year 75 of Chaldaean (i.e., Seleucid) calendar Dios 14 at dawn Mercury was ½ cubit above α Librae. The date corresponds to year 512 of Nabonassar Thoth 9/10 (= 29/30 October -236).

Year 67 of Chaldaean calendar Apellaios 5 at dawn Mercury was ½ cubit above β Scorpii. The date corresponds to year 504 of Nabonassar Thoth 27/28 (= 18/19 November -244).

*Almagest* XI 7. Year 82 of Chaldaean calendar Xanthikos 5 in evening Saturn was 2 digits below γ Virginis. The date corresponds to year 519 of Nabonassar Tybi 14 (= 1 March -228).

The parts of the relevant Diaries that contained these observations have not survived, and the Normal Star Almanacs for 82 SE (LBAT 1003 and 1004) do not contain this information in their preserved part.

Clearly these three observation reports of planets' passings by of Normal Stars reached Hipparchus through yet another source, which gave dates in years of the Seleucid era and months of the Macedonian calendar. Again, Hipparchus has converted them into years of Nabonassar and months in the Egyptian calendar.

### *3. Normal Star Almanacs*

The earliest discussion of Normal Star Almanacs was by Epping [1889a] pp. 17-108. He presents in tabular form the "Lunar Six" and remarks concerning eclipses taken from BM 34033 (LBAT 1055) for 189 SE (= -122/1) on pp. 18-20, the same from BM 34078 (LBAT 1051; Epping knew only a part of this tablet, Sp 175) for 188 SE (= -123/2) on p. 21-22, and the same from BM 34032 (LBAT 1059) for 201 SE (= -110/09) on pp. 23-24. The dates of the planets' passing by the Normal Stars given in these tablets were utilized in his attempt to identify the Normal Stars on pp. 117-133; the data on the planets' Greek-letter phenomena are found on pp. 135-150 (Θ on pp. 135-136, Φ and Ψ on pp. 136-139, and Γ and Ω on pp. 140-150), and the dates of Sirius phenomena on p. 150. A transliteration and a translation of LBAT 1055 appears on pp. 153-159 and of LBAT 1059 on pp. 160-167. Rarely do the dates of the "Lunar Six" in these Normal Star Almanacs agree with the reports found in the extant Diaries for -123, -122, and -110. However, the Diaries for -123 and -110, though they record far fewer planetary phenomena than do the Normal Star Almanacs, are in agreement with them (except for wording) when they overlap.

Kugler [1907] published another group of three Normal Star Almanacs in their entireties. These are the following:

BM 34080 (LBAT 1010) for 104 SE (= -207/6) on pp. 88-89 (transliteration) and 116 (computations). The little that survives of the Diary for -207 does not contradict the data on this tablet.

BM 41599 (LBAT 1022) for 120 SE (= -191/0) on pp. 90-91 (transliteration) and 111 (computation), and Tafel VII (copy). The fragments of the Diaries for -191 do not overlap with this tablet.

BM 34758 (LBAT 1057) for 194 SE (= -117/6) on pp. 100-105 (transliteration) and 112 (computation), and Tafel VI (copy); Kugler had only two of the three fragments of this tablet. The Diary for -117 agrees in some particulars, disagrees in others:

| Diary | Normal Star Almanac |
|---|---|
| XII 5 or 6. Φ of Saturn 4½ cubits below α Gem. | XI x. Saturn 4 cubits below α Gem. |
| | XII 10. Φ of Saturn in Gemini |
| XII [10. Venus] 4½ cubits [below α] Arietis | XII [9]. Venus 4½ cubits [below α] Arietis |
| XII 29. Solar eclipse (after) 5 months; omitted | XII 29. 54° (= 3;36 h) after sunset, solar eclipse. |

There was no solar eclipse visible at Babylon on 17 March -116 according to computation; and, of course, the Normal Star Almanac dates it after sunset! It is clearly computed.

Kugler [1924] pp. 464-553 provides a long commentary on Normal Star Almanacs and Almanacs, which he calls collectively Astronomical Calendars or Ephemerides. He knows of eight Normal Star Almanacs: two of those published in Epping [1889a], namely LBAT 1055 and LBAT 1059, and all three that he published in Kugler [1907]; in addition, following Strassmaier, he joins Sp II 177 + 61+ to Sp 175 (which latter Epping had published) to form the almost "complete" BM 34078 + 34588 (LBAT 1051 + 1052), transliterated and translated on pp. 532-534, and he publishes two additional Normal Star Almanacs:

BM 45696 (LBAT 1020) for 111 SE (= -201/0), excerpted in translation on pp. 531-532. None of these excerpts occur in what remains of the Diary for -201.

BM 34076 (LBAT 1038) for 172 SE (= -139/8), transliterated and translated on pp. 533-534. Nothing remains of the Diary for -139. Kugler divides these eight examples into two groups: an older one, consisting of those for 104 SE, 111 SE, and 120 SE, whose heading is simply "measurements for year x," and a younger one, consisting of those for 172 SE, 188 SE, 189 SE, 194 SE, and 201 SE, whose heading is "measurements of the 'attainments' of the planets for year x." The "attainments" are supposed to be the planets' entries into zodiacal signs, though according to Sachs [1948] p. 280 only the Normal Star Almanacs from Uruk record the planets' entries into zodiacal signs; and all those that Kugler knew come from Babylon and do not record the planets' entries into zodiacal signs.

Schaumberger [1935a] provided a translation of and notes on LBAT 1020 on pp. 368-371, and a copy on Tafel X; of LBAT 1051 + 1052 on pp. 371-373, and a copy on Tafeln XI and XII; and notes on LBAT 1038 on p. 374, and a copy on Tafeln XIII and XIV. Goetze [1947] published a copy of a tablet in the Columbia University Library (LBAT 1007) for 96 SE (= -215/4).

Sachs [1948] pp. 281-282 knew of thirteen Normal Star Almanacs; he has added to those known to Kugler and the one copied by Goetze MLC 1860 (LBAT 1004) for 82 SE (= -229/8), from Uruk; MLC 1885 (LBAT 1025) for 133 SE (= -178/7), also from Uruk; Istanbul U. 180(3) + 193a + 193b (LBAT 1030a) for 150 SE (= -161/0), also from Uruk; and Istanbul U. 194 (LBAT 1031) for 151 SE (= -160/59), also from Uruk. Sachs [1955] pp. xxi-xxiii lists 122 tablets of Normal Star Almanacs, of which some have been joined and others are duplicated. The earliest dated one known to us is BM 48104 for 19 SE (or for an earlier regnal year 19), the latest dated one BM 32247 for 234 SE, both unpublished; 51 fragments in LBAT are undated.

The Normal Star Almanacs contain, as outlined in Sachs [1948], either the "Lunar Three" or the "Lunar Six", with the first occurring more likely in earlier examples, the second in later ones; the Greek-letter phenomena of the planets; the planets' passing by of the Normal Stars, which gives its name to this class of almanacs; the dates of solstices, equinoxes, and Sirius phenomena; and records of solar and lunar eclipses.

According to all those who have studied the Normal Star Almanacs and the Almanacs, *all* of the data they record are predictions; see Sachs [1948] p. 287 and Neugebauer [1975] p. 555. Since the earliest dated Normal Star Almanac is for -292/1 and the earliest dated Almanac possibly for -261/0, one might hypothesize that the dates of the occurrences of the Greek-letter phenomena of the planets were computed from the Ephemerides, but it is demonstrated above and in II B 4 below that the evidence is against this. Computations of the dates of their passings by of Normal Stars and entries into zodiacal signs could have been made with velocity schemes associated with the subdivisions of synodic arcs (see II C 2.3 and II C 4.2b, and cf. the method employed in Huber [1958]), and computations of the dates of their Greek-letter phenomena and their passings by of Normal Stars with the Goal-year periods (see II B 5). However, an unpublished study by Hunger has shown that there is almost always a difference—sometimes considerable—between dates derived from Goal-year Texts and those recorded in Normal Star Almanacs. The computations of solar and lunar eclipse-possibilities would have relied on the "Saros" cycle and related texts for times and magnitudes (see II C 1.1).

### *4. Almanacs*

The Almanacs, according to Sachs [1948] pp. 277-280, contain the "Lunar Three", the Greek-letter phenomena of the planets, the dates of the planets' entrances into the zodiacal signs, the positions of the planets at the beginnings of months, the dates of solstices, equinoxes, and Sirius phenomena, and reports of solar and lunar eclipses.

Kugler [1907] was the first to publish examples of Almanacs, which he calls the second class of "Ephemerides."

BM 33873 (LBAT 1123) for 129 SE (= -182/1), transliterated and translated on pp. 92-95, computations on p. 111, and copy on Tafel VII. The planetary phenomena generally do not match those of the fragments of Diaries for -182.

| Diary | Almanac |
|---|---|
| I around 14. Ξ of Venus end of Aries | I 22. Ξ of Venus beginning of Aries |
| II 23. Ω of Mercury in II; not observed | II 21. Ω of Mercury in II |
| XI around 4. Ψ of Jupiter 4 digits before β Tauri, 2½ cubits low to South | XI [x]. Ψ of Jupiter in Taurus |

BM 31051 (LBAT 1134) for 178 SE (= -133/2), transliterated and translated on pp. 96-99, computations on p. 112, and copy on Tafel VIII. Again, there is little agreement between this Almanac and the Diary for -133.

| Diary | Almanac |
|---|---|
| V summary. Jupiter in Cancer, Venus at beginning of V in Leo. | V 1. Jupiter and Venus in Cancer. V 15. Venus reached Leo. |
| XI 21. Ξ of Mercury beginning of Pisces; ideal Ξ on 20th. | XI 20. [Ξ] of Mercury. |

BM 33797 (LBAT 1190) for 301 SE (= -10/9), transliterated and translated on pp. 104-105, computations on p. 113, copy on Tafel IX. There exists no Diary for -10.

Kugler [1924] published a new transcription, translation, and commentary on LBAT 1123 on pp. 483-485, and a new translation of and a commentary on LBAT 1134 on pp. 496-498. He also published a number of new Almanacs, on pp. 470-513.

BM 46046 (LBAT 1172) for 236 SE (= -75/4), transliterated on p. 471. There is no disagreement with the fragment of a Diary for -75.

| Diary | Almanac |
|---|---|
| IX summary. Jupiter and Mars in [Gemini] | IX beginning. Jupiter and Mars in Gemini. |
| IX [x]. Γ of [Mercury] in Sagittarius | IX 7. Γ of Mercury in Sagittarius. |

BM 34722 (LBAT 1169) for 236 SE (= -75/4), on pp. 471-472. There is no overlap with the fragment of the Diary for -75.

BM 45698 (LBAT 1174) for 236 SE (= -75/4), transliterated, translated, and commented on on pp. 472-480. Adds nothing to the overlap of LBAT 1172 with the fragment of a Diary for -75.

BM 45953 (LBAT 1167) for 234 SE (= -77/6), transcribed on p. 486. The results of comparison with the Diary for -77 are mixed (some of the following Almanac entries are derived from the other Almanacs for -77/6).

| Diary | Almanac |
|---|---|
| I around 5. Ξ of Mercury in Taurus. | I 6. [x] of Mercury. |
| II 13. Ω in Gemini; not observed. | II [x]. Ω. |
| II summary. Jupiter and Mars in Pisces, [Venus] in Aries, Mercury in Gemini, Saturn in [x] | II 1. Jupiter and Mars in Pisces, Venus in Aries, Mercury in Gemini, Saturn in Libra |
| 16. Venus [reached] Taurus | II 16. Venus reached Taurus. |
| 26(?). Mars reached Aries | II 28. Mars reached Aries. |
| III around 8. Ψ of Saturn, 2½(?) [....] | III 8. Ψ of Saturn in Libra. |
| III 18. Γ of Mercury in Gemini; (ideal) Γ on 16th. | III 16(?). Γ of Mercury in Gemini. |
| III summary. 12. Venus reached Gemini | III 12. Venus reached Gemini |
| IV around 2. Σ of Mercury end of Gemini | IV 1. Σ of Mercury end of Gemini |
| IV around 23. Φ of Jupiter 1 cubit 8 digits behind η Piscium | IV 23. Φ of Jupiter in Aries. |
| IV summary. 7. Venus reached Cancer | omitted (in break ?) |
| 16. Mars reached Taurus | IV 17. Mars reached Taurus |
| V 13. Ξ of Mercury in Virgo; (ideal) Ξ on 12th. | V 10. Ξ of Mercury in Virgo. |

| | |
|---|---|
| V summary. 2. Venus reached Leo around | V 2. Venus reached Leo. |
| 13. Σ of Venus in Leo around 29. Ω of | V 12. ŠÚ |
| Mercury end of Virgo. | omitted (in break ?) |
| 29. Saturn | V 29. Saturn reached Scorpius |
| VI 2. Ω of Mercury end of Virgo; not observed. | omitted (in break ?) |
| VI 21. Θ of Jupiter | VI [x]. Θ of Jupiter in Aries. |
| VII summary. 13. Mercury reached Libra | VII 13. Mercury reached Libra |
| around 30. Σ of Mercury in Libra around 4. | omitted |
| [Ω] of Saturn in [x] | omitted (in break ?) |
| VIII around 2. Ξ of Venus in Scorpius; not observed | omitted |
| VIII around 8. Θ of Mars; not observed | VIII 8. Θ of Mars |
| VIII around 12. Γ of Saturn in Scorpius | omitted (in break ?) |
| VIII 20. Ψ of Jupiter ½ cubit in fr[ont ...] | VIII 20. Ψ of Jupiter in Pisces |
| XI around 28. Σ of Mercury in Aquarius; not observed | XI 28. Σ of Mercury. |
| XI summary. [x]. Venus reached Aries | omitted (in break ?) |
| 15. Mercury reached Aquarius | XI [x.X] reached Aquarius |
| XII 10. Φ of Saturn 3 cubits above [α Scorpii] | XII 10. Φ of Saturn in Scorpius |

The other fragments of Almanacs for 234 SE (= -77/6) published by Kugler are:

BM 33633 (LBAT 1161), transliterated on p. 486.

BM 46021 (LBAT 1168), transliterated on p. 487; since joined to LBAT 1167.

BM 34667+34668 (LBAT 1164+1165), transliterated on pp. 487-488.

These were all combined by Kugler into one Almanac, which he called "Tafel Σ", transliterated, translated, and commented on on pp. 488-492.

Other Almanacs published in Kugler [1924] are:

BM 34232 (LBAT 1122) for 128 SE (= -183/2), translated and commented on on pp. 492-493. There is no overlap with the surviving fragments of the Diary for -183 except for part of the summary of month VIII, where the two sources are not in disagreement.

BM 34121 (LBAT 1127) for 158 SE (= -153/2), translated and commented on on pp. 494-495. There is no overlap with the fragment of the Diary for -153.

BM 34051 (LBAT 1134) for 178 SE (= -133/2), translated and commented on on pp. 496-498. The results of comparison with the Diary for -133 are again mixed.

| Diary | Almanac |
|---|---|
| VI [x]. Σ of [Mercury] in Leo | VI 12. Σ of Mercury in Leo. |
| XI 20. Φ of Mars | omitted (in break ?) |
| XI 21. Ξ of Mercury beginning of Pisces; (ideal) Ξ on 20th. | XI 21. [Ξ] of Mercury beginning of Pisces |
| XII around 28. Θ of Mars | XII 22. Θ of Mars. |
| XII summary. Around 16 Ω of Mercury end of Pisces. | XII 18. Ω of Mercury. |

BM 34949 (LBAT 1139) for 185 SE (= -126/5), translated and commented on on pp. 498-499. There are some entries overlapping with those in the Diary for -126.

| Diary | Almanac |
|---|---|
| VIII [around] 13. Γ of Mercury in Scorpius | VIII [x]. Γ of [Mercury] in Scorpius |
| XII 14. lunar eclipse (after) 5 months, omitted. | XII 14 lunar eclipse (after) 5 months, omitted. |

BM 35039 (LBAT 1158) for 220 SE (= -91/0), translated on pp. 499-500. There are no remains of a Diary for -91.

BM 35149 (LBAT 1157) for 220 SE (= -91/0), translated and commented on on pp. 500-501.

BM 34991 + 35335 (LBAT 1154 + 1155) for 209 SE (= -102/1), transliterated and commented on on pp. 502-504; since joined with LBAT 1156. Nothing is left of any Diary for -102.

BM 34042 (LBAT 1153) for 209 SE (= -102/1), transliterated and commented on on pp. 502-505.

BM 33784 + 33790 (LBAT 1191 + 1192) for 303 SE (= -8/7), translated and commented on on pp. 505-506. No Diary survives for -8.

BM 34345 (LBAT 1129) for 160 SE (= -151/0), translated and commented on on pp. 507-513. None of the entries corresponds to any of those in the fragment of a Diary for -151.

Schnabel [1925a] published an Almanac from two fragments in Berlin, VAT 290 + 1836 (LBAT 1196), for 305 SE (= -6/5). There are no remains of a Diary for -6. Schaumberger [1925] pointed out that this Almanac records the "conjunction" of Saturn and Jupiter in Pisces in -6 that Kepler had suggested might have been the "star" that led the Wise Men to Bethlehem; unfortunately for this theory, the Almanac notes that both planets were in Pisces in months IV, V, VI, and VII, but does not refer to their actual conjunction which occurred on 24 May -6. Schaumberger [1926], together with Schoch, noted that the Babylonians recorded every instance in which any two planets (including Saturn and Jupiter) were in the same zodiacal sign, and that such "conjunctions" had no meaning for them. Pieces of another Almanac for 305 SE (= -6/5), BM 34659 (LBAT 1194), were published by Schaumberger [1935a] (Sp II 142) and [1943] (Sp II 795). The story of the attempt to identify the Jupiter-Saturn conjunction of -6 with the "Star of Bethlehem" and the role that the Almanacs for -6 played in this attempt is ably told by Sachs and Walker [1984], wherein a unified text, translation, and plates of LBAT 1193, 1194, 1195, and 1196 will be found.

Schaumberger [1935a] published notes on LBAT 1172 on pp. 359-361 and a copy on Tafeln I and II; notes on LBAT 1161, 1164 + 1165, and 1167 + 1168 on pp. 361-362 and copies on Tafeln III (LBAT 1161 and 1168), IV (LBAT 1164 + 1165), and V (LBAT 1167); translation of LBAT 1127 on pp. 362-363 and a copy on Tafel V; translation of LBAT 1122 on p. 363 and a copy on Tafel VI; translation of and notes on LBAT 1157 and 1158 on pp. 363-364 and copies on Tafel VI; translation of LBAT 1139 on p. 365 and copy on Tafel VII; notes on LBAT 1129 and copy on Tafel VII; notes on LBAT 1153 and 1154 + 1155 on pp. 366-367 and copies on Tafeln VIII (LBAT 1154 + 1155) and IX (LBAT 1153); notes on LBAT 1191 + 1192 on pp. 367-368 and copies on Tafeln VIII (LBAT 1192) and IX (LBAT 1191).

Sachs [1948] 277-280 listed twenty-three Almanacs, whose dates range from 128 SE (= -183/2) to 305 SE (= -6/5); they are LBAT 1122, 1123, 1124 (MLC 2195 for 147 SE (= -164/3), from Uruk), 1127, 1129, 1134, 1139, 1153, 1154 + 1155, 1157, 1158, 1161, 1164 + 1165, 1167, 1168, 1169, 1172, 1174, 1190, 1191, 1192, 1194, and 1196. Sachs [1955] pp. xxiii-xxiv adds to this list fifty dated and eleven undated exemplars. The earliest one is apparently BM 33746 (LBAT 1117), which may be for 50 SE (= -261/0) or possibly for 283 SE (= -28/7); the next earliest is BM 40101+ 55536 (LBAT 1118 + 1119) for 92 SE (= -219/8). The latest dated Almanac is a tablet in Dropsie College (LBAT 1201) for 385 SE (= 74/5 A.D.). This Almanac and five others from the first century A.D.—namely, MM 86.11.354 (not in Sachs [1955]) for 342 SE (= 31/2 A.D.), DT 143 (LBAT 1197) for 317 SE (= 36/7 A.D.), BM 45982 (LBAT 1198) for 355 SE (= 44/5 A.D.), BM 40083 (LBAT 1199) and its duplicate, BM 40084 (LBAT 1200), for 372 SE (= 61/2 A.D.)—were translated, transliterated, and copied by Sachs [1976]. An undated Almanac was published in Hunger [1976a] p. 101 (no. 99).

The fact that these Almanacs contain data for an entire year indicates that they are either excerpted from observational texts or they are computed. That the latter hypothesis is correct is clear from the completeness of the record and the absence of references to weather conditions or to ideal dates. See above p. 161. In addition, the Almanacs datable to the first century A.D. contain frequent errors in naming the zodiacal sign in which a phenomenon is expected to occur.

Though the data in the Normal Star Almanacs and the Almanacs are computed, the few cases where their dates of planetary Greek-letter phenomena overlap with dates of these phenomena in the Ephemerides indicate that the latter are not their source. We present a few examples.

| | |
|---|---|
| LBAT 1038 for 172 SE (= -139/8). | |
| III 13. Θ of [Jupiter]. | ACT 604 (System A) III 10;27,40 Θ of ♃. |
| | ACT 620 (System B) III 12;26 Θ of ♃. |
| V 2. Γ of Mercury in Cancer | ACT 302 (System $A_1$) V 1 Γ of Mercury at Cancer 24°. |
| LBAT 1134 for 178 SE (= -133/2). | |
| II 21. Σ of Mercury | ACT 302 [II 8] Σ of Mercury [at Aries 25;12] |
| III 1. Θ of Saturn | ACT 702 (System B) III 3;6,15 Θ of Saturn at [Scorpius] 29;5[3,45] |
| III 3. Ξ of Mercury at end of Gemini | ACT 302 [III 8] Ξ of Mercury [at Gemini 22] |
| III 16. Ω of Jupiter at beginning of Cancer | ACT 610 (System A') III [20] Ω of Jupiter at Cancer 1;58,7,30 |
| V 20. Γ of Mercury in Leo | ACT 302 [V 16] Γ of Mercury [at Leo 2;22,30] |
| VI 12. Σ of Mercury in Leo | ACT 302 [VI 12] Σ of Mercury [at Leo 26;41,30] |
| VIII 6. Ξ of Mercury in Scorpius | ACT 302 [VIII 4] Ξ of Mercury [at Scorpius 13;46,40 |
| VIII 23. Ω of Mercury at beginning of Sagittarius | ACT 302 [IX x] Ω of Mercury [at Scorpius 27;46,40] |
| X 20. [Σ] of Mercury | ACT 302 [X 22] Σ of Mercury [at Sagittarius 2;22,50] |
| XI 21. Ξ of Mercury at beginning of Pisces | ACT 302 [XI 24] Ξ of Mercury at [Pisces 0;]2[6,40] |
| XII 18. [Ω] of Mercury | ACT 302 [XII x] Ω of Mercury [at Pisces 22;26,40] |
| XII 19. Ψ of Jupiter in Cancer | ACT 610 XII 24;30,40 Ψ of Jupiter at [Cancer] 16;45. |

## *5. Goal-Year Texts*

The Goal-Year Texts, as described by Sachs [1948] pp. 282-285, present for a single given year the planetary data (dates of the Greek-letter phenomena and of the passings by of the Normal Stars) for a certain number of years previous to the given year for each planet. These "Goal-Year Periods" are as follows:

| | |
|---|---|
| Saturn | 59 years (≈ 2 rotations and 57 synodic periods) |
| Jupiter | 71 years (≈ 6 rotations and 65 synodic periods) |
| Jupiter | 83 years (≈ 7 rotations and 76 synodic periods) |
| Mars | 79 years (≈ 42 rotations and 37 synodic periods) |
| Mars | 47 years (≈ 25 rotations and 22 synodic periods) |
| Venus | 8 years (= 8 rotations and ≈ 5 synodic periods) |
| Mercury | 46 years (= 46 rotations and ≈ 145 synodic periods) |

There is also a column of lunar phenomena, that is, the occurrences of eclipses and the "Lunar Six," for the year 18 before the given year and (see Sachs [1955] p. xxv) the four lunar intervals in the middles of the last six months of year 19 before the given year.

Sachs [1955] p. xxv also pointed out that of the Goal-Year periods the two for Jupiter (71 years and 83 years) bring a close return of the dates of Greek-letter phenomena and of the dates of passings by of the Normal Stars respectively; the same is true of the two for Mars (79 years and 47 years).

The Goal-Year periods with modifications are recorded by Ptolemy in *Almagest* IX 3 (note that by "solar year" he means a tropical year of 6,5;14,48 days).

| Planet | Synodic periods | Solar years | Rotations |
|---|---|---|---|
| Saturn | 57 | 59 + 1¾ days | 2 + 1;43° |
| Jupiter | 65 | 71 – 4 9/10 days | 6 – 4 5/6° |
| Mars | 37 | 79 + 3;13 days | 42 + 3 1/6° |
| Venus | 5 | 8 – 2;18 days | 8 – 2¼° |
| Mercury | 145 | 46 + 1 1/30 days | 46 + 1° |

Note that the corrections to the rotations are simply the corrections to the solar years multiplied by the mean daily motion of the Sun:

$$1;45 \times 0;59 = 1;43,15$$
$$4;54 \times 0;59 = 4;49,6$$
$$3;13 \times 0;59 = 3;9,47$$
$$2;18 \times 0;59 = 2;15,42$$
$$1;2 \times 0;59 = 1;0,58$$

After Ptolemy, these Goal-Year periods reoccur in the astronomies of Greece, India, the Islamic world, Western Europe, and Central Asia; see Pingree [1998] p. 135.

Since the nineteen-year cycle implies that one year equals 12;22,6,20 synodic months, and since one month according to lunar System B contains 29;31,50,8,20 days, one year contains 12;22,6,20 × 29;31,50,8,20 = 6,5;14,49,... days. This is very close to Ptolemy's value of 6,5;14,48 days which he used in computing his table of Goal-Year periods. Therefore, assuming his lengths of the synodic arcs were nearly correct, the days that he adds to or subtracts from the solar years in each Goal-Year period are close enough for us to use them. Ptolemy even computes for us the number of days in each planet's Goal-Year period:

| | | |
|---|---|---|
| Saturn | 21,551;18 | = 5,59,11;18 days |
| Jupiter | 25,927;37 | = 7,12,7;37 days |
| Mars | 28,857;43 | = 8,0,57;43 days |
| Venus | 2,919;40 | = 48,39;40 days |
| Mercury | 16,802;24 | = 4,40,2;24 days |

If we divide each of these numbers by 29;31,50 we will get the approximate number of days, in the mean, by which the calendar date will shift between the forebear and the Goal-Year.

| | | |
|---|---|---|
| 5,59,11;18 days | ≈ 729 months + 23;31 days | ≈ 730 months – 6 days |
| 7,12,7;37 days | ≈ 877 months + 29;19 days | ≈ 878 months ± 0 day |
| 8,0,57;43 days | ≈ 977 months + 6;11 days | ≈ 977 months + 6 days |
| 48,39;40 days | ≈ 98 months + 25;40 days | ≈ 99 months – 4 days |
| 4,40,2;24 days | ≈ 589 months + 8;54 days | ≈ 589 months + 9 days |

How the Babylonians made the necessary adjustments is not known to us, but see II C 2.1.

The first publication of Goal-Year Texts was by Epping-Strassmaier [1890/1891]. BM 92682 (LBAT 1297) is transliterated and translated in ZA 5, 353-363 with computations on 363-366; there is a copy in ZA 6, 229-230. LBAT 1297 was later joined to BM 34053 (LBAT 1296). There are six sections: the first is for Jupiter in 153 SE (= -158/7), 83 years before 236 SE (= -75/4); the second is for Venus in 228 SE (= -83/2), 8 years before -75; the third is for Mercury in 190 SE (= -121/0), 46 years before -75; the fourth is for Saturn in 177 SE (= -134/3), 59 years before -75; the fifth is for Mars in 157 SE (= -154/3), 79 years before -75; and the sixth is again for Jupiter, but in 165 SE (= -146/5), 71 years before -75. There are a few overlaps with Diaries:

| | |
|---|---|
| section 2 (Venus) | Diary for -83 |
| III 23. reached Leo | III 23. Venus reached Leo |
| section 3 (Mercury) | Diary for -121 |
| III 20. Ξ in Cancer | III 17. Ξ of Mercury in Cancer |
| section 4 (Saturn) | Diary for -134 |
| VIII 4. Ω in Scorpius | VIII 4. Ω of Saturn in Scorpius; not observed |
| section 6 (Jupiter) | Diary for -146 |
| X 1. above μ Gem 8 digits | X 1. Ψ of Jupiter 6 digits before μ Gem, 8 digits high to the North [....] |
| XI 3. above μ Gem 8 digits | |

In the same article is transliterated and translated part ($S^+$ 76-11-17,1949) of BM 32222 (LBAT 1237) in ZA 6, 89-98 with notes and computations on pp. 98-101 and a copy on p. 231. This tablet has six sections: the first is for

Venus in 110 SE (= -201/0), 8 years before -193; the second is for Mercury in 72 SE (= -239/8), 46 years before -193; the fourth is for Saturn in 59 SE (= -252/1), 59 years before -193; the fifth is for Mars in 39 SE (= -272/1), 79 years, and in 71 SE (= -240/39), 47 years before -193; and the sixth is for the Moon in 100 SE (= -211/0), 18 years before -193. With it we can compare the Diaries for -193 as well as for the forebear years.

| | |
|---|---|
| section 1 (Venus; no Greek-letter phenomena recorded) | |
| II 3. above η Gem 1 cubit 4 digits | II 1. above η Gem 1 cub. 4 dig. (Diary -193) |
| II 5. above μ Gem 1 cub. 4 dig. | II 3. above μ Gem 1 cub. 4 dig. (Diary -193) |
| II 21. below β Gem 2½ cub. | III 1. below β Gem 2 cub. (Diary -193) |
| III 3. above δ Cnc 2/3 cub. | III 1. above δ Cnc 5/6 cub. (Diary -193) |
| IX 4. below β Lib 2 cub. | IX 4. below β Lib 2 5/6 cub. (Diary -193) |
| X 2. above α Sco 3 cub. | X 3. above α Sco 3 cub. (Diary -201) |
| X 14. above ϑ Oph 2 cub. | X 14. above ϑ Oph 2 cub. (Diary -201) |
| | X [13?]. above [ϑ Oph] 2 cub. (Diary -193) |
| XI 22. below β Cap 1½ cub. | XI [20]. [below β] Cap 2 cub. (Diary -193) |
| XII 9. above γ Cap 1 cub. 8 dig. | XII 5. above γ Cap 1 cub. 8 dig. (Diary -193) |
| XII 11. above δ Cap 1 cub. 8 dig. | XII [9]. above δ Cap 1 cub. 8 dig. (Diary -193) |

| | |
|---|---|
| section 2 (Mercury) | Diary for -193 |
| II 14. Γ in Taurus; not observed | II 14. Γ; [omitted] |
| II 28. Σ in Taurus; not observed | II 2[8]. Σ; omitted |
| VIII 20. Ξ in Sagittarius; not observed | VIII [x]. Ξ in Sagittarius; (ideal) Ξ on 20th |
| VIII 29. Ω in Sagittarius; not observed. | VIII around 29. Ω in Sagittarius; not observed |
| section 4 (Saturn) | |
| VIII 18. [Ω] | VIII 9. Ω in Scorpius. |

In the case of Mercury in LBAT 1237 the corresponding dates in column 2 and in the Diary for -193 are all computed. The Mercury ephemeris ACT 300 (System $A_1$) gives the date of Γ in -193 as II 9;21,5, and notes that it will be omitted.

In the same article was published, in ZA 6, 217-225, a partial translation with notes and computations of CBS 17 (LBAT 1295). There are five sections preserved: the first is for Jupiter in 154 SE (= -157/6), 71 years before -86; the second is for Jupiter in 142 SE (= -169/8), 83 years before -86; the third is for Venus in 217 SE (= -94/3), 8 years before -86; the fourth is for Mercury in 179 SE (= -132/1), 46 years before -86; and the fifth, after a break, is for the Moon in 207 SE (= -104/3), 18 years before -86.

Kugler [1907] pp. 84-87 published in transliteration and translation BM 34579 (LBAT 1251); there is a copy on Tafel V. To this was later joined BM 36006 (LBAT 1252). This tablet is divided into seven sections: section one is for Jupiter in 69 SE (= -242/1), 71 years before -171; section two is for Jupiter in 57 SE (= -254/3), 83 years before -171; section three is for Venus in 132 SE (= -179/8), 8 years before -171; section four is for Mercury in 94 SE (= -217/6), 46 years before -171; section five is for Saturn in 81 SE (= -230/29), 59 years before -171; section six is for Mars in 61 SE (= -250/49), 79 years before -171; and section seven is for Mars in 93 SE (= -218/7), 47 years before -171.

| | |
|---|---|
| section 3 (Venus) | Diary for -179 |
| VIII 15. [....] above γ Cap | VIII 15. [x]+1 digits [above γ Cap] |
| XI [x]. Ω at end of Pisces | XI around 29. Ω in Pisces. |
| section 4 (Mercury) | Diary for -171 |
| I 16. Σ in Aries. | [I] 16. Σ in Aries. |
| V [x]. Σ near ε Leo | V 18. Σ at beginning of Leo. |

Kugler-Schaumberger [1933a] published transcriptions and translations of and notes on three Goal-Year Texts; Kugler-Schaumberger [1933b] published computations relevant to the same three and copies of them.

Part of BM 32154 (LBAT 1216), namely $S^{+}$ 76-11-17,1881, was transliterated, transcribed, and annotated in Kugler-Schaumberger [1933a] pp. 4-7; there are computations in Kugler-Schaumberger [1933b] pp. 100-105 and a copy on pp. 113-114. There are eight sections at least fragmentarily preserved: the first is for the Moon in [63] SE (= -248/7), 18 years before -230; it records a lunar eclipse (after) five months on I 13 (= 19 April -248), which occurred before sunset in Babylon; a solar eclipse on I 28 (= 4 May -248); a lunar eclipse on VII 15 (= 13 October -248), which did not occur at Babylon; and a solar eclipse on VII 29 (= 27 October -248), also not visible at Babylon; the second section is for Jupiter in 10 SE (= -301/0), 71 years before -230; the third is for Jupiter in year 4 of Antigonus (= -313/2); the fourth is for Venus in 63 (read 73) SE (= -238/7), 8 years before -230; the fifth is for Mercury in 35 SE (= -276/5), 46 years before -230; the sixth is for Saturn in 22 SE (= -289/8), 59 years before -230; the seventh is for Mars in 2 SE (= -309/8), 79 years before -230; and the eighth is (probably) for Mars in 34 SE (= -277/6), 47 years before -230.

| | |
|---|---|
| section 2 (Jupiter) | Diary for -230 |
| I 19. Γ in Taurus | I [20]. Γ in Taurus, (ideal) Γ on 18th |
| section 4 (Venus) | Diary for -230 |
| I 11. Ξ in Taurus | I 13. Ξ in Taurus |
| [VI 13]. 1 cubit 4 digits above α Sco | VI 1+[x]. 1 cubit 4 digits above α Sco |

BM 32286 (LBAT 1218) was transliterated, translated, and annotated in Kugler-Schaumberger [1933a] pp. 7-8; there are computations in Kugler-Schaumberger [1933b] pp. 105-107 and a copy on pp. 114-115. In this fragment only the first section, for Jupiter in 19 SE (= -292/1), 71 years before -221, part of the second, for Jupiter in 7 SE (= -304/3), 83 years before -221, and part of the last, for the Moon in 72 SE (= -239/8), 18 years before -221, survive.

Part of BM 34625 + 34741 + 55543 (LBAT 1219 + 1220 + 1221) was transliterated, translated, and annotated in Kugler-Schaumberger [1933a] pp. 8-12; there are computations in Kugler-Schaumberger [1933b] pp. 107-112 and a copy on pp. 115-116. There are eight sections on the tablet (not all known to Kugler and Schaumberger): the first is for Jupiter in 20 SE (= -291/0), 71 years before -220; the second, for Jupiter, is practically gone; the third is for Venus in 83 SE (= -228/7), 8 years before -220; the fourth is for Mercury in 45 SE (= -266/5), 46 years before -220; the fifth is for Saturn in 32 SE (= -279/8), 59 years before -220; the sixth is for Mars in 12 SE (= -299/8), 79 years before -220; the seventh is for Mars in 44 SE (= -267/6), 47 years before -220; and the eighth is for the Moon in months VI to $XII_2$ of [72] SE (= -239/8), 19 years before -220.

Sachs [1948] pp. 282-285 knew only the seven Goal-Year Texts discussed above. In Sachs [1955] were added at least 73 dated texts and 62 undated ones. The earliest is Musée de Rouen HG-x (LBAT 1213), whose Goal-Year is 76 SE (= -235/4), the latest BM 34631 (LBAT 1305), whose Goal-Year is probably 352 SE (= 41 A.D.). In addition there are tablets which excerpt data for the two planets Mercury (Goal-Year period 46 years) and Mars (Goal-Year period 47 years) as a first step toward constructing a Goal-Year text. Such tablets are BM 55517 (LBAT 1372), with Mercury and Mars data for 7-9 SE (= -304/3 to -302/1), so that the Goal-Years were 54-55 SE (= -257/6 and -256/5); BM 46231 (LBAT 1373), with Mercury data for 54-56 SE (= -257/6 to -255/4) and Mars data for 54-55 SE (= -257/6 and -256/5), so that the Goal-Years are 101-102 SE (= -210/09 and -209/8); and BM 35339 + 46235 + 46242 (LBAT 1374 + 1375 + 1376), with Mercury data for 83-85 SE (= -228/7 to -226/5) and Mars data for 82-84 SE (= -229/8 to -227/6) and 50-52 SE (= -261/0 to -259/8), so that the Goal-Years are 129-131 SE (= -182/1 to -180/79). Sachs also remarks that BM 41958 (LBAT 1384), which contains Mercury data, may be from a Goal-Year Text.

There is no doubt that the data in the Goal-Year Texts were derived from annual observation reports like the Diaries and, in part, from computations. There are few overlaps of data in them with the data in our Diaries, however; but to a large extent the lack of coincidences is due to the fact that our Diaries are nowhere near as complete in their records as was needed to produce a Goal-Year Text. This leads us again to the hypothesis that there were other Diaries available in Babylon than those which we have recovered examples of.

If we now look at Normal Star Almanacs, Almanacs, and Goal-Year Texts together, we will see that no one of them derives from another nor, because of their completeness, from our Diaries.

I. Normal Star Almanacs contain computed "Lunar Three" *or* "Lunar Six"; computed and complete Greek-letter phenomena of planets; computed and numerous planets' passings by of the Normal Stars; computed and complete dates of solstices, equinoxes, and Sirius phenomena; and computed eclipses.

II. Almanacs contain computed "Lunar Three"; computed and complete Greek-letter phenomena; computed dates of the planets' entries into zodiacal signs; positions of the planets at the beginnings of months; computed dates of solstices, equinoxes, and Sirius phenomena; and computed eclipses.

III. Goal-Year Texts contain observed and computed complete dates of Greek-letter phenomena and passings by of the Normal Stars; the "Lunar Six"; and observed and computed eclipses.

## *6. Planets*

A number of tablets present the dates and zodiacal signs for the occurrences of the Greek-letter phenomena of individual planets; sometimes their passings by of Normal Stars are also included, or lunar phenomena. Some of these data represent observations (derived from the Diaries or other observational records) and some represent computations made later than the dates of the phenomena themselves. It is likely that many of these tablets were compiled for the purpose of comparing computations according to a system to the observations in the earlier Diaries; others, derived from observational records, may have been intended to provide data for constructing mathematical systems. Any computed values included in this latter class would have been intended to provide the correct number of occurrences of a phenomenon in a longer period.

An unpublished tablet found by John Britton at Harvard (see Swerdlow [1998] pp. 86-87) contains the earliest assemblage of observations of the Greek-letter phenomena of a single planet; the surviving fragment provides data on the occurrences of Γ and Ω of Mars in years 2 to 10 of Šamaš-šumu-ukīn (-665 to -656). Some of the information is tabulated in the lower seven lines of the table in Swerdlow [1998] p. 202.

The tablet containing the next oldest observations is BM 76738 + 76813. This tablet was announced by Walker [1982] pp. 20-21 as containing observations of Γ and Ω of Saturn, often noted as occurring near named Normal Stars, for years 1 to 14 of Kandalanu (-646 to -633). Walker has kindly provided us with a transliteration. Several phenomena were not observed. Usually no day is mentioned for them; in one case where the day is recorded, it is said to have been "calculated." A table indicating the years (in modern notation), months, and days of the 21 observations and the zodiacal sign in which the Normal Star was if it was named is to be found in Swerdlow [1998] p. 198. The range of the synodic periods—20 to 27 days—suffices to show that these are real observations. The collection was certainly put together sometime after the dates it records, and was presumably intended to be used in analyzing the motion of Saturn.

A tablet from Uruk recording observations of the Greek-letter phenomena (Γ, Φ, Ψ, and Ω) of Saturn for the years 28 to 31 (-576/5 to -573/2) of Nebukadnezar II was published by von Weiher [1993] p. 111 (no. 171) and explained by Hunger [1999]. It gives not only the dates (regnal year, month, and day) of each occurrence, but also usually the "Normal Star" near which the phenomenon occurred, though there is some question about which stars this tablet is in fact referring to. Only one phenomenon is stated to be not observed, an occurrence of Φ in year 28 $XII_2$ (-575 March 24 to April 21) in Sagittarius; the tablet notes neither the day nor the nearby Normal Star, and therefore employs no computation. Nor is it remarked in half of the cases how many cubits or fingers Saturn is from the Normal Star, nor whether it is before or behind, above or below. These deviations from normal procedures in recording information in the Diaries and the completeness of the record make it doubtful that this tablet was derived from a Diary unless the tradition of Diary-keeping in Uruk, for which we have later evidence, was different from that of Babylon.

BM 33066 (LBAT 1477) is known as "Cambyses 400" because of its position in Strassmaier [1890]. Epping [1890b] explained the lunar and planetary data in Cambyses 400; Oppert [1891] recognized the identity of the lunar eclipse of year 7 of Cambyses IV 14 (= 16 July -522) with that of year 7 of Cambyses, 225 of Nabonassar, VII 17/18 recorded by Ptolemy in *Almagest* V 14. Kugler [1903] claims that errors with regard to the calendar of the year 7 of Cambyses committed in the section on lunar phenomena on the obverse indicate that the material on a tablet of Cambyses' time was later reworked and inscribed on BM 33066; he also translates and comments on the planetary data and comments on the two lunar eclipses recorded on the reverse. Kugler [1907] pp. 61-75 transliterated and translated the entire tablet; an improved version by Hunger will be available shortly.

The text of the tablet consists of four parts. The first, on the obverse, gives the "Lunar Six" for months I to $XII_2$ of year 7 of Cambyses (-522/1); since the year 7 had no $XII_2$ (this is actually month I of year 8) and because of other errors, Kugler concluded that the tablet is a later copy, with changes introduced, of an original. He believed that the data were all computed since the set is complete whereas in the real world some of the 78 intervals reported would not have been observed because of adverse weather; of course, this argument does not preclude a mixture of observations and calculations. How such computations might have been made we do not know.

We have already noted above that the earliest Diary that we have that contains all of the "Lunar Six" is that for -372, though some of them appear in the Diary for -567. There are several tablets recording "Lunar Six" listed in Sachs [1955]; the oldest is BM 34075 (LBAT 1431) for the years -322/1 to at least -318/7. Based on this evidence, a fully developed theory for predicting the "Lunar Six" in existence in the late sixth century B.C. seems improbable; but, on the other hand, we have very little material from this early period in any case.

The second part contains a paragraph each recording the dates and nearby constellations of Greek-letter phenomena of Jupiter (Γ, Φ, Ψ, Ω), Venus (Γ, Σ, Ξ, Ω), Saturn (Γ, Ω), and Mars (Γ, Φ, Ψ, Ω) in year 7 and year 8 (year 9 for Mars) of Cambyses. Since the non-existent month $XII_2$ is mentioned before year 8 in each of these paragraphs and since there was no year 9 of Cambyses (this was year 1 of Darius I), it is clear that this section, though it presumably contains a mix of observations and computations because of its completeness, also passed through the hands of the compiler of part one. Fragmentary data for Mercury (Γ, Σ, Ξ, Ω) are found on the right edge of the tablet.

Part three excerpts observations for year 7 of Cambyses of the planets passing by each other. Added is a statement concerning Venus that Kugler [1907] p. 71 took to mean "reached its greatest elongation," but this interpretation seems unlikely; Kugler [1903] pp. 225-226 had computed that on the date in question, VI 24 (= -522 September 23/4), Venus was about 6° from the Moon in the morning. The dates are given almost in the order of the calendar: IV 1, VI 24, VII 23, VII 29, VII 12, VII 11, VIII 2, and X 5. This suggests that they were derived from a Diary, but that the redactor of the tablet inadvertently missed the earlier conjunctions in month VII and added them at the end.

Finally, the fourth part gives detailed reports concerning two lunar eclipses, that for IV 14 (= -522 July 16, which is the eclipse also recorded, from a Babylonian record, by Ptolemy), and that for X 14 (= -521 January 10). The evidence points to the tablet's having been based in part on Diaries or other observational texts for -522/1 to -520/19 and in part on computations, and to its having been executed in the late fifth or early fourth century B.C.

CBS 11901 (LBAT 1478), a tablet from Nippur, was referred to by Weidner [1914a] p. 6, who dated it to about -1500; Weidner [1914b] pp. 9-17 presented a partial transliteration and a long and irrelevant discussion. Weidner's discussion was irrelevant because of the false date that he had assigned to the tablet; Kugler [1914] pp. 233-242 demonstrated that the phenomena it records occurred in -424.

The text presents in column 1, for the period from month IV to month IX of that year, the "Lunar Three" (i.e., the 1 or 30 that indicates whether the previous month had 30 or 29 days, the day of the Full Moon, and the day of the Moon's last visibility), and in column 2 the days on which Γ, Σ, Ξ, and Ω of Venus and Mercury occurred, on which Γ and Ω of Saturn, Jupiter, and Mars occurred, on which solstice(s) and equinox(es) occurred, on which Γ of Sirius occurred, and on which solar and lunar eclipses occurred. The dates of the Greek-letter phenomena of the planets are shown by Kugler to be based on observations, as were undoubtedly the days for the Lunar Three; and we have seen above that the schemes for computing solstice, equinox, and Sirius phenomena dates did not exist in the fifth century B.C. Therefore we conclude that CBS 11901 was compiled from a Diary or other observational text for -424; we do not know what connection it had with Nippur.

The eclipses are both for month VII, the lunar on VII 15 (= 9 October -424) and the solar for VII 28 (= 23 October -424). Since the solar eclipse of 23 October -424 was not visible in Mesopotamia, it must have been predicted. The difference between the two eclipses is signalled in the text by its giving the time since sunset, 10° (= 0;40 h), till the beginning of the lunar eclipse, but no details concerning the (predicted) solar eclipse are to be found. As noted above, the earliest reference to a predicted solar eclipse that has survived in our extant Diary fragments is that for 25 September -302.

BM 32209 (LBAT 1411) + 41854 (LBAT 1412) lists very incompletely conjunctions of the Moon with Mars from -422/1 to -416/5 and for -412/1 on the obverse, and of the Moon with Saturn from -422/1 to -420/19, for -416/5, for -403/2 and -402/1, and for -399/8 on the reverse. We can compare some of the Mars set with the Diary for -418.

| Diary | BM 32209 |
|---|---|
| $GE_6$ 9 $MURUB_4$ *sin ina* IGI AN 3 SI | MU 5 BAR 7 $MURUB_4$ *sin ina* IGI |
| *i sin ana* ULÙ SIG | AN 3 SI *i sin ana* ULÙ SIG |

Though the two sources disagree on the day of the month, BM 32209 clearly has copied the Diary. It is easy to believe that the other entries were also compiled from Diaries; but the purpose of the compilation eludes us. The scribe also carefully notes the intercalary months; they follow the nineteen-year cycle.

In the case of the well-known Jupiter observation tablet, BM 34750+ (its part BM 35328 (LBAT 1394) was published by Kugler [1907] pp. 80-85 and Tafel IV) the scribe informs us in a colophon: "Appearances of Jupiter which are from year 18 (= -386/5) of Arses, who is called Artaxerxes (II), the king, to month IV of year 13 (= -345/4) of Ochus, who is called Artaxerxes (III), the king. Written (as) a copy of clay tablets and wooden tablets [of] Diaries belonging to ...., the son of Mušallim-Bēl. Tablet of ...., the son of Marduk-šāpik-zēri. Hand of Bēl-šunu, the son of ...." Presumably the "clay tablets and wooden tablets of Diaries" are the "short" Diaries from which the multi-month Diaries were compiled. Unfortunately, in the periods actually covered by the surviving fragments of BM 34750—-386 to -383, -364 to -358, -356 to -355, -350 to -348—the only overlapping Diary that we possess is an incomplete one for -384 VII to XII; it offers no data to be compared with BM 34750+.

That tablet records not only the dates of the observed or computed phases (Γ, Φ, Θ, Ψ, and Ω) of Jupiter (see Swerdlow [1998] p. 200) and its passings by of Normal Stars and other planets, but also, after the name of each month, a "30" or a "1" indicating that the previous month had 29 or 30 days respectively. All of this information, of course, was available in the Diaries.

Our attention now shifts to the beginning of the Seleucid era, about the time that the mathematical systems of the ACT material were being devised. BM 35501 (LBAT 1403) + 35532 (LBAT 1402) + 35713 (LBAT 1410) gives observations and computations of the Greek-letter phenomena of Jupiter and its passings by of the Normal Stars and the other planets, excluding the Moon, from SE 11-16, 20-25, and 48-50 (= -300 to -294, -291 to -285, and -263 to -261), and the same phenomena for Mercury from SE 46 to 50 (= -265 to -261). We can compare the data of this tablet with the remaining Diaries for -300, -299, -294, -291, -289, -288, -287, -286, -264, -263, -262, and -261.

Jupiter

| Diary | BM 35501+ |
|---|---|
| -300 VIII 19. Jupiter, while receding to the west, was [...] below [α Gem] | [VIII?... Jupiter] 3 cubits [below] α [Gem] |
| -289 [II ...] Γ of Jupiter in "Bull of Heaven" | II 7. [Γ] of Jupiter in ♉. |
| III 9. [Jupiter] 2 cubits above α Tau | III 9. Jupiter 2 cubits above α Tau |
| -288 VII 21. Φ of Jupiter in ♋ | [VII x.] Φ of [Jupiter] in ♋ |
| -287 [I 20?] first part of night. Venus 1½ cubits above Jupiter | I [1]8, first part of night. Venus [...] above Jupiter |
| -262 V 20. Γ of Jupiter in ♌; it was small, rising of Jupiter to sunrise: 11;40°; (ideal) Γ on 19th. | V 20. Γ in ♌; [..]; (ideal) Γ on 19th. |

Mercury

| Diary | BM 35501+ |
|---|---|
| -261 II 8 first part of night. Mercury [... above η Gem] | II 8? [...] above η Gem |
| IX 20. Ξ of Mercury in ♑; it was bright and high, sunset to setting of Mercury: 15°; (ideal) Ξ on 19th or 18th. | [IX] 20. ⌜Ξ⌝ |

There can be little doubt that the compiler of BM 35501+ used Diaries, but not only those available to us.

BM 32590 (unpublished) contains on the obverse observations of occurrences of the Greek-letter phenomena (Γ, Φ, Θ, Ψ, and Ω) of Mars from SE 27 to 35+ (= -284 to -276+) and on the reverse observations of Mars' passings by of Normal Stars from SE 11 to 16 (= -300 to -294). We compare Diaries for -300 and -299 for the passings by and -284, -283, -281, -278, and -277 for the Greek-letter phenomena.

| Diary | BM 32590 |
|---|---|
| -285 VII 15. Ψ of Mars 2/3 cubit below η Psc | [VII] 15. Ψ of Mars [...] η Psc |

While this is the only overlap in our material, it remains most probable that BM 32590 was compiled from Diaries.

BM 32231 + 32273 (unpublished) lists on the obverse the Greek-letter phenomena (Γ, Σ, Ξ, and Ω) of Mercury and its passings by of Normal Stars for SE 14 to 17 (= -297 to -293), and on the reverse a fragmentary account of Mars' passings by of Normal Stars in SE 15 (= -296/5). The only Diary available for comparison is that for -294; there are no overlaps.

A 3405 (LBAT 1479), a tablet from Uruk, is briefly described by Sachs [1948] pp. 286-287. It gives the Greek-letter phenomena of the planets and lunar eclipses for the period from SE 60 to 70 (= -251/0 to -241/0); each year has a section within which each month has a subsection. This text uniquely provides the degree of longitude of the planets at their phases and of the Moon at its eclipse; it is highly likely that these longitudes were computed by means of ACT-type systems. It also mentions, for the lunar eclipses, the eclipse magnitude and whether the Moon was near the ascending or descending node; this information could be derived from the lunar Ephemerides. Sachs believes that the tablet was copied about 50 years after the events it describes.

Another tablet from Uruk, A 3456 (LBAT 1377), records the Greek-letter phenomena (Γ, Σ, Ξ, and Ω) of Mercury and its passings by of Normal Stars and the other planets from SE 116 to 132 (= -195/4 to -179/8), with occasional notices of computed equinoxes and solstices. It was published by Hunger [1988] and the dates of the Greek-letter phenomena were tabulated by Swerdlow [1998] pp. 205-207 and 209-210. Swerdlow [1998] p. 107 regards these as observations, but notes that A 3456 contains material not normally found in Diaries. Indeed, when we compare the Babylonian Diaries with this tablet, there are both agreements and disagreements; and, of course, in both computations are mingled with observations.

| Diaries | A 3456 |
|---|---|
| -195 IX 16. Ω | Ω in ♑; not observed |
| IX 24. Γ in ♑ | Γ; rising of Mercury to sunrise: 19°; (ideal) Γ on 24th. |
| XII 13. Ξ in ♈; bright and high; sunset to setting of Mercury: 15°; (ideal) Ξ on 11th. | [...]; bright; sunset to setting of Mercury: 15°(?); (ideal) Ξ on 11th. |
| -194 III [3?]. Σ ...; omitted | Σ omitted |

| | |
|---|---|
| -193 II 2[8?] Σ; omitted | II 27. Σ; omitted |
| VIII [x]. [Ξ]; sunset to setting of Mercury: 15°; (ideal) Ξ on 20th. | VIII 22. Ξ in ♐; sunset to setting of Mercury: 15°; (ideal) Ξ on 20th. |
| VIII 29. Ω in ♐; not observed. | VIII 29(?). Ω in ♐; not observed. |
| X [x]. Σ in ♑; not observed. | X 27. Σ in ♑; not observed. |
| XII [x]. [Ξ] in ♓ | XII 1. Ξ in ♓; sunset to setting of Mercury: 15°; (ideal) Ξ on XI 26(?). |
| -192 II 18. Σ in ♊; bright and high; sunset to setting of Mercury: 15°; (ideal) Σ on 16th. | [II x.] Σ in ♊; bright; sunset to setting of Mercury: 15°; (ideal) Σ on 17th. |
| II 20. Mercury 2½(?) cubits below β Gem | Mercury 2 cubits 10 digits [below β] Gem |
| -191 IV 22. [Γ] | IV 27. Γ in ♌; rising of Mercury to sunrise: 16°; (ideal) Γ on 22nd. |
| VIII [x.] Γ at end of ♎; bright and high; rising of Mercury to sunrise: 21°; (ideal) Γ on [x] | VIII 8. Γ at end of ♎; bright and high; rising of Mercury to sunrise: 20°; (ideal) Γ on 2nd. |
| -189 VII 10. [Mercury] 2/3 cubit [above Mars] | VII 10. Mercury 5/6 cubit above Mars. |
| -186 XI 14. Ξ in ♍(!) 2½ cubits behind Venus; bright and high; sunset to setting of Mercury: 16°; (ideal) Ξ on 12th. | XI 14. Ξ in ♓; bright; sunset to setting of Mercury 16°; (ideal) Ξ on 11th. |
| XII [1]. Ω at end of ♓. | XII 1. Ω in ♓. |
| -183 II 3. Mercury [...] above γ Gem | II 4. [...] above γ Gem |
| IV [1. Γ in ♊; ...;] (ideal) Γ on [III] 26 | III 26. Γ in ♊; not observed |
| -182 X 17. Γ in ♑ 1 cubit in front of Moon; bright; rising of Mercury to sunrise: 16°; [...] | [X ...]. Γ [...]; rising of Mercury to sunrise [...]; (ideal) Γ on 24th. |

These examples suffice to demonstrate that the scribe of A 3456 used a different set of Diaries—presumably originating in Uruk—from the Babylonian Diaries available to us, though the conventions and even many of the computations of the two schools were identical or very similar.

A strong indicator of the fact that the scribe of A 3456 did not use an Ephemeris is that the computed dates on that tablet and the entries in A 3424 + 3436 (ACT 300), an Ephemeris from Uruk for -193/2 to -168/7, are not in agreement. Nor do the entries in ACT 300 agree with those in BM 45980 (ACT 301), an Ephemeris from Babylon, where they overlap (from -178/7 on), even though both belong to System $A_1$; both A 3456 and ACT 300 belong to local traditions in Uruk.

| A 3456 | ACT 300 |
|---|---|
| -187 [VIII x] Γ; (ideal) Γ on 10+. | [IX 3]. Γ in ♐ |
| -186 [VI x. Σ] in ♍; (ideal) Σ on 2nd | VI 5. Σ in ♍ |
| VIII 25. Γ in ♏: (ideal) Γ on 24th | [VIII] 28. Γ in ♏ |
| X 8. Σ in ♑; not observed | X 14. Σ in ♑ |
| -185 I 26. Σ; omitted | I 25. Σ in ♈ |
| V 6. Γ at end of ♋; (ideal) Γ on 4th | [V] 3. Γ in ♋. |
| VIII 21 Γ in ♏; (ideal) Γ on 19th | [VIII] 22. Γ in ♏ |
| -184 IX 2. Σ at beginning of ♐; not observed | VIII 25. Σ. |
| -183 VIII 11. Σ in ♏; not observed | VIII 15. Σ. |
| XI 2. Γ in ♑; not observed | [XI] 3. Γ in ♑ |
| XII 12. Σ in ♓; not observed | X[II 13]. Σ. |
| -182 [III] 24. Γ in ♊; not observed | [III 22]. Γ in ♊ |
| [VIII x.] Σ; (ideal) Σ on 1st. | [VIII 7]. Σ. |
| [X x. Γ]; (ideal) Γ on 24th. | [X 25.] Γ in ♑ |
| XII 7. [Σ] in ♒; not observed | [XII 9.] Σ. |
| -181 [II] 18 [Γ] at end of ♉; [not] observed. | [II 17]. Γ [in ♉]. |
| [VI] 3. Γ | [V 30]. Γ [in ♍] |
| IX [1]7. [Γ] | [IX 19]. Γ [in ♐]. |
| -180 II 20. Γ; omitted | [II 6]. Γ [in ♉] |
| [V x]. Γ in ♌; (ideal) Γ on 23rd. | [V 24]. Γ [in ♌]. |

It is important to remember that the restorations of the dates in ACT 300 are absolutely certain since they are the result of mathematical calculations. This does not prove that the authors of the Diaries never used the Ephemerides, but it does show that there was some distance between the observers and compilers of Diaries and the men who computed the Ephemerides.

IM 44152 (LBAT 1409), which is perhaps from Uruk, is a collection of the dates and longitudes (by zodiacal signs) of the Greek-letter phenomena (Γ, Φ, Θ, Ψ, and Ω) of Jupiter for SE 160 to 172 (= -151/0 to -139/8). The data for -149 to -139 are tabulated by Swerdlow [1998] pp. 200-201. We have, to compare with this, Diaries for -149, -148, -146, -144, -143, -142, -141, and -140.

| Diary | IM 44152 |
|---|---|
| -146 X [18, 19, or 20] Ψ 6 fingers before μ Gem | X 19. Ψ in ♊ |
| -143 IV 22. Sunset to setting of Jupiter: 13° | IV 25. Ω in ♍ |
| VI 1. Γ in ♍; rising of Jupiter to sunrise: 12;30°; (ideal) Γ on V 28 | V 28. Γ in ♍ |
| -141 I 12. Θ | I 13. Θ |
| $VI_2$ 20. Ω at beginning of ♏ | $VI_2$ 11. Ω in ♏ |
| VII 22. Γ in ♏; small; rising of Jupiter to sunrise: 11;30°; [...] | VII 20. Γ in ♏ |

| | |
|---|---|
| XI [20 to 24] Φ 1 cubit before ϑ [Oph] | XI 23. Φ in ♏ |
| -140 I 27. Θ | I 27. Θ |
| IX [7. Γ in ♐; bright and high; rising of Jupiter to sun]rise: [...]; (ideal) Γ on 5th. | IX 6. Γ in ♐ |

Clearly these dates are derived from different sources. The Diaries we can consult are Babylonian; IM 44152 may have come from Uruk. If it did, the situation is the same as in the case of A 3456, and we have additional information in IM 44152 concerning the Diaries composed in Uruk.

Finally, Sachs [1955] describes the tablet BM 34368 (LBAT 1481) + 41596 (LBAT 1482) and BM 42147 (LBAT 1483) as having "planetary data, arranged in tabular form, for several years."

### *7. Eclipses*

It is likely, though it cannot be directly demonstrated, that the reports of consecutive lunar eclipses arranged in eighteen-year sections were compiled as a sort of "index" to the Diaries, that is, as a device for checking the accuracy of eclipse theory. Obviously, they were not recorded on the surviving tablets until after the last eclipse on each was observed; for these are observation reports, including much information that could not have been computed for the past or predicted for the future.

We do not know, therefore, the date of the fragment, BM 41985 (LBAT 1413), which contains reports of observations of lunar eclipses in -747/6 and -746/5. However, the next tablet described in Sachs [1955], BM 32238 (LBAT 1414), records lunar eclipses from -730/29 to -316/5, and refers before breaking off to year 2 of Antigonus (-314/3); clearly the tablet was inscribed at the end of the fourth century. Other such collections are: three fragments of a single tablet, BM 45640 (LBAT 1415), BM 35115 (LBAT 1416), and BM 35789 (LBAT 1417), for -702/1 to -359/8; BM 32234 (LBAT 1419) for -608/7 to -446/5; BM 38462 (LBAT 1420) for -603/2 to -575/4; BM 41536 (LBAT 1421) for -563/2 and -562/1; BM 34787 (LBAT 1426) for -441/0 to -421/0; BM 34684 (LBAT 1427) for -424/3 to -406/5; BM 37652 (LBAT 1429) for -382/1 to -362/1; BM 37043+37107 (unpublished) for -327/6; BM 48090 (LBAT 1432) for -279/8 and -278/7; BM 34236 (LBAT 1436) for -194/3 to -159/8; BM 34963 (LBAT 1334) + BM 35198 (LBAT 1435) + BM 35238 (LBAT 1443) for about -88 to -86 together with other lunar data; and two undatable tablets, BM 45667 (LBAT 1451) and BM 41800 (LBAT 1452); the latter is possibly to be dated to -283.

Although the broken nature of most of these fragments often prevents us from knowing the earliest or, more importantly for the purpose of dating, the latest eclipses recorded on them, it is likely that most were compiled in the fifth and fourth centuries when eclipse theory was being developed.

Alongside these detailed reports of lunar eclipses there exist lists of the dates (years and months only; the days, which must always be the 13th, 14th, 15th, or 16th, were not recorded) of observed lunar eclipses arranged in eighteen-year groups. Thus BM 32363 (LBAT 1418) covers the period from at least -646/5 to -574/3; BM 36910 (LBAT 1422) + BM 36998 (LBAT 1423) + BM 37036 (LBAT 1424) from -489/8 to -379/8; and BM 37094 (LBAT 1425) from -454/3 to -435/4. They were presumably connected with the establishment and testing of the predictive "Saros Canon," which dates from some time earlier than -384. See II C 1.1.

We also have reports of one, or sometimes two, lunar eclipse(s) for dates late in the Seleucid period. BM 33643 (LBAT 1437) describes one that occurred on SE 121 XII 15 (= 28 February -189); BM 42053 (LBAT 1439) one on SE 124 IV (listed as penumbral in Meeus-Mucke [1983] under -187 Aug 1) and one on SE 127 VIII (= 24 November -184), which is dated VIII 15 in the Diary for -184; BM 41129 (LBAT 1440) one on SE 157 XII 14 (= 21 March -153); BM 33982 (LBAT 1441) one on SE 183 VIII 13 (= 5 November -128); BM 45845 (LBAT 1442) one on SE 192 II 14 (= 1 June -119), which is dated II 13(?) in the Diary for -119; BM 42145 (LBAT 1444) one on SE 231 I 14 (= 21 April -80), which is dated II 13(?) in the Diary for -80; BM 33562A (LBAT 1445), which was published by Strassmaier [1888] pp. 147-148 and Epping [1889b], one on SE 232 I 13 (= 11 April -79); and BM 42073 (LBAT 1446) one for SE 232 VII(?) 14 (= 5 October -79). The last three tablets contain detailed reports on lunar eclipses that do not appear in Kudlek-Mickler [1971], but are listed as partial in Meeus-Mucke [1983] and in Stephenson-Steele [1998]: BM 41565 (LBAT 1447) with its duplicate, BM 45628 (LBAT 1448), one on SE 244 X 15 (= 19 January -66); BM 34940 (LBAT 1449) one on SE 245 X 13/4 (= 8 January -65); and BM 32845 (LBAT 1450) one on SE 246 IX 14 (= 28 December -65).

BM 36754 (LBAT 1430) records the dates (year and month) of a sequence of solar eclipses arranged in groups of eighteen years from -347/6 to -258/7. According to Stephenson-Steele [1998] thirty-one solar eclipses were visible at Babylon between -347 and -257, whereas BM 36754 apparently lists many more. This suggests that the eclipses in BM 36754 were computed—perhaps by a scheme related to the "Saros Cycle" such as those discussed in II C 1.1b. BM 71537 (unpublished) seems to contain reports of observations beginning in the Achaemenid period and ending in -312/1.

There is, finally, one tablet recording a single solar eclipse. BM 33812 (LBAT 1438) describes the eclipse of SE 121 XII 29 (= 14 March -189). It is noted in the Diary for -190 as having happened at the end of month XII.

## C. Theoretical Texts

### *1. Early Theoretical Texts, Solar and Lunar*

#### *1.1. The "Saros" Cycle*

The "Saros" cycle (so-called because Edmund Halley applied this name to an eclipse cycle in 1691; see Neugebauer [1938]) can be expressed as the period-relation (see Neugebauer [1975] pp. 502-505):

223 synodic months ≈ 242 draconitic months ≈ 239 anomalistic months ≈ 241 sidereal months ≈ 18 years.

The most accurate of these approximations is that a "Saros" cycle exactly equals 223 (3,43) mean synodic months. Then, if a solar year is approximately 6,5;15 days and 12;22,8 synodic months,

3,43 syn. m. ≈ 18;1,45 years ≈ 18 years + 10;40 days

This period-relation was first discovered in a cuneiform text by Epping-Strassmaier [1893], who had found it on BM 34597 (LBAT 1428); they provide a table summarizing its data on pp. 176-177 with regnal years, years before 0 in our calendar, and the numbers of the Babylonian months. This text was published in a copy by Strassmaier [1895] and in a partial transcription with corrections by Aaboe [1972] p. 114.

Subsequently Aaboe-Britton-Henderson-Neugebauer-Sachs [1991] published photos, transliterations, and translations of two tablets mentioned in II B 7, BM 36910+36998+37036 (LBAT 1422+1423+1424), from -490 to -374, on pp. 4-8; of BM 37044 (LBAT 1425) for -454 to -434, on pp. 10-11; and, again, of BM 34597 (LBAT 1428), for -400 to -271, on pp. 12-16. From the physical characteristics of BM 34597 the authors conclude that it originally began at least as early as -490, the initial year of LBAT 1422+1423+1424, and continued to -257. Because these tablets give regnal years in the Achaemenid and Macedonian periods and years in the Seleucid period after year 6 of Antigonus, they were obviously copied after -374, -434, and -257 respectively.

All three texts belong to the same general scheme. Each column contains 38 eclipse-possibilities in 18 years; the eclipse possibilities are arranged in five groups with intervals of 6 months between entries; between each group is a period of five months in which there are no eclipse possibilities. The number of entries in each group is 8, 8, 7, 8, and 7 respectively. The entries are simply the names of the Babylonian months; the regnal year or the year in the Seleucid Era is given to the left of the first month of that year to appear in the list. The year is marked to indicate that it has an intercalary Ulūlu ($VI_2$) or an intercalary Addaru ($XII_2$). In LBAT 1422+1423+1424 the intercalations are the actual ones rather than schematic; in LBAT 1425 also the intercalations are the actual ones. In LBAT 1428 the intercalated years after -497 coincide with those ordained by the nineteen-year cycle. The authors present a reconstruction of the "Saros canon" from -526 to -257 on p. 21, and argue from the occurrence of the phrase "5 months" at the "correct" boundaries in this scheme (extended backwards) in eclipse-reports from -746 to -422 that the eighteen-year cycle in precisely this form goes back to the middle of the eighth century B.C.; all the more surprising, then, if this were true, that the authors of the *Reports* and *Letters* in the seventh century would look for an eclipse in three or four consecutive months. The earliest eclipse-reports seem to have been compiled in the fifth or fourth century B.C.

On p. 19 the authors give a table of lunar eclipses visible at Babylon between -746 and -238. From this it is easy to see how many were missed because they occurred during the day at Babylon; 22 between -743 and -725, 18 in the next column, and so on. The scheme was constructed by counting the eclipses that were not observed.

Britton [1993] argues that the "Saros" cycle and probably also the nineteen-year cycle were known before approximately -525; that the "Saros" cycle was reformed and the nineteen-year intercalation cycle was established between approximately -525 and -475; and that the system of measuring longitudes by the twelve zodiacal signs had been introduced by approximately -450. He justifies these conclusions by citing the following texts:

I. Records of two depositions concerning incidents that occurred at Larsa and Uruk on 13 III 8 Cyrus (= 15 June -530) as discussed and interpreted by Beaulieu-Britton [1994]. The incidents involved the beating of a kettle-drum at the gates of the temples Ebabbar and Eanna on the occasion of a lunar eclipse. Since no lunar eclipse actually was visible in Mesopotamia on 15 June -530, the authors infer that one was predicted, and argue that the prediction was made on the basis of the "original" "Saros" cycle, and employ the reconstructed "Saros" cycle from -746 to -509 (including a shift downwards of one eclipse-possibility for the five-month interval at the beginning of the eighteen-year cycle that began in -526, a shift that produced the "reformed" "Saros" cycle) to show that it would indeed predict an eclipse on 13 III 8 Cyrus. We would note that the same result would come from using the 41-month cycle (visible lunar eclipse in XI 4 Cyrus = 21 February -533) or the 88-month cycle (visible lunar eclipse in II 1 Cyrus = 4 May -537); for the 41-month cycle, the 47-month cycle, and their combination, the 88-month cycle, see below.

II. The "Saros Canon" texts mentioned above. We would note that the oldest of these was copied after -434.

III. BM 36599+36941 and BM 36737+47912, discussed in II C 1.1a and 1b below, which was compiled after -456 because it presents calculated and observed data for the 38 eclipse possibilities from -474 to -456, follows the "reformed" "Saros" cycle. This same text gives the longitudes of the eclipse-possibilities measured in zodiacal signs, degrees, and minutes. We note that this inclusion could be explained on the hypothesis that the tablet was compiled late in the fifth century on the basis of computations combined with eclipse reports, later than the period -456 to -438 suggested by Britton. The alleged reference to a zodiacal sign in the Diary for -463 cited by Britton on p. 69, fn. 11, is, in fact, to a constellation, while that alleged to occur in the Diary for -453 is inconclusive.

Pannekoek [1917] proposed a series of steps by which the Babylonians could have accumulated the knowledge needed to construct the "Saros Canon". The first was to notice, from observations of when the Sun and the Moon "balance each other" in the middle of the month, that if the Full Moon rose before sunset, a lunar eclipse might occur that night; if it rose after sunset, there would be no eclipse that night (we have reworded Pannekoek's statement to form a simpler criterion than his). After one lunar eclipse was observed, another might be seen six months later. In fact, five or six lunar eclipses separated by six months each would follow each other; the series would begin with one or two partial eclipses, in the middle would be a few total eclipses, and at the end one or two partial eclipses again. Before and after such a series would be an interval of "six or twelve" months without an eclipse. A collection of records of five such series would show that each contains seven or eight six-month periods followed by a five-month period, and the reports of the sixth period would be similar to those of the first. It is this similarity in the character of the eclipses (magnitude and direction of impact) between the eclipses in series five series apart that Pannekoek claims to have convinced the Babylonians of the validity of the cycle; he emphasizes the fact that the return in anomaly as well as the return with respect to the node in eighteen years causes this similarity.

Schnabel [1924a] argues from predictions of eclipses in the *Reports* that the "Saros" cycle was being used in the Sargonid period. While it is not impossible that it was used by some, many statements in the *Reports* indicate that their authors were not well informed concerning eclipse possibilities.

Neugebauer [1937c] pp. 245-248 suggests that from very early times (we have seen this in *Enūma Anu Enlil* and in the *Letters* and *Reports*) it was realized that solar eclipses occurred at "New Moons" and lunar eclipses at Full Moons. He further assumes that it was realized that an eclipse occurred only when the distance of the Moon from the ecliptic was small; Neugebauer does not say this, but the evidence for this would lie in the records of the Moon's passings by of the Normal Stars so many cubits and digits above and below. The Diaries where those data are recorded we hypothesize to go back to -746. Therefore, there may have existed records of the Moon's passings by of the Normal Stars earlier than -750, though none has been found. Neugebauer's suggestion continues with the linkage of the eclipse period with an intercalation cycle, in order to allow predictions of the months in future solar years in which eclipses would occur (this intercalation scheme—i.e., presumably the nineteen-year cycle—would have been used by astronomers; the civil authorities did not follow this cycle until much later, as is reflected in the early part of the "Saros Canon" which refers retroactively to the actual calendaric dates used by the civil authorities) and with the hypothesis that the Babylonians constructed a simple linear zigzag function of lunar latitude in which $p = \frac{223}{242} = 0;55,17,21,...$, $M = 5°$, $m = -5°$, and $\mu = 0°$. Since $P = \frac{p}{1-p} = \frac{0;55,17,21}{0;4,42,39} = 11;44,11,45,...$, $d = \frac{2(M-m)}{P} = \frac{20}{11;44,11,45} = = 1;42,14,... \approx 1\frac{7}{10}\,{}^{\circ}\!/_{m}$. Then, starting from a latitude of 0°, the next latitude of 0° will occur in $\frac{10°}{1\frac{7}{10}\,{}^{\circ}\!/_{m}} \approx 5;53$ months. This means that in 223 months there are 38 eclipse possibilities. It must be noted that no tablet containing such a linear zigzag function has been discovered.

Van der Waerden [1940] points out that the eighteen-year cycle for predicting eclipse-possibilities is used in Goal-Year Texts of the second century B.C. He then refers to BM 35402 (LBAT 1593), which had been published by Kugler [1907] p. 48, for a lunar period of 684 years, and to part (Sp 184) of BM 34085 (LBAT 1599), which had been quoted in Kugler [1907] p. 53, for this period of 684 years as an eclipse period. He dates both tablets in the fifth century B.C. The period of 684 years he derives in the following way: 47 synodic months = $51\frac{1}{237}$ draconitic months $\approx 51\frac{1}{180}$ drac. mos. From this, 8460 syn. mos. = 9181 drac. mos., and $8460^{m} = 684^{y} \times 12\frac{7}{19}\,{}^{m}\!/_{y}$. One could more easily derive this from the nineteen-year cycle, which we know was used in the fifth century B.C., and an observation that an eclipse repeats in 47 synodic months. Then $47^{m} \times \frac{19y}{235m} = \frac{893}{235}y$; and to convert $\frac{893}{235}y$ into an integer number of years we multiply it by 180 to obtain 684 years. Van der Waerden notes that the 47 synodic month rule will work for the solar eclipse predicted in CBS 11901 (LBAT 1478)—the solar eclipse of 23 October -424 (= VII 28 of 40 Artaxerxes I) which was not visible in Babylon. According to the "Saros Canon" there was an eclipse possibility 48 synodic months previously, in month IX of 36 Artaxerxes I, which was also not visible in Babylon. But van der Waerden notes that 141 (47 × 3) months previously, on 31 May -435 (= II 28 of 29 Artaxerxes I), there was an almost total solar eclipse visible at Babylon. Van der Waerden then turns to the lunar eclipse predicted for III 15 of Nebukadnezar (= 4 July -567) found in the Diary for -567; he points out that 47 months earlier, on VI 12 of 33 Nebukadnezar (= 14 September -571) a lunar eclipse was visible in Babylon.

Finally, van der Waerden cites Schaumberger [1935] pp. 252-253 and 317, as having argued for the use of the eclipse period of 47 synodic months.

Schaumberger argues on the basis of a Report (no. 4 in Hunger [1992]), from [Issar-šumu-ereš] to Esarhaddon, which alludes to a lunar eclipse on III 14, "which is not at its predicted time." Schnabel [1924c] had already shown that this was the lunar eclipse of 10 June -668 (actually the early morning of 11 June). Schaumberger argues that it was probably the lunar eclipse of 47 synodic months earlier, on 23 August -672, that provided the prediction even though it was not visible in Babylon because it occurred before sunset. Van der Waerden [1940] p. 113 notes that there was also a visible lunar eclipse that occurred six months earlier than 11 June -668, on 17 December -669. On the preceding page, he discusses a Letter from Mār-Ištar whom Schaumberger [1935] pp. 315-317 believed to be the astrologer of Assurbanipal; Mār-Ištar claims in this letter to have looked in vain for a solar eclipse on III 28, 29, and 30. Schaumberger dated this solar eclipse possibility on 17 June -632, which was preceded by a solar eclipse 47 synodic months earlier, on 29 August -636. These data are repeated by van der Waerden; later Schott-Schaumberger [1942] combined Report no. 4 concerning the lunar eclipse of 11 June -668 and several Letters from Mār-Ištar (nos. 347, 351, and 362 in Parpola [1993]) and the previously mentioned Letter (no. 363) which actually reports a solar eclipse possibility on 24, 25, or 26 June -668. The conjunction occurred on 25 June at 9:16 PM at Babylon, and so at night when a solar eclipse could not be seen, at a longitude of 86°, while the longitude of the descending node was 58° so that no solar eclipse was possible. But a solar eclipse had occurred one month earlier, on 27 May -668 (the Babylonian observers looked several months in succession), but it was over before sunrise and thus was not seen in Babylon. It may have been the eclipse of 27 May that was predicted from the preceding solar eclipse, that of 15 April -675, just 88 synodic months previously, since 88 synodic months is also a useful eclipse period (see Schaumberger [1935] p. 252). But in no case does the prediction support the use of the 47 month period, except that 88 is the sum of two shorter eclipse periods, 41 and 47 synodic months.

The periods of 41 and 47 synodic months are good eclipse periods because they represent respectively the series of 7 eclipse possibilities (6 intervals of 6 months and one of 5) and 8 eclipse possibilities (7 intervals of 6 months and one of 5) that make up an 18-year eclipse cycle. Since the Babylonians recorded the time of night (for lunar eclipses) or of day (for solar ones) in either watches or time-degrees, it was possible for them to compute the excess or deficiency of the time of an opposition or conjunction with respect to an integer number of days in any one of these periods. This presumably explains the method by which some authors of Letters in the seventh century B.C. were able to predict the watch in which an eclipse would occur (e.g., Letter 78 in Parpola [1993] p. 58; cf. Letter 71 on p. 53; and Report 487 in Hunger [1992] p. 269). The use of time-degrees for the time of impact and eclipse-middle already existed in the early fourth century

B.C. (see the Diary for -370), and probably considerably earlier. Once one was fairly certain of the mean value of the number of days and time-degrees between eclipses separated by 5 or 6 months, or better by 41 or 47 months, simple division would produce a fairly accurate number of days and time-degrees in a mean synodic month.

The Babylonians' success in predicting lunar eclipses was investigated by Steele-Stephenson [1997]. They compiled a list of 35 predictions for which a time is recorded between -730 and -76, and found that 19 were "successful," 12 were "near misses", and just 4 were failures. Of these eclipses, one was of the eighth century B.C., two each of the seventh, the sixth, and the fifth, six of the fourth, seven of the third, thirteen of the second, and two of the first.

### *1.1a. Column Φ*

Neugebauer [1955] vol. 1, pp. 44-45, in describing column Φ of System A of the Moon, notes that it has the same period as the unabbreviated column F, which represents lunar velocity, and also the same positions of maximum and minimum values as in column F. He also notes that there are procedure texts (for the relation of Φ and F see ACT 200, section 5; and ACT 204, section 5; for that of Φ and G see ACT 200, section 14; ACT 205; ACT 206; ACT 207, ACT 207a, ACT 207b, ACT 207c, ACT 207ca, ACT 207cb, ACT 207cc, ACT 207cd, and ACT 207d, section 1; and for that of all three see ACT 204a and ACT 208) for computing F and G from Φ. Note that column G gives the excess of a synodic month over 29 nychthemera. Both Φ and G are measured in large-hours (= H), which are actually time-degrees on the equator (6;0 H = 6,0°); the large-hours are used in order to avoid confusion with degrees on the ecliptic.

Neugebauer [1957], in reading BM 36705+36725 which Sachs had identified and joined, discovered that it states that "17,46,40 is the difference for 18 years." In Neugebauer [1955] vol. 1, p. 59 it had been demonstrated ($\varphi$ = 0;17,46,40 H, and $\varepsilon$ is an interval of time) that:

$$d_{\Phi} = \tfrac{1}{\varepsilon}\varphi$$

or, since $\varepsilon = \frac{3}{28} = \frac{1}{9;20}$ synodic months,

$$d_{\Phi} = 9;20 \times 0;17,46,40 = 2;45,55,33,20 \text{ H}.$$

The quantity $\varphi$ represents the difference between two values of Φ separated by 223 synodic months in which there are 16 periods, P. Φ is based on a linear zigzag function whose parameters are:

$$M = 2;17,4,48,53,20$$
$$m = 1;54,47,57,46,40$$
$$\mu = 2;7,26,23,20$$
$$\Delta = 0;19,16,51,6,40$$
$$d = 0;2,45,55,33,20$$
$$P = \frac{2\Delta}{d} = \frac{1,44,7}{7,28} = 13 + \frac{7,3}{7,28} = 13;56,39,6,\ldots$$

Then 16 P = 223 + $\frac{3}{28}$ synodic months. Neugebauer concludes that $\Phi$ is the difference in time, expressed in time-degrees (large hours), between the lengths of two synodic months separated by one "Saros."

Neugebauer-Sachs [1967] pp. 192-199 published BM 36301, which contains three columns. Column I gives the months and longitudes of the Greek-letter phenomena ($\Gamma$, $\Phi$, $\Psi$, and $\Omega$) of Mars from, probably, -359 to -341 (see II C 2.3); column II gives the dates and longitudes of Venus for one synodic period, perhaps in -343/2 (see again II C 2.3; for both columns dates in the Seleucid period are possible); and the bottom of column II with column III give ten values of $\Phi_2$ ($\Phi_1$ is the value of $\Phi$ at conjunctions, $\Phi_2$ its value at oppositions), five obscure lines of which the first is "36 Kandalānu" (Kandalānu ruled Babylon for just 22 years, from -646 to -625), and eight numbers that are multiples of $\varphi$ (17,46,40). The eight values of $\Phi_2$ occur in lines of a full table of $\Phi_2$ that are distant from each other by multiples of 14 (increased in one case by 1, diminished in two by 1). This difference of multiples of 14 lines is clearly connected with the fact that

$$P_\Phi = \frac{1,44,7}{7;28} = 13;56,39,6,\ldots \approx 14$$

However, the order of the values of $\Phi_2$ is not that of the order of the lines in a complete table of $\Phi_2$, so that the purpose of this column remains mysterious. However, if the reference to Kandalānu bears any relation to the date of this column, the $\Phi$ function was already developed in the late seventh century B.C. in Babylon. This early date is, of course, not certain, nor, indeed, probable.

BM 37149, published in Neugebauer-Sachs [1967] pp. 199-200, is a fragment giving seven values of "velocities" which are each a multiple of 11,6,40. This number, half of 22,13,20, represents the difference between values of $\Phi$ one half month apart (e.g., $\Phi_2$ - $\Phi_1$).

A text of unknown date, BM 36744+37031, published by Neugebauer-Sachs [1969] pp. 94-96, connects in some unclear fashion the "Saros" (18 years) with column $\Phi$ (1,57;47,57,46,40°, which is the minimum value of $\Phi$).

The maximum value of $\Phi$ is 2,17;4,48,53,20°, but there are versions of $\Phi_1$ and of $\Phi_2$ which truncate the maximum at 2,13;20° (which all entries greater than 2,13;20° are replaced by) and the minimum at 1,58;31,6,40° (which all entries lower than 1,58;31,6,40° are replaced by). Such a text is BM 36822+37022, apparently for -397, which is published by Aaboe-Sachs [1969] pp. 3-11. This text will be discussed more fully in the section on System A of the lunar Ephemerides (II C 4.2a.1). An example of the truncated version of $\Phi_1$ for an even earlier date is provided by the second text in Aaboe-Sachs [1969] pp. 11-16, which is preserved in duplicate copies: BM 36599+36941 and BM 36737+47912. This gives lunar functions for the eighteen years from -474 to -456; the original tablet was, of course, compiled after the last date. Entries are for the 38 solar eclipse possibilities between those two dates. The longitudes of the conjunctions are given according to a scheme whereby alternate eclipse possibilities, regardless of whether or not they follow by six months or five, are diminished by 5° and 5;30° respectively, while the zodiacal signs progress by six signs or by five depending on whether the interval is six months or five. Since there are 33 six-month and 5 five-month intervals, the increase in zodiacal signs of solar motion is 223 zodiacal signs, while the decrease in degrees is 38 × 5;15° = 199;30°. The longitudinal increase of the Sun, therefore, is 223 zodiacal signs diminished by 6 zodiacal signs and 19;30°, or 12 rotations and 10;30°. This is a crude approximation presumably derived from:

1 year = 12;22,8 synodic months.

From this it follows that:

223 months = 18;1,44,41,... years = 18 years and 10;28,6,...°

Already, then, in the fifth century B.C. the Babylonians had both Φ in its truncated form and System A's relation of a solar year to synodic months. This text will also be more fully considered in our discussion of System A of the Moon (II C 4.2a.1).

*1.1b. Predictions of Solar Eclipses*

For now we will instead pause to consider column VI of BM 36599+36941 and BM 36737+47912, in which are preserved remarks, apparently derived from Diaries or some other observational texts, concerning the 38 solar eclipse possibilities. Unfortunately, the only Diary that we have for this 18-year period is a fragment of one for -463, and it preserves neither month II nor month VII. However, we can easily correlate the solar eclipses actually observed at Babylon as reported in Kudlek-Mickler [1971] with these reports:

| Kudlek-Mickler p. 29 | BM 36599+ and 36737+ |
|---|---|
| 20 March -469. Magnitude 0.62 | Month XII, (eclipse) which (was expected to) pass by; 1½ di[gits.] |
| 2 July -465 | Broken |
| 26 December -465 | Broken |
| 30 April -462 | Broken |
| 2 August -457 Mid-eclipse at 12.39 P.M. | I did not watch; (expected) at 50° (= 3;20 h) left to sunset. |

In neither the first nor last case where the data coincide is the Babylonian statement close to what modern calculations produce.

But that the Babylonian astronomers understood that the visibility of solar eclipses and their magnitudes depend on factors that we would attribute to the draconitic as well as to the anomalistic and synodic months is indicated by the predictions in column VI with appropriate remarks of the times and magnitudes of solar eclipses that were not visible at Babylon. They still, of course, had no control over or conception of the complex problem of the path of the Moon's shadow across the surface of the earth.

The theoretical foundations for a mathematical scheme that would provide predictions of the magnitudes of solar (or lunar) eclipses based on mathematically determining the distance of the mean Sun from a lunar node were discussed by Aaboe [1972]. He suggests a saw-function based on the fact that, as in System A, the mean longitudinal increment for the Sun in a synodic month is $\frac{6,0°}{12;22,8m}$ = 29;6,19,... °/m, and the longitudinal decrease for the lunar node in a mean synodic month is 1;33,55 °/m. Therefore, in a mean synodic month the longitudinal difference between the mean Sun and the lunar node is very close to 30;40°. From this we can compute that, if a mean conjunction occurred at one of the Moon's nodes, 6 months later the Sun will be 6 × 30;40° or 184° from the same node. Aaboe further suggests that, in order to use integers, one can multiply all numbers by 0;45; this converts 30;40 into 23, 180 into 135, and 4 into 3, which means that the mean Sun will cross the node 23 times in 135 months. The saw-function that allows us quickly to determine which months contain a conjunction at which a solar eclipse is possible and which contain one at which a solar eclipse is not possible has the parameters:

$$M = 135$$
$$m = 0$$
$$d = 23$$
$$P = \frac{M}{d} = 5;52,10,\ldots \text{ months},$$

where P is the mean interval between eclipse possibilities. In order to improve the predictive power of the scheme, one can drop the saw-function by half of the 30;40°, or by 15;20°, so that the initial conjunction occurs just 15;20° before the longitude of the node; and, for convenience's sake, one can construct a saw-function of the eclipse possibilities themselves, with

$$M = 20$$
$$m = 0$$
$$d = 3$$
$$P = \frac{M}{d} = 7;40,$$

where 7;40 is the number of eclipse-possibilities separated by six months preceding one separated by five months. This is close to the "Saros" in which in 18 years there are three series of 8 eclipse possibilities and two of 7 since

$$3 \times 8 + 2 \times 7 = 38$$

while $7;40 \times 5 = 38;20.$

The "Saros" implies a saw-function in which

$$M = 223$$
$$m = 0$$
$$d = 38$$
$$P = \frac{M}{d} = 5;52,6,\ldots \text{ months},$$

and a derived saw-function in which

$$M = 38$$
$$m = 0$$
$$d = 5$$
$$P = \frac{M}{d} = 7;36 \text{ possibilities.}$$

Aaboe is able to point to a saw-function in one Babylonian tablet, BM 34705+34960 (= ACT 93), an undated piece from Babylon. The second column of this tablet gives a saw-function in which

$$M = 39,0$$
$$m = 21,0$$
$$d = 3,4$$
$$P = \frac{\Delta}{d} = \frac{18,0}{3,4} = \frac{135}{23}$$

The period is exactly the same as is that of the saw-function described above based on integer parameters. On this tablet, next to each crossing of the node by the Moon, is a number which probably indicates the eclipse-magnitude. Aaboe also notes that an exact replica of the period relation, 23 eclipse possibilities in 135 synodic months, is embedded in the Chinese Triple Concordance system proposed by Liu Hsin at about the beginning of the current era; this eclipse cycle was called the Phase Coincidence Cycle (see Sivin [1969] pp. 11-14 and 69). And Aaboe proposes that a mysterious Demotic papyrus, P. Carlsberg 31, that had been published in Neugebauer-Parker [1960/1969] vol. III, pp. 241-243, should be emended to refer to months rather than years. This emendation produces a saw-function in which

$$M = 129$$
$$m = 0$$
$$d = 22$$
$$P = \frac{M}{d} = 5;51,49, \ldots \text{ months.}$$

M is less than 135 by six months and d is less than 23 by one eclipse possibility.

The text found in the duplicates BM 36599+36941 and BM 36737+47912 is taken up again by Neugebauer [1975] vol. 1, pp. 525-527. He shows that column III, unexplained in Aaboe-Sachs [1969], is a column of solar eclipse-magnitudes at intervals of 12 months. The parameters of the zigzag function describing the eclipse magnitude are (c is the maximal eclipse magnitude):

M = 1,58;30 digits
c = 18 digits
d = 46;30 digits
$\Delta = 2(M + c) = 4,33$ digits
$2\Delta = 9,6$ digits

In a 12-month period the increase or decrease in the eclipse-magnitude will be: 12d = 9,18 digits ≡ 12 digits mod. $2\Delta$.

This works very well for the sequence of numbers opposite alternate eclipse-possibilities in each series of eight or seven possibilities (the maximum is 18).

series 1): 5, 17, 7, -5 and 6, 18, 6.
series 2): 0, 12, 12, 0 and 5, 17, 7, -5.
series 3): 4;40, 16;40, 7;20, -4;40 and 7, 17, 5, -7.
series 4): 4, 16, 8, -4 and 9, 15, 3.
series 5): 2, 14, 10, -2 and 2, 14, 10, -2.

Neugebauer could not explain the different values for the increase or decrease in eclipse-magnitudes in five and six month intervals. However, he did show that the entries in column IV bear a simple relation to the eclipse-magnitudes (Ψ) in column III:

| Ψ | col. IV (d=3) | Ψ | col. IV (d=1) | Ψ | col. IV (d=3) |
|---|---|---|---|---|---|
| 1 | 19 | 9 | 41 | 16 | 50 |
| 2 | 22 | 10 | 42 | 17 | 53 |
| 3 | 25 | 11 | 43 | 18 | 56 |
| 4 | 28 | 12 | 44 | | |
| 5 | 31 | 13 | 45 | | |
| 6 | 34 | 14 | 46 | | |
| 7 | 37 | 15 | 47 | | |
| 8 | 40 | | | | |

He suggested that the 56 corresponding to a maximal eclipse magnitude of 18 digits may be read as 0;56 of total obscuration. Against that interpretation is the pattern by which the numbers in col. IV increase.

Moesgaard [1980] pp. 77-81 discussed columns III and IV in the context of his general theory of the origins and development of Babylonian lunar theory. He defines a wave (w) as $\frac{\text{1 rotation of the Moon}}{17.5}$ = 20;34,17, ...° because there are 35 points where the Moon crosses the ecliptic in 19 years, and it crosses each one (approximately) twice. Then the distance, k, between two crossing points will be $\frac{6,0°}{35}$ = 0;30 w = 10;17,8, ...°. Moesgaard explains the entries in column III as the values of a "belt phase" function, which is the "wave-function" (what is left after subtracting the number of w's corresponding to the rotations and the k's at an epoch from the number of w's in a synodic month multiplied by the number of synodic months since epoch; the w's in a synodic month he computes to be about 18.9148042). Unfortunately, he does not demonstrate by computation that the entries in column III correspond to the values produced by his "belt phase" function. Moesgaard's explanation of column IV is more persuasive; he claims that its numbers represent the duration of the eclipses expressed in time-degrees, so that the maximum, 56, is equal to 3;44 h. This still does not explain the change in differences from 3 to 1 and back to 3 in the sequence of increasing numbers in column IV reconstructed by Neugebauer.

A column—numbered III—of BM 36599+36941 and BM 36737+47912 that had not been understood by Aaboe-Sachs [1969] is explained by Britton [1989]. He begins by showing that it is possible, from the actual lunar eclipse occurrences recorded in an eclipse-table from -746 to -710 (446 months) to derive at least five eclipse cycles:

1) M = 47, d = 8, P = $\frac{47}{8}$ = 5;52,30
2) M = 88, d = 15, P = $\frac{88}{15}$ = 5;52,0
3) M = 135, d = 23, P = $\frac{135}{23}$ = 5;52,10,26,...
4) M = 223, d = 38, P = $\frac{223}{38}$ = 5;52,6,18,...
5) M = 358, d = 61, P = $\frac{358}{61}$ = 5;52,7,52,...

We have seen that there is evidence that the Babylonians used the first four of these as well as perhaps another not mentioned by Britton:

0) M = 41, d = 7, P = $\frac{41}{7}$ = 5;51,25,42,...,

which is not very good at all.

From among these possible periods he constructs several combinations numbered by the number of series (a) of seven or eight eclipse possibilities, which is the same as the number of five-month eclipse intervals in them. Thus, number 3 above is called C(3) since in it a = 3, and number 4 is called C(5) since a = 5 in it. Similarly, there is a C(6) which is double C(3); in it M = 270 months, d = 46 eclipse possibilities, and a = 6 series. From this period he constructs an "eclipse magnitude" function, Ψ*, which is the difference between the eclipse limit for that period, c(a), and the distance of the eclipse center from the node, η; since 3d/8 < c(a) ≤ 3d/7, we know that c(a) for C(6) lies between 17 ⅛ and 19 $\frac{5}{7}$—say, at 18 (measured in units of 0;40°). Then, when η = 0, Ψ* = 18. And, since the difference of η for one month is n(a) and that for six months is just a, the difference for 6 months in C(6) is 6 and for 12 months it is 12. The entries in column III are not directly generated from Ψ*(6) or Ψ(6); several adjustments were made for reasons discussed by Britton. However, this seems to be the correct solution to the meaning of this column. The units in which the entries in column III are measured are digits according to Britton, who supports Moesgaard's interpretation of column IV. However, he concludes that both columns III and IV were originally intended for lunar eclipses, and then were applied by analogy to solar eclipses.

In Aaboe-Britton-Henderson-Neugebauer-Sachs [1991] pp. 68-71 was published an additional fragment, BM 36580, of BM 36737 containing parts of columns II and III. The evidence from this fragment was incorporated into his discussion of column III by Britton [1989]. Another important early text on eclipse magnitudes was published in this volume by Aaboe et alii, pp. 43-62; this is on the reverse of BM 36651+36719+37032+37053. This text gives the date (regnal years and months), the longitude of the Moon at opposition, and the "eclipse magnitude"; it covers two "Saros" Cycles, from -416 till -380.

The longitudes are computed by subtracting 10;30° from 12 zodiacal signs for twelve month intervals, and the same amount from 11 zodiacal signs for eleven month intervals; since there are 14 twelve month intervals and 5 eleven month intervals in a "Saros" Cycle, the latter again equals 14 × 12 zodiacal signs + 5 × 11 zodiacal signs – 19 × 10;30° = 223 zodiacal signs – 199;30° = 18 rotations + 7 zodiacal signs – 199;30° or, with rotations of the Sun, 18 years – 10;30°. The text lists eclipse possibilities at the ascending node separately from those at the descending node. The values at the ascending node link up with those in the duplicates of the "Solar Saros" published by Aaboe-Sachs [1969], while the values at the descending node suffer from consistent errors. Moreover, the six month longitudinal increments in column II of the longitudes are based on a solar progress of 29;15° a month from ascending to descending node, one of 29° a month from descending to ascending node. Precisely the same monthly solar velocities are found in BM 36400, published in Aaboe et alii [1991] pp. 63-67. This tablet gives longitudes of the Moon at conjunction in each month from -264 to -258, so that these rather crude approximations continued to be used until at least the middle of the third century B.C.

BM 36651+36719+37032+37053, published in Aaboe et alii [1991], contains a fragment of a column of "eclipse magnitudes", Ψ, computed according to the rules of Ψ(6), in which:

$$M = 270$$
$$d = 46$$
$$P = \frac{M}{d} = 5;52,10,26,...;$$

this is the function from which column III of the "Solar Saros", Ψ*, was derived. But this tablet's values of Ψ are modified so that they fit the "Saros" period of 18 years and in order to reflect the effect of both lunar and solar anomaly. This scheme, the authors point out, has relations with both System A and System B, but is prior to both.

Steele [1997] compiled a list of 61 predictions of solar eclipses for which a time is given between -357 and 37 A.D. He divided these into two classes: Class A consists of 28 eclipses that were visible at the latitude of Babylon, but not necessarily at its longitude, Class B consists of 33 eclipses that were not visible at Babylon or any other locality on Babylon's parallel of latitude. The predictions of Class A had an average error in predicted time of almost two hours, those of Class B an average error of almost four hours.

*1.1c. Origin of Φ.*

We return briefly to consideration of column Φ. Brack-Bernsen [1990] (see also Brack-Bernsen [1993]) showed that the sum of the "Lunar Four", which are the observations of time-intervals measured in time-degrees from among the "Lunar Six" made in the middle of the month and called ŠÚ (moonset to sunrise on the morning before opposition), NA (sunrise to moonset on the next morning), ME (moonrise to sunset on the evening before opposition), and $GE_6$ (sunset to moonrise on the evening after opposition), produce a graph, Σ, that is remarkably similar to the linear zigzag function of Φ. From Σ she constructs a linear zigzag function, $\hat{\Sigma}$, which has the period of Φ (13;57) but a different amplitude. From this she concludes that

$$\Phi = \hat{\Sigma} + K$$

where K lies between 98° and 102°. She then shows that, while for Φ

$$6247 \text{ syn. mos.} = 6695 \text{ anom. mos.} = 448 \times P_{\Phi},$$

in terms of the "Saros" with 223 synodic months,

$$28 \text{ “Saroi”} + 3 \text{ syn. mos.} = 28 \times 16 \times P_{\Phi}$$

where 28 × 16 = 448, and claims that the complex structures of Σ repeat with the periodicity of 6247 months. She thus concludes that Φ did not arise from the "Saros", but that both Φ and the "Saros" arose from Σ.

This argument is expanded in Brack-Bernsen [1997], especially chapters 10-16 (pp. 69-133). She points out that ŠÚ and NA can be transferred to the western horizon by substituting the opposite-point of the Sun for the Sun itself, and similarly ME and $GE_6$ can be transferred to the eastern horizon. Then ŠÚ + NA is the setting-time of the two arcs of elongation of the Moon from the Sun in the evening and ME + $GE_6$ is the rising-time of the two arcs in the morning. Since the oblique ascension of the rising arc is the oblique descension of the setting arc, their sum cancels out, and the motion of the Moon relative to the Sun in sphaera recta without any influence from the local terrestrial latitude is properly represented by ŠÚ + NA + ME + $GE_6$ = Σ. The Babylonians must have arrived at the decision to use Σ purely empirically, if at all.

Each of the "Lunar Four" depends on four variables: the time from opposition to sunrise or sunset, the velocity of the Moon, the longitude of the Moon, and the latitude of the Moon. Either ŠÚ + NA or ME + $GE_6$ depends on just the Moon's velocity and longitude, but with the longitude as the dominant factor; in Σ the Moon's velocity is the dominant factor. This is the link to Φ and its relative F, the lunar velocity function. The period of Σ and of $\hat{\Sigma}$ is 13;56,39,6,... = $\frac{1,44,7}{7,28}$, which is the period of Φ. Then $\hat{\Sigma} + K = \Phi$, where K is about 100°.

Brack-Bernsen traces the practice of observing at least some of the "Lunar Four" back to Tablet 14 of *Enūma Anu Enlil* and to II ii 43 - II iii 12 of MUL.APIN, and she notes that the "Lunar Six" are recorded on BM 33066, "Cambyses 400" (see II B 6). What she does not note is that some of the 78 intervals reported on this tablet for the "Lunar Six" must be computations or estimates; that the tablet is a revised copy of an original; and that some of the "Lunar Six" appear in the Diary for -567, though the Diary for -372 is the first known to have all six. She does, however, refer to BM 34075 (LBAT 1481), which records "Lunar Sixes" from -322/1 to -318/7. This evidence, scanty as it is, places the collection of observations of the "Lunar Four" in the sixth century B.C., perhaps the last half though conceivably somewhat earlier. This would fit nicely with the theory—if it is correct—that the "Saros" cycle and Φ both go back to the late fifth century B.C.

In the next part of her book (pp. 89-121) Brack-Bernsen presents and analyzes the data concerning the "Lunar Six" or "Lunar Four" found in "Cambyses 400" (pp. 99-103); on the reverse of BM 34034 (LBAT 1285), which lists ŠÚ+NA and ME+$GE_6$ for months VII to XII of -136/5 and the "Lunar Six" from the end of month $XII_2$ of that year to the end of month XII of the next year, -135/4 (pp. 90-92, 98-100, and 102); and other Goal-Year Texts (pp. 104-121). On pp. 123-126 she explains the discussion of ŠÚ+NA and ME+$GE_6$ found on lines 29-30 and 36-38 of the obverse of TCL 6 11, a tablet from Uruk. She concludes by claiming—apparently correctly—that Φ is derived from $\hat{\Sigma}$, which in turn is derived from the empirical data which forms Σ; what she confesses an inability to explain as yet is the constant difference, K, between $\hat{\Sigma}$ and Φ.

It had been a problem recognized by Aaboe [1969] and by Brack-Bernsen [1980] that, though central to column Φ is lunar anomaly, the length of the "Saros" (which contains an almost integer number of anomalistic months) depends more on solar anomaly than on lunar anomaly; for the longitudes of the bounding conjunctions of a "Saros" move through the ecliptic once in every 35 or so "Saroi", and with that motion the effect of solar anomaly changes markedly. Britton [1990] attempts to eradicate this problem by noting that column Λ (which we will discuss in II C 4.2a.1) gives the length of the twelve months preceding a syzygy and column Φ gives the length of the 223 months following that same syzygy, so that the sum of Λ and Φ is the length of 235 months, which is 19 solar years. In that period (though not in the "Saros" itself) the effect of the solar anomaly disappears.

*1.1d. Time-degrees*

We have seen in the discussion of the *ziqpu* star lists (II A 5), of the GU text (II A 6), and perhaps of the early observational text of Mercury (II B 1.2) that the use of time-degrees goes back to the early centuries of the first millennium B.C.; the *bēru* and UŠ of the earlier texts, such as Tablet 14 of *Enūma Anu Enlil* and MUL.APIN, refer to time measured by the water-clock, whereas the *ziqpu* star lists and the GU text measure it by meridian transits of the *ziqpu* stars or other stars that fall on the same meridians. This technique is applied to lunar eclipses in Letters 134 (Parpola [1993] p. 108) and 149 (Parpola [1993] pp. 113-114). We include the observational text of Mercury with these because the later practice of recording the time from planet-rise to sunrise or from sunset to planet-set (see II B 2.1) utilizes time-degrees. Time-degrees, as just noted in II C 1.1c, provide the measurements of the "Lunar Six" and "Lunar Four" which were already being made in the sixth century B.C., as also the measurements of the times of lunar eclipses (see II B 2.5) from the seventh century on (see II A 6).

There exist several broken tablets apparently giving the rising-times of twelfths (dodecatemoria) of zodiacal signs in terms of simultaneous meridian crossings of *ziqpu* stars; the fact that they divide the 360° of the ecliptic into twelve zodiacal signs shows that this method must be dated after ca. -400. Most of the texts were published by Schaumberger [1955]: A 3427 obverse 1-25 (Scorpius and Sagittarius) on pp. 238-242; U 196 (Capricorn) on pp. 242-243; and BM 34713 (LBAT 1499) reverse 10-30 (Aries and Taurus) on pp. 244-247. Schaumberger also published, on pp. 247-251, a text (A 3414 + U 181 a,b,c,d) that details the setting and rising of *ziqpu* stars and arcs close to them. Horowitz [1994] pp. 97-98 believes that the small fragment BM 77242, which he publishes there, belongs to the same type, but it is too incomplete to provide certainty. Such texts allowed Babylonian astronomers to convert ecliptic arcs as they rose (or set) into time-intervals.

In the Diaries the times of various astronomical phenomena are given by the culmination of *ziqpu* stars; this method is especially applied to lunar eclipses. The earliest attested occurrence is in month IX of -214.

*1.1e. Lunar Latitude*

The problem of lunar latitude is obviously closely bound to the "Saros" which contains very close to 242 draconitic months; indeed, without this return to the same node the system of predicting eclipses would not work. We should now consider sections 1 and 4 of BM 41004 (published by Neugebauer-Sachs [1967] pp. 200-205), which deal with lunar latitude. This was copied by Iddin-Bēl, the son of Marduk-šāpik-zēri, who refers to year 5 of Philip (= -318/7) in MNB 1856.

This text begins with a simple scheme in which the Moon has a maximum latitude of 5° North and South of the ecliptic and in which its elongation from the node it has passed increases by 1;40° a month—that is, 20° a year. If the Moon starts near the Stars at a latitude of 5° N, it will reach its descending node near the King in 90°; it will reach a latitude of 5° S near the Head of Scorpion after proceeding 90° further; and it will reach its ascending node behind the Goat-Fish after another 90° of motion. In the second year the longitudes at which the Moon reaches maximum, mean, and minimum latitudes will have progressed by 20°, so that they will occur at 20° behind the Stars, behind Cancer, near Libra, and passing by the Goat-Fish. In the third year these positions will be near the Ribbon of the Fishes, behind Gemini, near Virgo, ... (the fourth position is omitted by the text).

The text then changes the width of the Moon's path from 10° to 6 cubits (= 12°), and the period of return to the same node at the same longitude is given as 18 years and the size (diameter) of the Moon as 12 fingers (= 1°). Finally, we are told that the Moon returns to its same longitude in 82 integer days; for 82 equals 3 × 27 1/3, where 27 1/3 is an approximation to the length of a sidereal month.

Some of this information is repeated in section 4, where it is stated that the Moon returns to its same place among the Normal Stars in 19 years. This is probably not a reference to the nineteen-year cycle as it is understood by Moesgaard [1980] p. 51 since the following sentence refers to the repetition of a lunar eclipse if one occurred at the beginning of the period. If one chose the right eclipse in the "Saros" there would be at least the possibility of an eclipse in eighteen years and two six-month intervals; but if one used the "Saros" itself, eighteen years would lead one always to an eclipse possibility, though not necessarily to an actual eclipse. Therefore, because of this and because of the context of the passage, we believe that Neugebauer-Sachs were correct in regarding the text's 19 as an error for 18. The context is provided by the following sentences which state that the latitude of the Moon (literally, its being high or low with respect to a Normal Star) is repeated after the period, now emended correctly to 18 years (19 years does not contain an integer number of draconitic months).

### *1.2. The Nineteen-year Cycle*

The famous nineteen-year cycle, well known from the accounts in Greek sources of the Athenian astronomer Meton and from its use in regulating the Jewish calendar, has been recognized for quite some time now as the equivalent of the sum of a "Saros" of 18 years and ca. 10;30 days (which equal 223 synodic months) and 12 synodic months (which equal 1 year diminished by ca 10;30 days). The resulting cycle is:

19 years = 235 synodic months.

This can also be expressed by the statement that in nineteen solar years there are seven intercalated synodic months.

Any observation-based or computation-based method of guaranteeing that, say, the vernal equinox always falls in a synodic month bearing a particular name will necessarily, if successful, involve the intercalation of seven synodic months in nineteen years. What distinguishes the Babylonian nineteen-year cycle is the distribution of these intercalated months. If we begin a nineteen-year cycle in 0 SE (= -311/0) the intercalations will occur in the following years of any cycle before or after it (* means an intercalated month XII, ** an intercalated month VI):

2*, 5*, 8*, 10*, 13*, 16*, 19**

If we begin counting months from $VI_2$ of year 19, omitting all the intercalary months, the intervals between intercalations are:

30, 36, 36, 24, 36, 36, 30,

which cannot be an accidental pattern.

The questions remain: When was this scheme devised? When was it adopted by a civil authority? The evidence suggests that the civil authorities had still not adopted the cycle at the end of the reign of Cambyses (-525) (see Neugebauer [1975] vol. 1, p. 354), and we have seen above that the dates in the "Saros Canon" follow the computed cycle after -497. There still remain a very small number of documents dated after -497 in the civil calendar that are not in accordance with the nineteen-year cycle, but they are extremely rare exceptions that presumably can be explained away. So the question remains how many years before -497 the astronomers—e.g., those who used the "Saros" to predict eclipses—used the cycle.

The answer to that question cannot be given. It is possible to make predictions for so many months in the future, though easier to keep track if one can name that month. The use of the nineteen-year cycle among astronomers, then, might possibly go back to the eighth century B.C. as did the Diaries, but we can neither prove nor disprove this hypothesis.

As a consequence, the literature on the Babylonian intercalation cycle has concentrated on compiling lists of intercalary months attested in inscriptions and on tablets—including, of course, the "Saros Canon" after it was discovered. We record a few of these investigations here: Mahler [1891]; Strassmaier [1892]; Mahler [1894] and [1898]; Kugler [1909]; Sidersky [1916]; Kugler [1924] pp. 408-438; Meissner [1924]; Parker-Dubberstein [1956]; Brinkman-Kennedy [1983]; and Kennedy [1986].

### *1.3. Solstices, Equinoxes, and Phases of Sirius*

Weidner [1914a] tries to prove that the Babylonians knew of the precession of the equinoxes because their dates for the occurrences of the equinoxes and solstices in the last six centuries B.C. were approximately correct, and so were they in CBS 11901 (LBAT 1478), which Weidner claimed to have been written in about -1500. However, Kugler [1914a] pp. 233-242 demonstrated that the date of CBS 11901 is -424/3.

Schnabel [1923] pp. 233-237 claims that Kidenas/Kidinnu discovered precession along with System B of the Moon at Sippar in -313; Kugler

[1924] pp. 582-621 and 627-630 sharply criticized Schnabel's conclusions. Schnabel [1927] replied with further arguments in favor of his theory, but now dated Kidinnu to -378. Schnabel's thesis was finally decisively refuted by Neugebauer [1950].

The method employed by the Babylonians to compute the dates of solstices and equinoxes in the last few centuries B.C. was explained by Neugebauer [1947a], who showed concerning the fragment of a tablet from Uruk, U 107+124, that it contains the dates of the summer solstices in the years -167 to -153 and that these dates are computed from the nineteen-year cycle; for the epact in that cycle (i.e., the difference between twelve mean synodic months and a solar year) is 11;3,9,28, ... tithis, while that in the Uruk solstice-scheme is 11;3,10 tithis. This interpretation was confirmed by Neugebauer [1948], who adds the dates of many other computed solstices and equinoxes found on numerous tablets. Among previously unpublished tablets he cites A 3456 (LBAT 1377); MLC 1860, 1885, and 2195; and two from Uruk in Istanbul, U 193a and U 194. The simple rule is that successive summer solstices differ by 11;3,10 tithis and successive colures by 93 tithis.

Sachs [1952] showed that the dates of the phases of Sirius (Γ, Θ, and Ω) are related to this solstice and equinox scheme by fixed intervals between the summer solstice and Γ of Sirius (+21 tithis), and Θ (+191 tithis), and Ω (+327 tithis). In examples of dates of the phases of Sirius in fifth and fourth century B.C. Diaries he finds deviations from the Sirius scheme, while in a text (BM 32327+32340 (LBAT 1432b)) he finds that the dates of Θ and Ω agree with the Sirius scheme, while those of Γ are earlier by 1 tithi. Next, he points out that each of the colures in the Uruk solstice scheme and Θ and Ω in the Sirius scheme can fall in either one of two months, while Γ always falls in month IV. From this he conjectures that, when the nineteen-year cycle was introduced, the distribution of the intercalary months was designed to keep the dates of Γ of Sirius in month IV. We agree that one of the consequences of the distribution of the intercalary months was to keep the date of Γ in month IV, but, as we have shown previously, that distribution reflects a pattern of numbers that must be deliberate.

For tablets displaying this scheme see II B 2.4 above, where it is conjectured that the use of these schemes goes back to -368, which falls nicely in the range proposed by Sachs, less then a century after -380, though that fact is not free from accident.

Aaboe-Sachs [1966] pp. 9 and 11-12 publish a text, BM 36810+36947 and BM 36811, that lists the dates of summer solstices from years 2 to 12 (apparently of Artaxerxes III, and so -356 to -345), and, in a list at right angles to the first, for 15 additional years. Using the table given above in II B 2.4 it is easy to see that these 26 years fit the solstice scheme perfectly for the years from -356 to -331; note that -331 is the last year in a nineteen-year cycle.

Neugebauer-Sachs [1967] pp. 183-190 published a considerably older text, found on BM 36731, that gives the dates of winter solstices, vernal equinoxes, and Ω and Γ of Sirius for the years -617/6 to -586/5. The connection of the dates of the colures, which are very difficult to determine, with those of the phases of Sirius makes good sense since the dates of the phases are easily observed. But the scheme underlying the dates is more difficult to understand; it seems that the difference between successive dates of colures and phases is 12 months 11 "days" and 1,36 time-degrees (= 1;36 H). Since 1,36° equals 0;16 tithis (1,36 : 6,0 = 0;16), the epact will be 11;16 tithis, far higher than the nineteen-year cycle's epact of 11;3,10 tithis. However, the difference 1,36 is mentioned in a lunar procedure text (ACT 200 sect. 11), so that the interpretation of 1,36° as 0;16 tithis seems to be correct, and this text becomes the earliest cuneiform text to refer to tithis. Neugebauer-Sachs suggest that the difference 12 months 11;16 tithis could mean

$$12^{m} \times 29;30^{\,d/m} = 354 \text{ days.} \quad 0;16^{\tau} \approx 0;15,45^{d}$$

$$1 \text{ sidereal year} = 12 \text{ months} + 11 \text{ days} + 0;16 \text{ tithis} = 6,5;15,45 \text{ days.}$$

However, it seems odd, if the author of the text used tithis which are defined as thirtieths of a mean synodic month, that he would compute with months containing 29½ days and use the 11 both as tithis (in the epact) and as civil days (in the length of a sidereal year). We feel that we do not really understand this text.

Hunger [1991] (copies in von Weiher [1993] 168 and 169) publishes two tablets from Uruk that give the computed dates in months and "days" (these must be tithis since there are always 30 of them in a synodic month) of the summer solstices. The first (W 22925) covered 37 years of which the first corresponded to the year 44 of Artaxerxes II (-360/59). The value of ε is precisely 11 tithis. The second text (W 22801+22805) covered years 14 to 21 of Nabopolassar, 1 to 43 of Nebukadnezar II, 1 to 2 of Amel-Marduk, and 1 to 4 of Neriglissar—i.e., from -612/1 to -556/5. In what remains of this tablet, while ε is usually 11 tithis, there are two occasions when it is 12 tithis. These two years correspond to -607/6 and -568/7; since they are 40 years apart, in this very early tablet ε = 11;3 tithis.

In Neugebauer-Sachs [1967] pp. 190-192 is published a fragment of a table, BM 36838, listing the dates in months and tithis of the Γ of the Stars (Pleiades) and of the Γ of an unidentified constellation or star in the years from 64 SE (= -247/6) to 86 SE (= -225/4). Γ of the Stars occurs 1 month 15 tithis (= 45 tithis) after the Vernal Equinox, Γ of the unknown constellation 16 tithis after the Vernal Equinox.

## *2. Early Theoretical Texts, Planetary*

### *2.1. Period Relations*

There is a tablet (BM 36823 (LBAT 1393)) that arranges observations of the phases of Jupiter for -525/4 to -489/8 in 12-year periods and another (BM 32299 (LBAT 1388) + 45674 (LBAT 1387)) that arranges those of Venus for -462/1 to -417/6 in 8-year periods. Such arrangements probably were made much later than the recorded observations. We have also noted the Goal-Year Texts and the Goal-Year Periods (II B 5). Here we will discuss a few other texts that refer to planetary period relations.

Kugler [1907] pp. 45-48 published and discussed (copy on Tafel II) BM 45728, which gives the following periods: 27 days (a sidereal month) for the Moon, 8 years for Venus, 6(?) years for Mercury, 47 years for Mars, 59 years for Saturn, and 27 years for Arrow. The periods for Venus, Mars, and Saturn are Goal-Year periods, and that for Mercury would be if the scribe had written 46 instead of 6. Kugler suggests that the 27-year period of the fixed star, Arrow, refers to a primitive intercalation cycle (19 + 8 years) that he would date most probably before -532, but at least before the fourth century B.C. It is surprising to see Goal-Year Periods this early; Kugler's dating of the tablet is questionable.

On pp. 48-53 Kugler published and discussed (copy on Tafel III) the end of BM 35402 (LBAT 1593), where we find long and short periods for the planets.

Jupiter: long period: 344 years = 83 × 4 + 12 years = 71 × 4 + 5 × 12 years
short period: 63 months + 10 "days" = 1900 tithis.
Venus: long period: 6400 years = 8 × 8 × 100 years
short period: 63 months and 20 "days" = 1910 tithis.
short period: 7 days, 14 days, 21 days (presumably the periods of invisibility at inferior conjunction)
Mars: (short period): 65 months = 1950 tithis
(short period): 6 months and 20 "days" = 200 tithis (Ω to Γ in MUL.APIN)
long period: 284 years = 79 × 3 + 47 years, its ACT period.
Saturn: (short period): 10 months = 300 tithis
long period: 560 years = 59 × 9 + 29 (59 × 4 + 29 = 265, its ACT period).
Moon: long period: 684 years = 19 × 36 years.

This fragment, then, shows a knowledge of both Goal-Year Periods (Jupiter and Venus) and ACT periods comprised of Goal-Year Periods and shorter periods (Mars and Saturn).

Neugebauer-Sachs [1967] pp. 200-203 and 206-208 publish a transliteration of BM 41004 and a translation of and commentary on sections 2 and 3 of that tablet that gives periods of Jupiter, Venus, Mars, Saturn, and Mercury. This tablet was copied by Iddin-Bēl, the son of Marduk-šāpik-zēri, who seems to have been active at the end of the fourth century B.C.

Jupiter: 12 years – [7 days]
12 years – 1 month (incorrect)
12 years + 5°
71 years (Goal-Year Period)
83 years – 7 days (Goal-Year Period)

Since 71 years = 12 × 6 – 1 years and the motion of Jupiter in 12 years is one rotation + 5°, in 6 periods of 12 years it is 6 rotations + 30°; 30° is the approximate motion of Jupiter in one year. This justifies the period of 71 years. The period of 71 years plus that of 12 years – [7 days] justifies the period of 83 years – 7 days. Clearly 71 years produces a return relative to the Sun—i.e., a return of Greek-letter phenomena—while 83 years produces a return in passings by of Normal Stars—i.e., a return in longitude.

Venus: 12 (read 8) years – 4 days (Goal-Year Period)
16 years – 2 days
48 years + 2 days
64 years + 1 or 2 days
8 years – 4° (Goal-Year Period)
16 years – 2°

All except the first and the fifth period are incorrect. The first and the fifth periods produce a return of passings by of the Normal Stars.

Mars: 32 years – 5 days
47 years + 4 days (Goal-Year Period)
64 years + 4 days
126 years

These numbers are inconsistent. A period of 47 years is used in the Goal-Year Texts for a return of passings by of Normal Stars, one of 79 years for a return of Greek-letter phenomena, which should be an integer number of years; but 32 years – 5 days combined with 47 years + 4 days yields 79 years – 1 day. Further, 2 × (32 years – 5 days) gives 64 years – 10 days rather than 64 years + 4 days, and 2 × (47 years + 4 days) combined with 32 years – 5 days give 126 years + 3 days, not just 126 years.

Saturn: 59 years – 6 days (Goal-Year Period)
30 years + 9 days
12° in 1 year
30 years + 7;20°
147 years

Period 3 is a crude approximation leading to 1 rotation in 30 years. Period 4 tells us that in 30 years the motion of Saturn is 1 rotation + 7;20°, but period

2 indicates that one rotation occurs in 30 years + 9 days. There are obviously inconsistencies. Period 5, 147 years, is based on 3 × (59 years – 6 days) diminished by 30 years + 9 days; but this equals 147 years – 27 days.

| | |
|---|---|
| Mercury: | 13 years – 3 days |
| | 46 years – 1 day (Goal-Year Period) |
| | 125 years |

Period 3 is derived from 3 × (46 years – 1 day) diminished by 13 years – 3 days precisely.

Neugebauer-Sachs also refer to four additional texts, namely the two published by Kugler [1907], an unpublished parallel to the second (BM 40113), and BM 34560 (LBAT 1515). The main differences found in this last text are:

| | |
|---|---|
| Jupiter: | 12 years + 6° |
| | 71 years – 5° |
| Saturn: | 59 years – 5(?) days |
| Mercury: | 46 years. |

Some of these period relations and others similar to them appear in Procedure Texts.

*2.2. Planetary Latitudes*

Neugebauer-Sachs [1967] pp. 208-210 publish BM 37266, a tablet also copied by Iddin-Bēl, the son of Marduk-šāpik-zēri. It deals on the obverse with the latitudinal motion of a planet and on the reverse with the monthly velocities of a planet; on the obverse are two sections: one for, probably, Jupiter, and the other explicitly for Saturn, while the planet whose velocity is discussed on the reverse is probably also Saturn.

The first latitude scheme posits a high in Scorpius and a low in Pisces with the rate of increase and decrease being 8 digits or 0;40° for every 30° of longitudinal increase. This leads to the following reconstructions:

| | | | |
|---|---|---|---|
| Aries | –1;20° | Libra | +1;20° |
| Taurus | –1;20° | Scorpius | +1;20° |
| Gemini | –0;40° | Sagittarius | +0;40° |
| Cancer | 0° | Capricorn | 0° |
| Leo | +0;40° | Aquarius | –0;40° |
| Virgo | +1;20° | Pisces | –1;20° |

This puts the ascending node in Cancer, which is correct for Jupiter, and the maximum latitude at ±1;20°, which is also very close to being correct for Jupiter.

The scheme for Saturn's latitude prescribes that it goes up in Virgo, Libra, and Scorpius, proceeds to one node, goes down in Pisces, Aries, and Taurus, and proceeds to the other node. In this description "it goes up" means that it is at its maximum northern latitude, "it goes down" at its maximum southern latitude. The text continues by stating that from Taurus "it directs itself to going up," from Scorpius "it directs itself [to going down]." The pattern, then, is identical to that in the first section, with the ascending node in Cancer as is correct for Saturn. However, the maximum latitude should be ±3°, which requires an increase and decrease of 12 digits or 1° for every 30° of progress in longitude.

*2.3. Subdivisions of Synodic Arcs*

The reverse of BM 37266, the tablet whose obverse contains schemes for the latitudinal motion of Jupiter and Saturn, offers the rudiments of the following scheme of velocities:

| | | |
|---|---|---|
| Γ | [40 (read 48?) per month] | |
| | 35 (read 36) per month | |
| | 24 per month | |
| | 12 per month | 10° |
| Ψ (read Φ) | 2/3 of 24 per month<br>1/3 of 12 per month | |
| Ψ | 12 per month | |
| | 36 per month | |
| | 40 (read 48?) per month | 10° |
| Ω | 50 per month | |
| Γ | 40 (read 48) per month | |

This is based on the theory that from Γ to Φ and from Ψ to Ω are each 10° in a period of 4 months; if the "corrections" we have offered are correct, the numbers denote digits since 48 digits = 4°, 36 digits = 3°, 24 digits = 2°, and 12 digits = 1°. If the retrogression also occupies a period of 4 months, the retrograde arc will be $4 \times 2/3 \times 24 + 4 \times 1/3 \times 12 = 80$ digits = 6;40° (note that 6;40° is the value for Φ to Ψ in System A of Saturn). The total synodic arc, then, exclusive of the period of invisibility from Ω to Γ, will be 20° – 6;40° = 13;20°. This is a very crude approximation to the synodic arc of Saturn.

Column I of BM 36301, published in Neugebauer-Sachs [1967] pp. 192-195, gives the years, months, longitudes, and names of the Greek-letter phenomena (Γ, Φ, Ψ, and Ω) for Mars; the top and bottom of the column are broken. The computational system that lies behind these data is similar to that of System A, but not identical. The column originally covered 34 occurrences of phenomena, that is, 8½ synodic periods, over 19 years; these years are either from year 45 of Artaxerxes II (= -359/8) to year 17 of

Artaxerxes III (= -341/0) or from year 4 of Alexander IV (= -312/1) to 17 SE (= -294/3).

The reverse of MNB 1856, published by Neugebauer-Sachs [1969] pp. 93-94, gives the month and day for Mars' entry into the zodiacal signs from Capricorn to Gemini and for the occurrences of Φ, Θ, and Ψ in year 5 of Philip (= -318/7). This tablet, like several others that we have discussed, was copied by Iddin-Bēl, the son of Marduk-šāpik-zēri.

Column II of BM 36301, published by Neugebauer-Sachs [1967] pp. 193-198, divides a synodic period of Venus from one occurrence of Ψ to the next into 23 segments; each entry consists of the name of the month, the number of the tithi, and the longitude of the planet. At the appropriate places, entry 3 and entry 13, are noted the occurrences of Φ and Ξ. It appears that the dates and longitudes of $\Psi^1$, Φ, Σ, Ξ, and $\Psi^2$ were computed from some arithmetical scheme.

| line | date | longitude | phenomenon |
|---|---|---|---|
| 1 | III | Cancer 15° | [Ψ] |
| 2 | IV | Cancer | |
| 3 | V 12 | Cancer 26° | Φ |
| 4 | VI 12 | Leo 26° | |
| 5 | VII 12 | Virgo 26° | |
| 6 | VIII 12 | Libra 26° | |
| 7 | IX 12 | Scorpius 26° | [A] |
| — | | | |
| 8 | IX 20 | Sagittarius 3;45° | |
| 9 | X 15 | Capricorn 3;45° | |
| 10 | XI 10 | Aquarius 3;45° | |
| 11 | XII 5 | Pisces 3;45° | |
| 12 | XII 29 | Aries 3;45° | [Σ] |
| — | | | |
| 13 | III 9 | Gemini 27;45° | Ξ |
| 14 | IV 5 (4) | Cancer 27;45° | |
| 15 | IV 29 | Cancer (Leo) 27;45° | |
| 16 | V 29 | Leo (Virgo) 27;45° | |
| 17 | VI 19 | Virgo (Libra) 27;45° | [B] |
| 18 | VI 22 | Scorpius 1;20° | |
| — | | | |
| 19 | VII 22 | Sagittarius 1;20° | |
| 20 | VIII 22 | Capricorn 1;20° | |
| 21 | IX 22 | Aquarius 1;20° | |
| 22 | X 22 | Pisces 1;20° | |
| 23 | XI 26 | [Aries 5;20°] | [Ψ] |

According to this scheme, Venus' synodic period is divided into six parts as follows:

| | | |
|---|---|---|
| Ψ → Φ | | |
| Φ → A | $120^{\tau}$ | 120° (1°/τ) |
| | $8^{\tau}$ | 7;45° (≈ 0;55°/τ) |
| A → Σ | $99^{\tau}$ | 120° (1;12°/τ) |
| Σ → Ξ | $70^{\tau}$ | 84° (1;12°/τ) |
| Ξ → B | $100^{\tau}$ | 120° (1;12°/τ) |
| | $3^{\tau}$ | 3;35° (≈ 1;12°/τ) |
| B → Ψ | $154^{\tau}$ | 154° (1°/τ) |
| | $554^{\tau}$ | 609;20° |

The retrograde arc, Ψ → Φ, at the beginning should increase the time by about 30 tithis and decrease the longitudinal progress by about 20°.

The date of this scheme could be one of several possibilities: year 25 of Artaxerxes I (= -439/8), year 33 of Artaxerxes I (= -431/0), year 41 of Artaxerxes I (= -423/2), and year 8 of Darius II (= -415/4). Another astronomical possibility is 147 SE (= -164/3), but it seems historically to be too late.

A similar scheme for Venus is found on BM 33552, published in Britton-Walker [1991], another text copied by Bēl-apla-iddina, the son of Mušallim-Bēl, in about -320:

| | | | |
|---|---|---|---|
| Φ → A | $128^{d}$ | 128° | 1°/d |
| A → B | $272^{d}$ | 340° | 1;15°/d |
| B → Ψ | $128^{d}$ | 128° | 1°/d |
| Ψ | $12^{d}$ | 0° | 0°/d |
| Ψ → Φ | $32^{d}$ | -20° | -0;37,30°/d |
| Φ | $12^{d}$ | 0° | 0°/d |
| Φ → Φ | $584^{d}$ | 576° | |

A second section of the tablet correlates the length of invisibility at inferior conjunction to the zodiacal sign in which it occurs:

| | |
|---|---|
| $1^{d}$ | Sagittarius (read Capricorn) to Pisces |
| $4^{d}$ | Aries |
| $8^{d}$ | Taurus |
| $12^{d}$ | Gemini |
| $15^{d}$ | Cancer to Virgo |
| $6^{d}$ | Libra and Scorpius |
| $2^{d}$ | Sagittarius |

A third and final section gives 260° as the arc from Φ to Σ, and the range of Venus' forward velocity as between $0;45^{\circ/d}$ and $1;15^{\circ/d}$.

Aaboe-Huber [1977] present a text, BM 37151+37249, that once contained a scheme for subdividing 5 synodic periods of Venus in the 8 years between, probably, -120 and -112. The tablet gives lists of dates, directions (East or West), longitudes, and Greek-letter phenomena (Ξ, Ψ, Ω, Γ, Φ, and Σ), but it is badly destroyed. Only parts of lines 2-13, 17-18, 21-25, and 40-58 remain. From these remnants one can see that a synodic period was divided into 11 subdivisions.

| Line | Date | Direction | Longitude | Phenomenon |
|---|---|---|---|---|
| 1 | | [West | Virgo 10 | Ξ] |
| 2 | | | Capricorn 19 | |
| 3 | | | Taurus 2 | |
| 4 | | | Taurus 24 | |
| 5 | | | Taurus 25;20 | Ψ |
| 6 | | [West] | Taurus 18;40 | Ω |
| 7 | | [Eas]t | Taurus 15;20 | Γ |
| 8 | | | Taurus 7;⸢40⸣ | Φ |
| 9 | | | Gemini 2;40(?) | |
| 10 | | | [Leo] 2[9;20(?)] | |
| 11 | | [Ea]st | [Capricorn] 28 [ | Σ] |
| 12 | | [We]st | [Aries] 1[0+x | Ξ] |
| 13 | | | [Leo] 2[0+x] | |
| ... | | | | |
| 17 | ]8 | West | | [Ω] |
| 18 | ]9 | East | | [Γ] |
| ... | | | | |
| 21 | | | [Aries] 11 [ ] | |
| 22 | ]4(?) | East | [Virgo] 10 [ | Σ] |
| 23 | ]1(?) | West | [Scorpius] 22 [ | Ξ] |
| 24 | | | [Pisces] 29;30° | |
| 25 | | | [Cancer] 11;30° | |
| ... | | | | |
| 40 | ]4(?) | East | | [Γ] |
| 41 | ]5(?) | | [Aquarius] 24 [ | Φ] |
| 42 | ]5(?) | | [Pisces] 19 [ | ] |
| 43 | ]5(?) | | [Gemini] 15;40(?) | |
| 44 | ] | East | Scorpius 17 | [Σ] |

| | | | | |
|---|---|---|---|---|
| 45 | ]10 | West | Aquarius 2 | [Ξ] |
| 46 | ]2(?) | | Gemini 9;30 | |
| 47 | ]1(?) | | Virgo 21;30 | |
| 48 | | | Libra 13 | |
| 49 | | | Libra 14;40 | Ψ |
| 50 | ]2(?) | West | Libra 10;40 | Ω |
| 51 | ] | East | Libra 5;20 | Γ |
| 52 | | | Virgo 28;20 | Φ |
| 53 | | | Libra 23;20 | |
| 54 | | | Capricorn 20 | |
| 55 | ]7(?) | East | Capricorn 26;30 | Σ |
| 56 | ] | West | Virgo 7;30 | Ξ |

The tablet is too broken to permit a reliable reconstruction of the scheme that underlies this text, though Aaboe-Huber tentatively propose the following (numbers of days (or tithis) without parentheses correspond to lines 1-12, those in parentheses to lines 45-56):

| | | | |
|---|---|---|---|
| $\Xi \rightarrow \Psi$ | 104 (102) | days | 1;15°/d |
| | 90 (90) | days | 1;8°/d |
| | 44 (43) | days | 0;30°/d |
| | 4 (5) | days | 0;20°/d |
| $\Psi \rightarrow \Omega$ | 20 (12) | days | -0;20°/d |
| $\Omega \rightarrow \Gamma$ | 10 (16) | days | -0;20°/d |
| $\Gamma \rightarrow \Phi$ | 23 (21) | days | -0;20°/d |
| $\Phi \rightarrow \Sigma$ | 30 (30) | days | 0;50°/d |
| | 80 (80) | days | 1;5°/d |
| | 119 (126) | days | 1;15°/d |
| $\Sigma \rightarrow \Xi$ | 62(?) (56) | days | 1;15°/d |

An alternative interpretation is offered in Britton-Walker [1991] p. 115.

Theoretical models similar to those described above are found in some of the Procedure Texts associated with ACT. Such models, reinterpreted to fit Indian norms (e.g., using nakṣatras instead of degrees to measure longitudes), appear in Sanskrit omen texts (saṃhitās) in a tradition that goes back to the period of the Achaemenid occupation of Gandhāra; see Pingree [1987a] and [1987b].

### *3. Proto-procedure Texts*

An early procedure text is BM 36722 and 40082, published in Neugebauer-Sachs [1969], pp. 96-111, which are parts of a single tablet copied by Iddin-Bēl, who may be the son of Marduk-šāpik-zēri, the son of Bēl-apla-iddina;

the last is known to have flourished in about -320. The contents relate to the Moon; the editors divide them into a table and seven sections.

The table and sections 1 to 4 are concerned with the determination of the *na* (the time between sunset and moonset on the first day of a month) and the KUR (the time between moonrise and sunrise on one of the last days of the month). The method employed in the table is to add a quantity dependent on the longitude of the Moon to the *na* or KUR of the previous month, subtract a quantity dependent on the longitude of the Moon when the previous sum exceeds one day's motion of the Moon; the additive and subtractive quantities related to longitudes of the Moon at the first degree of each zodiacal sign are tabulated. Since the Moon's anomaly and its latitude are ignored, these quantities apparently represent some crude estimate of the right ascensions, $\alpha$, of the differences in longitude, $\Delta\lambda$. Other parts of the table are concerned with two $\Delta\lambda$'s at opposition. Sections 1 to 4 provide examples for computing the *na* and the KUR from this table. However, section 4 includes rather obscure calculations of the length of the synodic month (column G of ACT texts) and the related function $\Phi$.

Section 5 deals with a computation relating to a lunar eclipse on VII 14; that this is purely schematic is indicated by the fact that the longitude of the Moon is given as Aries 15°—i.e., the conjunction occurred at Libra 0°.

Section 6 concerns lunar velocity, utilizing several functions—a truncated linear zigzag function, $F_0$, and another zigzag function, $\hat{\Gamma}$, of equal period but opposite phase with respect to untruncated F—that correspond to columns F and $\hat{G}$ in the ACT texts, though with different parameters. Also mentioned is 2,13,20, the value of descending $\Phi$ that corresponds to the constant minimum, 2,40, of G in the ACT texts. Section 7 presents a variant of the transformations that in the ACT texts connect $\Phi$ with G and $\hat{G}$. Though much remains obscure in this text, it clearly represents a stage in the development of the final ACT lunar theory.

A Procedure Text that may antedate the Seleucid era, but in any case represents a non-ACT approach to the problem of solar and lunar motion, is BM 36712 from Babylon, published in Sachs-Neugebauer [1956]. This text measures a year between two successive risings of the Arrow as composed of two parts: 6,4;30$^d$ + 0;40$^d$ = 6,5;10$^d$. In one rotation of the Sun, the text states, there are 13;22 rotations of the Moon, i.e.,

1 solar year = 12;22 synodic months.

The initial estimate of the length of a year comes from the next statement, that the Sun travels 80° in 81 days; for $\frac{360}{80}$ = 4½ intervals of 81 days, which yields a year of 81 × 4½ = 364½ $^d$. However, the text assumes that in 364½$^d$ the Sun has only traveled 359 1/3°; it therefore adds 2/3° to that quantity and 2/3$^d$ to the number of days. From this it computes that the mean daily motion of the Sun is 0;59,8,58,16,17,46,40° (compare 0;59,9$^{\circ/d}$ used in A 3417+... (ACT 185) and A 3406+... (ACT 186), both from Uruk) and that the mean daily motion of the Moon is 13;10,37,54,53,49,37,46,40°. Finally, the text assumes that a sidereal month equals 27;25$^{d}$, which should be but is not the quotient of $\frac{6,0^{\circ}}{13;10,37,54,53,49,37,46,40^{\circ/d}}$; that would be closer to a more correct 27;20$^{d}$.

## *4. Mathematical Astronomy of the ACT Type*

By the beginning of the Seleucid period, after a millennium and a half at least of observations of celestial phenomena and the invention of various mathematical models allowing at least rough predictions of future phenomena, the Babylonians were ready to devise consistent and powerful sets of rules (embodied in "Procedure Texts") that allowed them to generate tables (called by modern scholars "Ephemerides") permitting one to foretell the longitudes and times of the occurrences of lunar and planetary phenomena and the longitudes of these bodies in the intervals between these occurrences. The vast majority of the tablets on which the Procedure Texts and the Ephemerides are inscribed were published and explained by Neugebauer [1955], *Astronomical Cuneiform Texts* (whence the acronym ACT). Along with many articles on particular aspects of the ACT material which will be referred to in the course of this exposition, studies of the whole have appeared in van der Waerden [1974] pp. 205-283, Neugebauer [1975], vol. 1, pp. 368-540, and, for the planets, Swerdlow [1998].

### *4.1. A Survey of the Literature before ACT*

The beginning of the study of the ACT material goes back to Epping [1881] pp. 277-282 wherein J. N. Strassmaier refers to a number of tablets of the Seleucid and Parthian periods that he has seen and copied, including a Mercury text (BM 32599 = ACT 1050+824) and a Normal Star Almanac (BM 34033 = LBAT 1055); then Epping, on pp. 282-291, transcribes and interprets two columns—$K_1$ and $L_1$—of a tablet for New Moons from SE 208 to 210 = -103/2 to -101/0 (part of BM 34580, which is part of ACT 122). As he remarks on p. 291: "Der Anfang zur Entzifferung der astronomischen Tafeln Babylons ist gemacht."

ACT 122 belongs to Neugebauer's System B of the Moon. Fragments of this tablet containing parts of columns $F_1$, $G_1$, $H_1$, $J_1$, $K_1$, $L_1$, $M_1$, $N_1$, $O_1$, $P_1$, and $P_3$, and a fragment (BM 34066) of a tablet (BM 34066 +34277+34400+34488 = ACT 120) giving New and Full Moons for SE 179 = -132/1, which fragment contains parts of columns $K_1$, $L_1$, $M_1$, $N_1$, $O_1$, and $Q_1$, were studied by Epping [1889a] pp. 8-16 (the meanings of these columns will be described in II C 4.2). He recognizes the mathematical relations to each other of the numbers in adjacent columns, but not their astronomical meanings. The same two ACT tablets are discussed again by

Epping [1890a], who now introduces, on p. 236, a graph of a linear zigzag function.

The first ACT text of System A to be tackled was an excerpt of the entries in an Ephemeris of Full Moons at which lunar eclipses were possible (BM 45688 = ACT 60) in Epping-Strassmaier [1893]; the tablet provides columns $T_2$, $\Phi_2$, $B_2$, $C_2$, $E_2$, $\Psi_2$, $F_2$, and $\Sigma G_2$ for SE 137 to 160 = -174/3 to -151/0.

This situation with just three ACT texts having been examined persisted until Kugler [1900a] examined a large number of Ephemerides of the Moon, and recognized that they can be divided into two systems. Of his System I (Neugebauer's System B) he was able to study BM 34066 (part of ACT 120), BM 34037 (ACT 121), part of BM 34580 (part of ACT 122), BM 45694 (ACT 123), and part of BM 34069+34900 (part of ACT 126), and of his System II (Neugebauer's System A) BM 34575 (part of ACT 4), BM 34041 (part of ACT 5), BM 34582 (ACT 7), BM 34088 (part of ACT 9), BM 34619 (part of ACT 10), BM 35048 (part of ACT 11), BM 34604+34628 (part of ACT 13), BM 34600 (part of ACT 15), BM 34617 obverse (part of ACT 16), BM 34608 (ACT 51), BM 35688 (ACT 60), and a Procedure Text, part of BM 32651 (part of ACT 200).

Relying primarily for System A on ACT 60, which lists only oppositions which are lunar eclipse possibilities, Kugler [1900a] pp. 55-88 and 115-202 was able to explain most of columns T, Φ, B, C, E, Ψ, F, G, J, C′, K, and M; the one column whose function he (and most of his successors) failed to understand was Φ. Relying for System B on ACT 122, Kugler [1900a] pp. 9-53 and 88-114 explained the mathematical structure and the function of columns A, B, C, D, Ψ'', F, G, H, J, K, and L. He only had difficulty with column Ψ'', which he believed to provide the lunar latitude rather than the eclipse magnitude. In ACT 121 and 123 he analyzed columns ΔΨ', Ψ', and F', incorrectly with respect to the first two. Finally, he argued persuasively that System B was known to Hipparchus.

In an appendix Kugler [1900a] pp. 207-211 explained the structure of three types of Ephemerides for Jupiter—those called by Neugebauer Systems A, A', and B. He did not indicate the tablets upon which his investigations were based.

Kugler [1907] pp. 115-206 uses BM 34621 (ACT 602), BM 34571 (ACT 603), and BM 34771 (ACT 607) for System A of Jupiter; BM 45730 (part of ACT 610), BM 34570 (part of ACT 611 and of a Procedure Text, ACT 822), BM 35318 (part of ACT 612), and the Procedure Texts BM 33869 (ACT 810) and BM 45851 (part of ACT 813) for System A'; and BM 34594 (ACT 621), BM 34574 (ACT 622), and the reverse of BM 34587 (part of ACT 623) for System B. From ACT 622 he derived the parameter:

391 synodic periods of Jupiter = 427 sidereal years,

and the fact that 0° of the Babylonian zodiacal sign corresponding to Aries lay 4;36° West of the Vernal Equinox in -120.

For Saturn he relied on BM 34589 (part of ACT 704) and part of BM 33758 (part of ACT 704a) both of which, like all twelve ACT tablets of Saturn, belong to System B.

For Mercury he also knew only two tablets: part of BM 45980 (ACT 301) and part of BM 34585 (ACT 302), both of which belong to System $A_1$. From his analysis Kugler correctly determined the boundaries of the three zones of the ecliptic and their corresponding synodic arcs for Γ; from this he calculated that the mean synodic arc is $360° \times \frac{2673}{848} = 114;12,31,30°$.

And for Venus he found one tablet, part of BM 34128+34222 (part of ACT 410), representing System $A_1$, and representing System $A_2$ three tablets: BM 46176 (part of ACT 420 that he erroneously suggested to be part of ACT 421), BM 35118 (ACT 421), and, erroneously joined to it, BM 45777 (part of ACT 421a). Kugler discovered that in System $A_1$ 5 synodic periods of Venus = $8 \times 360° - 2;30°$ while in System $A_2$ 5 synodic periods of Venus = $8 \times 360° - 2;40°$.

Pannekoek [1916] recognizes that the dates of the occurrences of the phenomena in the Ephemerides of Jupiter published by Kugler [1907] are measured in months of which each consists of thirty equal units (which he calls "days", but we call "tithis"). He further asserts that the dates are derived from the longitudes by a single mathematical formula.

Sidersky [1916] pp. 75-94 discusses the structure of the part of BM 34580 (part of ACT 122) that Epping and Kugler had previously studied; he returns to this in Sidersky [1919], comparing the procedure for computing the first visibility of the Moon given by Maimonides in his *Sanctification of the New Moon* (see Neugebauer [1956]). Using also the part of ACT 120 published by Kugler, Sidersky attempted to discover the astronomical meanings of columns N, O, and Q; in this he was not successful (see Neugebauer [1955] vol. 1, pp. 81-82).

Schnabel [1923] pp. 123-130 argues that Κιδηνᾶς, who had already been identified with Kidin or Kidinnu, whose *tersītu* both ACT 122 (for -103/2 to -101/0) and 123a are said to be, was the inventor of System B of the Moon which both tablets follow. He begins with an anonymous commentator on Ptolemy's *Handy Tables* who stated in the early third century A.D. that the period-relation, 251 synodic months = 269 anomalistic months, is said to have been discovered by Κιδηνᾶς (see Jones [1990] p. 20); he then goes on to demonstrate, as had Kugler before him, that this and the other period relations of System B were known to Hipparchus.

In order to establish Kidinnu's priority to Hipparchus, Schnabel [1923] pp. 214-222 investigated several tablets in Berlin that he regards as both belonging to System B and antedating Hipparchus. VAT 7852 (ACT 161; see Schnabel [1923] pp. 242-243), which was copied at Uruk on 5 Duʾūzu

124 SE = 23 July -187, contains an auxiliary function for oppositions according to System B from 124 to 156 SE = -187/6 to -155/4; and VAT 7809 (ACT 101), which was copied at Uruk by Anu-aḫa-ušabši on 7 Ṭebētu <117> SE = 7 January <-193>, contains conjunctions according to System B for 118 and 119 SE = -193/2 and -192/1. In addition, Schnabel attempts unsuccessfully to date Kugler's SH 81-7-6,99 (BM 45694 = ACT 123) to 23 SE = -288/7; Kugler [1924] pp. 599-602 shows that the date is actually -76/5. Using Kugler's estimate of the difference between the Vernal Equinox and 0° of the Babylonian zodiacal sign corresponding to Aries in BM 34580 (part of ACT 122), 3;40°, he concluded that Kidinnu invented System B in -313; and, since Kugler [1900a] p. 10 misread the colophon of ACT 122 to refer to Sippar rather than to *Enūma Anu Enlil*, Schnabel located Kidinnu in Sippar. He also concludes that a second Babylonian astronomer mentioned in various Greek sources, Σουδίνης, was a follower of Kidinnu in the third century B.C.

Schnabel [1923] pp. 222-225 turns his attention to the third Babylonian astronomer referred to by name in the Greek tradition, Ναβουριανός. This name he recognizes in VAT 209 (part of ACT 18; see Schnabel [1923] pp. 244-245), which was found at Babylon and whose colophon states that it is the [*tersīt*]*u* of Nabû-x-man-nu (Schnabel restores the name to Nabû-ri-man-nu); it records conjunctions and oppositions according to System A for 263 SE = -48/7. Using the same methodology by which he dated Kidinnu to -313 and assuming that Nabû-ri-man-nu was the inventor of System A, Schnabel dates him to about -426.

Finally, Schnabel [1923] pp. 227-237 employs a long and spurious argument to claim that Kidinnu discovered the precession of the equinoxes.

Kugler [1924] pp. 524-530 discusses BM 45707 (part of ACT 654), which gives the daily longitudes of Jupiter from Kislimu 147 SE to Abu 148 SE = December -164 to August -163. Later in this volume he discusses (pp. 577-581) two tablets from Uruk that had been published in TCL 6 in 1922: AO 6477 (ACT 801), a Procedure Text concerning Mercury and Saturn, and AO 6481 (ACT 501), an Ephemeris giving the times and longitudes of Φ of Mars according to System A from 123 to 202 SE = -188/7 to -109/8. Finally, Kugler [1924] pp. 582-621 and 627-630 examines and refutes some of the claims with regard to Kidinnu made by Schnabel [1923].

Schnabel [1924c] interprets five tablets containing planetary Ephemerides and a Procedure Text; copies of four of these had been published in TCL 6 in 1922, and two of those four were discussed by Kugler [1924]: ACT 501 (reconstructed by Schnabel on pp. 106-107) and ACT 801. The other two tablets included in TCL 6 are AO 6476 (part of ACT 600), giving the dates and longitudes of Φ of Jupiter according to System A from 113 to 173 SE = -198/7 to -138/7, and AO 6480 (ACT 620), giving the times and longitudes of Θ of Jupiter according to System B from 127 to 194 SE =

-184/3 to -117/6. The fifth tablet, also from Uruk, is VAT 7819 (ACT 702), giving the dates and longitudes of Θ of Saturn according to System B from 123 to 182 SE = -188/7 to -129/8 (reconstructed by Schnabel on pp. 102-103).

Schnabel [1927] argues at length against Kugler [1924] pp. 582-621 that Kidinnu did discover the precession of the equinoxes; in the process he redates Nabû-ri-man-nu to -507 and Kidinnu to -378. He also transcribes on pp. 19-22 VAT 7844 (ACT 170), which gives auxiliary functions for conjunctions according to System B from 104 to 112 SE = -207/6 to -199/8; on pp. 28-30 VAT 7809 (ACT 101), a tablet that he had dealt with in Schnabel [1923] as had also Kugler [1924]; and on pp. 36-37 VAT 7821 (a part of ACT 185), giving daily longitudes of the Sun for 124 SE = -187/6.

Schaumberger [1935] pp. 375-394 restudies BM 34580 (part of ACT 122), a tablet previously studied by Epping [1889a] and [1890a] and by Kugler [1900a]. Schaumberger explains the last six columns: M, N, O, $P_1$, $P_3$, and $O_3$. In the course of solving the problems arising from these columns, he also explains columns $R_1$ and $Q_1$ in AO 6475 (a part of ACT 100), which gives conjunctions according to System B for 106 to 108 SE = -205/4 to -203/2.

Neugebauer [1936a] describes the mathematics of linear zigzag functions and the use of Diophantine equations to connect tablets recording the same function but separated from each other chronologically.

Neugebauer [1936b] pp. 523-529 determines, on the basis of ACT 200, a Procedure Text for System A of the Moon previously studied by Kugler [1900a], the lengths of the four seasons of a solar year on the assumption that a solar year contains 6,5;15 days. These values are:

| | Babylonian | Hipparchus - Ptolemy |
|---|---|---|
| Spring | 1,34;29,43 ≈ $94\frac{1}{2}^d$ | $94\frac{1}{2}^d$ |
| Summer | 1,32;43,24 ≈ $92\frac{3}{4}^d$ | $92\frac{1}{2}^d$ |
| Autumn | 1,28;35,21 ≈ $88\ \frac{7}{12}^d$ | $88\ \frac{1}{8}^d$ |
| Winter | 1,29;26,32 ≈ $89\ \frac{5}{12}^d$ | $90\ \frac{1}{8}^d$ |

On pp. 530-534 he investigates the lengths of daylight in System A, using Kugler's study of the same Procedure Text and of Ephemerides containing column C. Then (pp. 535-538) he derives the longitudes of the jumping points in System A's solar theory from two tablets also studied by Kugler, ACT 13 and 70. Turning his attention to System B, he investigates (pp. 538-543) the lengths of the seasons in that System, but determines that these values cannot be fixed precisely from the longitudes of the Sun at each syzygy provided by the Ephemerides; and (pp. 544-549) studies the lengths of daylight according to System B. Finally, he reconstructs on pp. 549-550 columns A, B, and C of ACT 123.

Neugebauer [1937b] pp. 38-53 investigates and explains the computation of column E, which gives the Moon's latitude, in System A; his analysis is based on the material available to Kugler [1900a] (parts or the whole of ACT 4, 9, 10, 11, 13, 16, 51, and 60) and to Schnabel [1923] (part of ACT 18), as well as the previously unpublished BM 34237 (ACT 17). In the course of his analysis he shows that behind column E lies an untabulated step-function with jumping points for conjunctions at Virgo 13° (ascending) and Pisces 27° (descending), and for oppositions at Pisces 13° (ascending) and Virgo 27° (descending). Then (pp. 53-59) he investigates column B, which employs a step-function with jumping points at exactly the same longitudes, and shows its relationship to column E. In the succeeding section of the paper (pp. 59-80) Neugebauer discusses some specific examples of column E as it appears and is used in several of the tablets listed above. Finally, on pp. 80-91 he defines the relationship between column Ψ, which denotes the magnitude of the eclipse, and column E.

Neugebauer [1937c] continues his investigations of Babylonian lunar theory, especially as it concerns the predictions of eclipses. He begins (pp. 199-215) by comparing the solar theories of System A (step-function) and System B (linear zigzag function) and defining their characteristics. As examples he uses for System A ACT 51, for System B ACT 122; he also refers to ACT 16 and 123. After a review of various situations that occur in linear zigzag functions (pp. 215-229), Neugebauer describes the relationship of lunar latitude (column E) to the anomalistic motion of the Sun (column B) in System A (pp. 229-259); he provides examples from ACT 9, 16, 18, and 60, as well as LBAT 1251. On pp. 259-273 he explains the theory of the magnitude of the eclipse (column Ψ) in System A; he also examines column Ψ of ACT 123 to attempt to obtain a value for the diameter of the Moon according to System B. An excursus on pp. 273-286 presents the Babylonian metrology of lengths, arcs, times, and weights, and their interrelations. In the following section (pp. 286-297) Neugebauer addresses the problem of determining the diameter of the earth's shadow from the value of the constant c (= 17;24 digits) which is used to derive Ψ from E. Since c represents the radius of the earth, its diameter is 2c = 34;48 fingers = 2;54°. Then on pp. 297-313 he describes the relation between the anomalistic motion of the Sun and the latitude of the Moon according to System B; he refers in the course of this discussion to A 3410 (ACT 104) from Uruk and ACT 121, 122, and 123. The remainder of the paper deals with mathematical aspects (the summing of linear zigzag functions (pp. 313-319) and the distribution of a linear zigzag function on a circle (pp. 319-321)), a comparison between the two Babylonian systems of the Moon (pp. 322-327), and a review of the whole article (pp. 328-339).

Van der Waerden [1940] p. 114 criticizes one statement made by Neugebauer [1937c], which was that the latitude function of System A was based on the period relation of the "Saros", namely:

223 synodic months = 242 draconitic months.

He demonstrates that, on the contrary, the ratio of draconitic to synodic months in System A—1;5,6,42,25—is closer to the modern value (1;5,6,42,18) than is that of the "Saros" cycle (1;5,6,43,35).

The same was pointed out in a different way by Pannekoek [1941] p. 10. He also questions (pp. 10-14) Neugebauer's interpretation of column E as referring to the latitude of the border rather than to that of the center of the lunar disk and (pp. 14-20) his contention that the so-called "Saros Canon" (LBAT 1428), which lists the lunar eclipse possibilities from -384/3 to -272/1 in eighteen-year groups, is derived from a tablet similar to ACT 60, in which are recorded the data for those oppositions which have the smallest values in column E. Pannekoek is undoubtedly right with regard to his criticism of the second idea since the "Saros" cycle was certainly known long before the development of System A, and Neugebauer [1945] accepts that column E records (measured in še or "barleycorns") the latitude of the Moon's center.

Van der Waerden [1941] pp. 25-32 reviews the three systems of Jupiter identified by Kugler [1907]. He then (pp. 33-39) computes, from Kugler's discussion of a Procedure Text for Jupiter, ACT 813, that the Babylonians based their computation of the time-interval of a synodic period of that planet on a solar velocity of 1° in 1;1,50 days; following Kugler, who corrects the data in the Procedure Texts where necessary, he presents the scheme of the time-intervals in tithis and the daily velocities of Jupiter in the three arcs of System A' between the occurrences of the Greek-letter phenomena in one synodic period; and examines the columns for Ψ and Ω of Jupiter according to System B in ACT 622. On pp. 39-40 van der Waerden describes the theory of System B of Saturn as found in ACT 704. In the last section of the original form of the paper he speculates that the ACT periods for Jupiter (36 sidereal rotations = 391 synodic periods) and Saturn (9 sidereal rotations = 256 synodic periods) are derived from the Goal-Year Periods.

Jupiter: 71 years = 65 synodic periods = 6 rotations - 5;33°

$71 \times 6 + 1 = 427$ years $= 65 \times 6 + 1 = 391$ synodic periods $= 6 \times 6$ sidereal rotations - 33;18° + $1 \times$ 33;45° ≈ 36 sidereal rotations

Saturn: 59 years = 57 synodic periods = 2 sidereal rotations + 0;50°

$59 \times 9 - 1 = 530$ years $= 57 \times 9 - 1 = 512$ synodic periods $= 9 \times 2 =$ 18 sidereal rotations + 7;30° - $1 \times 12°$ ≈ 18 sidereal rotations

$\frac{530}{2} = 265$ years $= \frac{512}{2} = 256$ synodic periods $= \frac{18}{2} = 9$ sidereal rotations.

This suggestion is not convincing.

After completing this much of his article, in April 1940, van der Waerden became aware of TCL 6 and Schnabel [1924a]; therefore, he was able to add pp. 42-48 to his original article in July 1940. In this appendix he adds remarks about System B of Saturn based on the Ephemerides ACT 702 and the Procedure Text ACT 801 and about System A of Jupiter based on the Ephemerides ACT 600 and 620; and he describes System A of Mars on the basis of ACT 501.

Following his determination of the schemes for the lengths of daylight in Systems A and B of the Moon in Neugebauer [1936b] and [1941] (in which article Kugler's Systems II and I were first called A and B), Neugebauer [1942] pp. 253-263 demonstrated that two whole families of schemes of rising-times (ἀναφοραί) of the zodiacal signs developed from these Mesopotamian parents in Greek and Latin astronomical and astrological literature.

Neugebauer [1945] returns to the problems of columns E and Ψ of System A that he had discussed in Neugebauer [1937b] and [1937c]. He here accepts the criticisms offered by Pannekoek [1941], and reinterprets the functions E and Ψ while adding a further function, Ψ', which is computed for all oppositions, not just for those at which there is an eclipse possibility. E is measured in barleycorns, Ψ (or Ψ') in fingers. Neugebauer also withdraws his previous attempt to compute the diameter of the Moon.

Neugebauer [1948] presents a mathematical solution to the problem of determining the interval between to separated pieces of a linear zigzag function; this is to be used for dating tablets containing the same linear zigzag function when the date of one of them is known.

Neugebauer [1950] demolishes step by step the arguments put forth in Schnabel [1927] in favor of the theory that Kidinnu discovered the precession of the equinoxes.

Neugebauer [1951] is based on the following tablets concerning Mercury: from Babylon BM 33618 (ACT 303), 34585 (ACT 302), 35277 (ACT 304), and 45980 (ACT 301 and 820a), and from Uruk A 3405 (not in ACT), 3409 (ACT 800a), 3424+3436 (ACT 300), and 3456 (ACT 198); AO 6477 (ACT 801); U 151+164 (ACT 800d) and 169 (ACT 800b); and Warka X 40 (ACT 800e). Thus, the increase in the number of available sources over those known to Kugler through Strassmaier's copies, both Ephemerides and Procedure Texts, naturally provides the basis for a much deeper understanding of Babylonian astronomy. This increase was due to Neugebauer himself, to Ehelolf (Berlin) and Kraus (Istanbul), and to Sachs working in London with Pinches' copies in the three years following the publication of Neugebauer's paper. In particular, what Neugebauer [1951] achieved was the discovery of the Babylonian rules for computing Σ and Ω of Mercury; Kugler had discovered the rules for Γ and Ξ decades earlier. He also demonstrates the grave inadequacies of Schoch's tables of the first and last visibilities of the planets and the Moon in Mesopotamia, published in Langdon-Fotheringham [1928].

Using some of his new material—namely, BM 35399 (ACT 201), which was found by Sachs among Pinches' copies (LBAT 91)—as well as a section (15) of ACT 200 that Kugler had not used, Neugebauer [1953] showed that the Babylonians not only had schemes of the rising-times of the zodiacal signs, but that they computed the oblique ascensions of ecliptic arcs.

On the eve of the publication of ACT Neugebauer [1954] presents a valuable survey of Babylonian planetary theory. Since we will cover all the material of this article in our discussion of ACT itself, excellent summary that it is, we will say no more about it save that, at the time of its writing (it was originally delivered as a lecture at the American Philosophical Society in Philadelphia on 23 April 1953), Neugebauer knew of eleven Ephemerides of Mercury (as in ACT), nine of Venus (as in ACT), seven of Mars (eight in ACT), about forty for Jupiter (forty-one in ACT), and eleven of Saturn (twelve in ACT), as well as twenty-eight Procedure Texts of the planets (thirty-eight in ACT, of which eleven are parts of Ephemerides); to these are to be added twelve unidentified planetary texts in ACT. Clearly, the fifty-three ACT type planetary texts discovered by Sachs in Pinches' copies, which he first saw in the summer of 1952, were incorporated into Neugebauer's *magnum opus* with remarkable rapidity.

### *4.2. ACT and After*

Neugebauer [1955] describes the two Babylonian lunar theories, System A and System B, and the lunar Ephemerides and Procedure Texts in volume I; and the various Babylonian planetary theories, Ephemerides, and Procedure Texts in volume II. The third volume contains transcriptions of all the texts, copies of some, and photographs of others. The general plan of the first two volumes of texts is as follows:

Volume I. Moon.

- ACT 1-99. System A and undetermined system. Ephemerides.
- ACT 100-199. System B. Ephemerides.
- ACT 200-299. Procedure Texts.

Volume II. Planets.

- ACT 300-399. Mercury Ephemerides.
- ACT 400-499. Venus Ephemerides.
- ACT 500-599. Mars Ephemerides.
- ACT 600-699. Jupiter Ephemerides.
- ACT 700-799. Saturn Ephemerides.
- ACT 800-899. Procedure Texts.
- ACT 1000-1099. Fragmentary and unidentified texts.

*4.2a. The Moon*

There are 171 lunar tablets in ACT; of these 58 Ephemerides and 29 Procedure Texts belong to System A, 77 Ephemerides and 2 (or perhaps just 1) Procedure Texts to System B. All except 1 or possibly 2 tablets of System A come from Babylon (the 1 or 2 from Uruk are the earliest full Ephemerides in ACT, datable to -187/6 to -185/4), while just 27 of the System B tablets come from Babylon, the rest from Uruk. The ranges of dates for the two Systems in the ACT material from the two cities is:

| | System A | System B |
|---|---|---|
| Babylon | -262/1 to -42/1 | -257/6 to -68/7 |
| Uruk | -187/6 to -185/4 | -207/6 to -150/49 |

But, if we restrict our count to full Ephemerides, the ranges change considerably for Babylon, minimally for Uruk:

| | System A | System B |
|---|---|---|
| Babylon | -170/69 to -42/1 | -135/4 to -75/4 |
| Uruk | -187/6 to -185/4 | -205/4 to -174/3 |

A much earlier Ephemeris of System A, dated -272/1, was discovered at Uruk and published by Hunger [1976a], pp. 100-101 (text 98). If Riley [1994] pp. 20-21 is right, ACT 123a, an Ephemeris from Babylon, is datable to the years -169/8 and -168/7. Moreover, there was certainly continuous interchange between the two cities. The most striking example is ACT 155, giving auxiliary functions from -207/6 to -187/6, which was copied at Uruk from a tablet from Babylon. Nevertheless, there is a clear preponderance of tablets belonging to System A coming from Babylon, while System B is better represented at Uruk than at Babylon. However, since we clearly have only a small fraction of all the ACT type tablets that were produced in antiquity, additional finds may change this picture drastically—and, of course, cause us to modify many more of our perceptions of Babylonian astronomy.

We have already seen (II C 1) that various elements of lunar and solar theory, especially those relating to eclipses, antedate the tablets included in ACT by centuries. Thus, the nineteen-year cycle goes back to about -500. The "Saros" possibly goes back to -746, though this is not very likely; a stronger argument can be made for the late sixth century B.C. It may antedate the nineteen-year cycle. Column Φ, which is closely entwined with the "Saros", may have already existed in the late seventh century B.C., and is more securely attested for the middle of the fifth century B.C.

*4.2a.1. System A of the Moon*

We will soon look more closely at the forebears of System A, but first we should describe what a full tablet of that System would contain (the obverse usually relates to conjunctions, denoted by a subscript 1 in Neugebauer's discussion, and the reverse to oppositions, denoted by a subscript 2). There are thirteen columns:

T. The dates in years of the Seleucid era and months.

Φ. The excess measured in large hours of the 223 months following the given syzygy over 6585 days.

B. The longitude of the Moon at syzygy assuming mean motion for the Moon, but varying the solar motion by means of a step-function that is characteristic of System A.

C. The length of daylight measured in large hours; derived from column B.

E. The latitude of the Moon measured in barleycorns.

Ψ. The potential magnitude of an eclipse measured in fingers.

F. Lunar velocity.

G. First approximation to the excess of the synodic month over 29 days on the assumption of a constant solar velocity of 30° per month; derived from column Φ.

J. Correction of G due to a variable solar velocity.

C'. Correction of the length of a synodic month due to the variation in the length of daylight; derived from column C.

K. Length of the month; sum of columns G, J, and C'.

M. Date of the syzygy (month, day, and large hours).

P. Duration of first or last visibility measured in large hours.

Aaboe-Sachs [1969] pp. 11-20 publish a text to which we have already referred (II C 1.1b), BM 36599+36941 and BM 36737+47912. There were six columns in this text relating to solar eclipse possibilities between 5 December -474 and 21 July -456. The second column preserves Φ, in its truncated version (i.e., values greater than 2;13,20 are replaced by 2;13,20), and the fifth gives the lunar velocity, F, using unabbreviated parameters but truncated at 15° per day, and the longitude of the conjunction. Thus we have here two variants of columns in System A. The values of the longitudes, as noted above, are not related to the step-function underlying column B, but rather to the relation of the "Saros" to solar motion:

$$223 \text{ months} \approx 18 \text{ rotations} + 10;30°.$$

This means that a solar year contains 12;22,7,50,... synodic months as opposed to System A's 12;22,8 synodic months.

Aaboe-Sachs [1969] pp. 3-11 present another text that has already been discussed, BM 36822+37022, which seems to date from -399 to -397. For each conjunction it lists $\Phi_1$, in the truncated version; F in its abbreviated form, also truncated; early approaches to C, G, and M computed differently from System A; and the longitudes of the conjunctions computed by means of a step-function differing from that used to compute column B of System A (see Aaboe [1966]). In this function the jumping points are Pisces 13;30° and Virgo 29° instead of Pisces 27° and Virgo 13°.

Pisces 13;30° → Virgo 20° $a_1 = 195;30°$, $w_1 = 28;20°$
Virgo 29° → Pisces 13;30° $a_2 = 164;30°$, $w_2 = 30°$

In System A, $w_1 = 28;7,30°$. Since $P = \frac{a_1}{w_1} + \frac{a_2}{w_2}$,

$$P = 6;54 + 5;29 = 12;23 \text{ months} = 1 \text{ solar year.}$$

Moreover, $\frac{w_1}{w_2} = \frac{28;20}{30} = \frac{17}{18}$, which is an awkward fraction to deal with; in System A $\frac{w_1}{w_2} = \frac{28;7,30}{30} = \frac{15}{16}$. This shows that, while step-functions existed by -400, the step-function for solar motion of System A had not yet been discovered. But a beginning had been made toward assembling the columns of System A.

A further text, from towards the end of the fourth century B.C. or beginning of the third (it was copied by Iddin-Bēl), was published in Neugebauer-Sachs [1969] pp. 96-111; this is BM 36722 and 40082. This is the "proto-procedure text" discussed in II C 3. It includes references to variant versions of columns Φ, F, and G.

But the earliest tablet belonging to System A is BM 40094, published in Aaboe [1969]. It contains columns $K_1$, $M_1$, $\Lambda_1$, $Y_1$, $\hat{C}'_1$, and $\tilde{K}_1$ for -318 to -315; on the basis of these remains of the right half of the tablet, Aaboe reconstructs the left half, which consisted of columns $T_1$, $\Phi_1$, $B_1$, $C_1$, $G_1$, $J_1$, and $C'_1$. While $29^d + G^H$ (H signifies large hours) is the approximate length of the preceding month and $6585^d + \Phi^H$ is the length of the succeeding 223 months, $354^d + \Lambda^H$ is the approximate length of the preceding 12 months, assuming constant solar velocity; Y is the correction for variable solar velocity, so that

$$354^d + \Lambda^H + Y^H = \text{the preceding 12 synodic months.}$$

These relations mean that Λ can be derived from G and Y from J. Column $\tilde{C}'$ is a correction of Λ for the variation in the length of daylight. Just as $C'_n = \frac{1}{2}(C_{n-1} - C_n)$, so $\tilde{C}'_n = \frac{1}{2}(C_{n-12} - C_n)$. And finally, just as K = G + J + C′, so $\tilde{K} = \Lambda + Y + \tilde{C}'$. Aaboe conjectures that $\tilde{K}$ (and its constituent parts) were introduced to serve as a check on M.

BM 40094 was joined to BM 45662 (ACT 128) in Aaboe-Hamilton [1979] pp. 23-28. This provided evidence for two more columns after $\tilde{K}$, a second column M derived from $\tilde{K}$ and a column $P_3$; the second column M confirms Aaboe's conjecture that the purpose of $\tilde{K}$ was to serve as a check on the first column M.

We have now established that System A was developing but did not yet exist in -400, while it was fully developed by -320. At this point we can consider the claim made by van der Waerden [1963] that System A was probably invented between 610 and 450 B.C., and System B between 500 and 440 B.C. Van der Waerden [1974] p. 246, taking BM 36599+36941; BM 36737; and BM 47912 to actually adhere to System A, repeated more or less his previous limits for its discovery; between -610 and -470, with a preference given to the period between -540 and -470. For System B he now (pp. 246-248) arrives at the limits -480 and -440. This date also, like that for System A, we shall show to be in error. Van der Waerden does not allow for developments; the whole theory must arise at once, so that he takes any step towards the theory to prove the existence of the whole theory. And he takes the convention of System A that the Vernal Equinox occurs when the Sun is at Hired Man 10° and that of System B that it occurs when the Sun is at Hired Man 8° to refer to points on the sidereal zodiac that he can identify; this, of course, is not the case. The difference in the longitudes can just as easily refer to a difference in the locations of the beginning of the constellation called Hired Man as to a difference in the locations relative to a fixed star of the Vernal Equinox point.

A group of auxiliary texts of undetermined date, though related to the methods of BM 40094, were published in Aaboe [1971]. BM 36699+36846 +37079+37886 (BM 36846 is ACT 207cd), and BM 36908 (pp. 6-11) are two pieces from a single tablet in which the first column gives values of Φ in its truncated form, the second the derived values of G, and the third a function measured in days that Aaboe calls G′. He conjectures that $G' = 29 + \frac{G-2;40}{6}$. Since 2;40 is the minimum value of G, the minimum of G′ is 29 days; its maximum is 29;22,57,20. This does not fit very well with Aaboe's conjecture that G′ gives the length of a lunation in days assuming constant solar velocity. BM 36793 (pp. 11-12) presents columns Φ and Λ′ where $\Lambda' = \Lambda + 0;3,24,7,...^{H}$. The meaning of Λ′ remains unclear. And BM 45930 (pp. 13-20) contains columns Φ and W where $177^{d} + W^{H}$ is the length of the preceding six synodic months assuming constant solar velocity. W can be derived from Φ. In an appendix (pp. 20-26) Aaboe re-interprets BM 46015 (ACT 55), showing that it contains columns F; G, Λ, and W mixed in one column; J, Y, and Z (the parallel correction for W) mixed in one column; C′; and K. These values are given for selected conjunctions between -131 and -109. Aaboe found a difficulty in one of the values for W; this was resolved in Aaboe-Hamilton [1979] pp. 15-16.

A group of quite early texts belonging to System A and relating to columns Φ and G were published in Aaboe [1968]. BM 36994 (p. 12) gives values of $\Phi_2$ for -305; BM 36824+37222 (pp. 12-14) lists $\Phi_2$ and $G_2$ on the obverse, $\Phi_1$ and $G_1$ on the reverse, for -276 to -274; BM 37203 (pp. 14-15) provides $G_2$ for -271; and BM 36700 (pp. 15-16) gives $\Phi_2$ and $G_2$ for -268.

The relations between Φ and G in the last three texts is not what one would expect from ACT. An undated tablet, BM 36311+36593 (pp. 16-30), provides columns Φ, Λ, and X, where X is the number of days that, added to $\Lambda^H$, equals the epact, ε; in other words,

$$1 \text{ solar year} = 354^d + \varepsilon^d = 354^d + \Lambda^H + X^d.$$

Aaboe-Hamilton [1979] pp. 33-34 publish BM 37021, which gives columns $K_1$ and $M_1$ for -266/5.

Of almost equal age with the first four texts published in Aaboe [1968] and the last mentioned one is BM 34934 (ACT 70), which gives columns $T_2$, $B_2$, and $E_2$ from -262/1 to -251/0; Neugebauer conjectures that it began originally with -266. But the oldest extant standard Ephemeris is A 3412+3420+3435 (ACT 1), a tablet probably from Uruk, giving columns $T_2$, $\Phi_2$, $B_2$, $C_2$, $E_2$, $F_2$, $G_2$, $J_2$, $C'_2$, $K_2$, and $M_2$, from -187/6 to -186/5. With this we move to a consideration of the parameters and computation of the standard (and variant) columns of System A in ACT, which is a remarkably consistent system.

Φ is a linear zigzag function in which:

$$M = 2;17,4,48,53,20^{H/m}$$
$$m = 1;57,47,57,46,40^{H/m}$$

(the truncated values, $M = 2;13,20^{H/m}$ and $m = 1;58,31,6,40^{H/m}$ do not occur in ACT texts)

$$\mu = 2;7,26,23,20^{H/m}$$
$$d = 0;2,45,55,33,20^{H/m}$$
$$\Delta = 0;19,16,51,6,40^{H/m}$$
$$P = \frac{2\Delta}{d} = \frac{1,44,7}{7,28} = 13;56,39,6,...^H$$
$$\Pi = 1,44,7 = 6247 \text{ months}$$

Since 6 H = 6,0 UŠ, if the sexagesimal place in each of these numbers is moved one place to the right, the result is the number of UŠ or time-degrees.

Neugebauer [1975] vol. 1, pp. 499-500, has shown that column Φ is tabulated from a column Φ* which is a linear zigzag function with the same maximum, minimum, and amplitude as Φ, but computed for successive tithis. Examples of column Φ* are found in ACT 80 and 81, both for -133/2. The difference in column Φ* is:

$$d^* = \frac{2\Delta + d}{30} = \frac{41;19,37,46,40}{30} = 1;22,39,15,33,20^H$$

The period, P, then, is an anomalistic month:

$$P = \frac{2\Delta}{d^*} = \frac{38;33,42,13,20}{1;22,39,15,33,20} = 27;59,33,...^d.$$

This value of an anomalistic month implies that:

6247 synodic months = 6695 anomalistic months.

The values of $\Phi_1$ reach their maxima at half an interval (i.e., half a mean synodic month) after the minima of $\Phi_2$. The rule for transforming $\Phi_1$ into $\Phi_2$ is given in BM 34079+35152+35324 (ACT 204), section 6, as interpreted in Aaboe [1968] pp. 8-9: add to $\Phi_1$ 0;20,39,48,53,20 (which is the difference of $\Phi$ for 15 tithis), or subtract it from $\Phi_1$; subtract 2;17,4,48,53,20 (M of $\Phi$) from the sum if it is greater than M or add 1;57,47,57,46,40 (m of $\Phi$) to the difference if it is less than m. The result is $\Phi_2$ for the current or the preceding month respectively.

Another Procedure Text, BM 32651 (ACT 200), section 5, discussed in Aaboe [1968] pp. 7-8, states that the unabbreviated parameters for the lunar velocity, F, are:

$$M = 15;56,54,22,30^{\circ/d}$$
$$m = 11;4,4,41,15^{\circ/d}$$
$$d = 0;42^{\circ/d}$$

Further, M (of F) occurs when $\Phi$ reaches its maximum, 2;17,4,48,53,20, and m when $\Phi$ reaches its minimum, 1;57,47,57,46,40. Then, the difference of F, 0;42, equals the difference of $\Phi$, 0;2,45,55,33,20, multiplied by 15;11,15. If one uses the truncated form of $\Phi$ (for which see Aaboe-Sachs [1969]) one obtains a truncated form of F:

| | $\Phi$ | F |
|---|---|---|
| M | 2;13,20 | 15 |
| m | 1;58,31,6,40 | 11;15 |

The same is repeated in section 5 of ACT 204, where we find (see van der Waerden [1974] p. 222):

$$F - 15 = 0;15,11,15\ (\Phi - 2;13,20).$$

The derivation of column G from the truncated version of $\Phi$ was already mentioned in several Procedure Texts; section 14 of ACT 200; VAT 1762 (ACT 204a); MLC 2205 (ACT 205); BM 35253 (ACT 206); BM 46116 (ACT 207a); BM 34134 (ACT 207b); BM 34737 (ACT 207c); BM 36004 (ACT 207ca); BM 41990 (ACT 207cb); BM 36438 (ACT 207cc)+37012+37026+37274+37319, now explained in Aaboe-Hamilton [1979] pp. 18-23; BM 36846 (ACT 207cd)+36699+37079+37886 and BM 36908, now explained in Aaboe [1971] (see above p. 224) pp. 6-11; and sections 1 and 7 of BM 35564+40081 (ACT 207d). Note that BM 33593 (ACT 207) was recognized by Aaboe-Hamilton [1979] pp. 9-10 to contain column W.

Neugebauer [1955] vol. 1, pp. 59-60 first explained how this works; the truncated function of $\Phi$ was hypothesized in van der Waerden [1966] pp. 147-153 and confirmed in Aaboe [1968], where also (p. 4) the relation of $\Phi$ to G is discussed again as also in Aaboe [1971] pp. 4-5. In brief, G is generated by a linear zigzag function, $\hat{G}$, with the following parameters:

$$M = 5;4,57,2,13,20^{H/m}$$
$$m = 2;4,59,45,11,6,40^{H/m}$$
$$\mu = 3;34,58,23,42,13,20^{H/m}$$
$$d = 0;25,48,38,31,6,40^{H/m}$$
$$\Delta = 2;59,57,17,2,13,20^{H/m}$$
$$P = \frac{1,44,7}{7,28} = 13;56,39,6,...^{H}$$
$$\Pi = 1,44,7 = 6247 \text{ synodic months.}$$

Then, when $\hat{G}$ lies between 2;53,20 and 4;46,42,57,46,40, $\hat{G} = G$; when $\hat{G}$ lies outside of these bounds, one uses a table (given, e.g., in Neugebauer [1955] vol. 1, p. 60, and Aaboe [1968] pp. 4-5, [1969] pp. 4-5, and [1971] pp. 8-9) that relates specific values of Φ as it is increasing or decreasing to specific values of G for cases where $\hat{G}$ lies outside the given bounds. The maximum of G is 4;56,35,33,20, which occurs when the increasing branch of Φ has the values 1;58,54,48,53,20 or 1;59,12,35,33,20, and the minimum of G is 2;40 when the decreasing branch of Φ has the value 2;13,20. In the truncated version of Φ the maximum is $2;13,20^{H}$, the minimum $1;58,31,6,40^{H}$. The maximum of G is offset from the truncated minimum of Φ.

Neugebauer [1975] vol. 1, pp. 506-511, describes how column G was computed from two modified columns of truncated Φ, R and S, such that the difference of G, line by line, equals the difference between R and S. With the parameters $\varphi = 0;0,17,46,40^{H}$ and $\varepsilon = \frac{3}{28}$ synodic months, where φ is the difference between the values for Φ separated by a "Saros" of 18 years or 223 synodic months and ε is the fraction of a synodic month in which Φ increases or decreases by φ—i.e., $\varphi = \varepsilon d = \frac{3}{28} \times 0;2,45,55,33,20 = 0;0,17,46,40$ (see II C 1.1a)—we have for the zigzag function behind columns R and S:

$$M = 7,30\varphi$$
$$m = 6,40\varphi$$
$$\mu = 7,5\varphi$$
$$d = \frac{\varphi}{\varepsilon}.$$

The result is that the difference between two values of G separated by 223 months is exactly equal to the difference between two successive values of Φ. This reflects the definitions of G and Φ:

$$G = 1 \text{ synodic month} - 29^{d}$$
$$\Phi = 223 \text{ synodic months} - 6585^{d}.$$

A Procedure Text, CBS 1493 (ACT 208), gives rules for computing G from F, a function that we have already seen to correlate with Φ; see Neugebauer [1955], vol. 1, pp. 59 and 61.

Finally, it had been shown in Aaboe [1969] that Λ can be derived from Φ, and in Aaboe [1971] that W can also be derived from Φ. There is a Procedure Text relating Φ to Λ, ACT 207d, sections 2 to 6 and 8. The relation of Φ to W is explained in Aaboe-Hamilton [1979] pp. 6-18, who refer to BM 34497 (ACT 1005) and, for Ephemerides including columns of W and V (where $148^{d} + V^{H}$ is the length of the preceding five synodic months assuming constant solar velocity), BM 45688 (ACT 60) column VIII, BM 77238 (ACT 61c) column I, and BM 46015 (ACT 55) column II.

We now see that the Babylonians operated with four basic time units in their lunar theory; each is the sum of a constant measured in integer days, a correction due to lunar anomaly, and a correction due to solar anomaly.

$$1 \text{ synodic month} = 29^d + G + J$$
$$6 \text{ synodic months} = 177^d + W + Z$$
$$12 \text{ synodic months} = 354^d + \Lambda + Y$$
$$223 \text{ synodic months} = 6585^d + \Phi$$

In the last case Φ represents the effect of lunar anomaly, and there is no extant correction for the solar anomaly. Aaboe [1969] pp. 11-15 hypothesized that there should have been a correction, S, to fill this gap. Brack-Bernsen [1980] computed the actual values of the corrections, showed that the magnitudes of the Babylonian functions (using truncated Φ instead of normal Φ) are about right, but that the effect of the hypothetical S on the length of the "Saros" is greater than that of truncated Φ.

It should be noted that two tablets from Babylon give the entries for Φ and other columns of System A for sequences of tithis in -133. BM 34606 (ACT 80) contains columns T*, Φ*, B*, and C* (the asterisk indicates that the time interval is a tithi), and BM 34803+34815 (ACT 81) columns T*, Φ*, B*, C*, E*, and F*.

Column B gives the longitudes of the Moon at conjunction and opposition on the assumption of constant velocity of the Moon, variable velocity of the Sun. The variation in solar velocity is generated by a step-function in which the jumping points are Virgo 13° (up) and Pisces 27° (down) for $B_1$, Pisces 13° (up) and Virgo 27° (down) for $B_2$. Using $B_1$, which relates directly to the Sun, we find the following parameters:

$$a_1 = 194° \qquad w_1 = 30^{°/m}$$
$$a_2 = 166° \qquad w_2 = 28;7,30°$$
$$\frac{w_1}{w_2} = \frac{16}{15}$$
$$P = \frac{194°}{30^{°/m}} + \frac{166°}{28;7,30^{°/m}} = 12;22,8^m = 1 \text{ solar year}$$
$$\Pi = 46,23 = 2783^m = 225^y.$$

The midpoint of the slow arc is Gemini 20°, the longitude of the solar apogee in the *Pauliśasiddhānta* of the *Pañcasiddhāntikā* and in the ardharātrikapakṣa of Āryabhaṭa.

If $s_1$ is an interval containing a jumping point (up) and $\sigma_1$ one containing a jumping point (down) while s and σ are the respective distances from the beginning of the interval to the jumping point, then $s_1 = 30 - \frac{s}{15}$ and $\sigma_1 = 28;7,30 + \frac{\sigma}{16}$. These rules are given in section 3 of ACT 200 in the form:

$$s_1 = s + 1;4\,(28;7,30 - s)$$
$$\sigma_1 = \sigma + 0;56,15\,(30 - \sigma).$$

Similar rules are employed when a synodic arc or time crosses over a zone-boundary in System A Ephemerides of the planets. Bernsen [1969], beginning with Aaboe's hypothesis that the division of the arcs on the ecliptic for step-functions pertaining to the Greek-letter Phenomena of the planets could be determined by examining the distribution of the longitudes of the occurrences of those phenomena (Aaboe [1965]), showed that indeed the distribution of the longitudes of conjunctions does depend almost exclusively on the longitude of the Sun (and very little on the velocity of the Moon), and that this distribution fits quite nicely the distribution resulting from column B. Goldstein [1980] made the following conjecture concerning how it was derived, without referring to Bernsen's demonstration (which, of course, does not prove that the Babylonians used any particular method of achieving a distribution consonant with reality). If the Babylonian astronomers began with the hypothesis that $\alpha_1 + \alpha_2 = 180°$, that $w_1 = 30^{\circ/m}$, and that $P \approx 12;22$ (for this period see BM 36712 discussed above), under these conditions, $w_2 = 28;16,...$, an irregular number and therefore not very useful in a step function. As Aaboe [1965] pointed out, the values of $w_1$ and $w_2$ in System A schemes are chosen to be regular numbers whose ratio to each other is that of two integers whose difference is one. The two regular numbers nearest to 28;16,... and above and below it are 28;26,40 and 28;7,30. The first number's ratio to 30 is $\frac{17,4}{18,0}$; Goldstein conjectures that this was rounded to produce the $\frac{17}{18}$ ratio attested in BM 36822+37022 where $w_1 = 28;20$ and $w_2 = 30$. The second number produces the very satisfactory ratio $\frac{15}{16}$. It remains, however, to correct the value of P since it was already known that a year contains more than 12;22 months—indeed, about 12;22,6,20 months. To achieve this correction it was only necessary to change the values of $\alpha_1$ and $\alpha_2$ from 180° to 194° and 166° respectively. The locations of the jumping points were chosen so as to reproduce the distribution of the longitudes of conjunctions.

Maeyama [1978] pp. 26-31 and [1981] pp. 263-267 and 326-332 attempts to derive the knowledge of the solar anomaly and the specific parameters of column B from observations of the variation in length of the "Saros". While this may quite well be the correct explanation of the discovery of the anomaly, Maeyama's derivation of the actual parameters is too complex to be plausible as an explanation of how the Babylonians proceeded. See also Brack-Bernsen [1997] pp. 41-46.

Column C gives the lengths of daylight for each syzygy. It is computed from the ratio 3 : 2 (i.e., $3,36^{\text{uš}} : 2,24^{\text{uš}}$) of the longest to the shortest daylight that we met with already in MUL.APIN. The Vernal Equinox occurs when the Sun's longitude is 10° of Hired Man—that is, of the first zodiacal sign. Where this point was positioned with respect to the fixed stars we do not know. The actual values of the lengths of daylight are found by adding or subtracting a given quantity for each degree in each 30° arc measured from the solstices; these given quantities are in the ratio to each other of 1 : 3 : 5, as are the differences between the UŠ in the lengths of daylight.

| | | | | | |
|---|---|---|---|---|---|
| ♑10° | 2,24uš | +0;8uš | ♋10° | 3,36uš | -0;8uš |
| ♒10° | 2,28uš | +0;24uš | ♌10° | 3,32uš | -0;24uš |
| ♓10° | 2,40uš | +0;40uš | ♍10° | 3,20uš | -0;40uš |
| ♈10° | 3,0uš | +0;40uš | ♎10° | 3,0uš | -0;40uš |
| ♉10° | 3,20uš | +0;24uš | ♏10° | 2,40uš | -0;24uš |
| ♊10° | 3,32uš | +0;8uš | ♐10° | 2,28uš | -0;8uš |

We have already noted the wide-spread use of this scheme and its relatives in Greek astrological and astronomical texts. The ἀναφοραί of each 30° arc on the ecliptic beginning with the Vernal Equinox are, according to this System A:

| | | |
|---|---|---|
| 1 | 20° | 12 |
| 2 | 24° | 11 |
| 3 | 28° | 10 |
| 4 | 32° | 9 |
| 5 | 36° | 8 |
| 6 | 40° | 7 |

This is true because

| | |
|---|---|
| 28°+24°+20°+20°+24°+28° = 144° | 10 |
| 24°+20°+20°+24°+28°+32° = 148° | 11,9 |
| 20°+20°+24°+28°+32°+36° = 160° | 12,8 |
| 20°+24°+28°+32°+36°+40° = 180° | 1,7 |
| 24°+28°+32°+36°+40°+40° = 200° | 2,6 |
| 28°+32°+36°+40°+40°+36° = 212° | 3,5 |
| 32°+36°+40°+40°+36°+32° = 216° | 4 |

Column C′, which provides a correction to the time of the syzygy due to the variation in the length of daylight, is derived directly from column C. If the entry in C for the syzygy of interest is C(n+1) and the preceding entry is C(n), then

$$C'(n+1) = \tfrac{1}{2}(C(n) - C(n+1)).$$

This is the change for the evening; the other half belongs to the morning.

Column E gives the latitude of the Moon's center at the time of the syzygy measured in še or barleycorns, where

72 še = 1° so that 1 še = 0;0,50°.

The maximum latitude is 7,12 še = 6°, though this is not intended as a real latitude, but is only a consequence of the method of computation. Nor does it affect any part of the Babylonian lunar theory, which is directed to the prediction of eclipses and of the appearances of the New Moon.

The method is to construct a linear zigzag function with the correct period, and then to modify this function in the central strip (the nodal zone, which extends 2,24 še = 2° North(+) and South(–) of the ecliptic) so that it increases or decreases at the appropriate rate. One can imagine an unmodified linear zigzag function for $E_c$ in which:

M = +6,0 še = +5°
m = –6,0 še = –5°
d = ±4 še for every degree of lunar elongation from its node.

Since there are two values of w in the step-function of column B, which generates the longitudes of the Moon—$w_1$ = 30° and $w_2$ = 28;7,30°—and the elongation of the Moon from its node, ω, equals the longitude of the Moon diminished by the longitude of the node, where the node moves -1;33,55,30° in a mean synodic month, there are two values of d:

$d_1$ in $\alpha_1$ = ±4 še × ω = ±4 še × 31;33,55,30° = ±2,6;15,42 še = 1;45,12,30°
$d_2$ in $\alpha_2$ = ±4 še × ω = ±4 še × 29;41,25,30° = ±1,58;45,42 še = 1;38,57,30°.

The jumping points are, of course, the same as for column B. The rules for computing the d when a jumping point is crossed are analogous to those used in column B.

The modification that changes $E_c$ into E is simply that, once the zigzag function reaches the nodal zone, the value of d doubles from ±4 še for a degree of elongation to ±8 še for a degree of elongation. Since the nodal zone is 4° wide within which the rate of increase of decrease is twice that of the zigzag function outside of the nodal zone, the period is preserved by increasing the maximum latitudes by +1° to +6° and by -1° to -6°. The rules for crossing the nodal zone are rather complicated; see Neugebauer [1955] vol. 1, pp. 49-50.

Aaboe-Henderson [1975] have computed that the node makes 1,2,37 = 3757 retrograde rotations in 4,0,0,0 = 864,000 synodic months, while the period of E = 15,0,0,0,0 years = 3,5,32,0,0,0 synodic months = 3,20,32,0,0,0 sidereal months = 3,21,20,24,22,11 draconitic months.

Two tablets from Babylon, BM 36636 (ACT 90) and BM 32351 (ACT 91), provide approximations to E based on an unmodified linear zigzag function in which:

M = +7,12
m = –7,12
Δ = 14,24
d = 2,27;15
P = 11;44,6,31,...
Π = 1,55,12.

The computation of column E is discussed in section 6 of ACT 200 (see Aaboe-Henderson [1975] pp. 208-211), section 4 of BM 33631 (ACT 200b), BM 34245+41608 (ACT 200c), section 3 of BM 34148 (ACT 200d), Rm 839 (ACT 200e), section 1 of BM 55530 (ACT 200i), and section 1 of BM 34079+35132+35324 (ACT 204).

Column $\Psi$ defines eclipse magnitudes measured in fingers where

$$1 \text{ finger} = 0;5° = 6 \text{ še}.$$

$\Psi$, then, is a modification of $\frac{E}{6}$—i.e., of the lunar latitude measured in fingers rather than barleycorns. The modification is:

$$\Psi = c \pm \frac{E}{6}$$

where c = 17;24 fingers and $\frac{E}{6}$ is added when the Moon is at its ascending node (i.e., passes from negative to positive values) and subtracted when the Moon is at its descending node. From the formula for $\Psi$ it is clear that 17;24 fingers is the greatest magnitude of an eclipse—the one that occurs when E = 0 (the Moon is on the ecliptic). Then, if E = 17;24 × 6 = 1,44;24 še, the limit of $\Psi$ of an actual eclipse is given by the formulae:

$$\Psi_M = c + \frac{1{,}44;24}{6} \text{ še} = 2c = +34;48 \text{ fingers}$$

But $\Psi_M$ lies outside of the nodal zone, whose bounds are ±2° = ±2,24 še. $\Psi$ is positive within the nodal zone, increasing from 0 to 17;24 as E increases from -17;24 fingers = 1;27° to 0°, and increasing from 17;24 to 41;24 as E increases from 0° to +2°. But between E = -2° and -1;27° at the ascending node and between E = +2° and +1;27° at the descending node, $\Psi$ is negative.

$$\left.\begin{array}{ll} \Psi = 17;24 + (-17;24) = & 0 \text{ fingers} \\ \Psi = 17;24 + (-24) = & -6;36 \text{ fingers} \end{array}\right\} \text{ascending node}$$

$$\left.\begin{array}{ll} \Psi = 17;24 - 24 = & -6;36 \text{ fingers} \\ \Psi = 17;24 - 17;24 = & 0 \text{ fingers} \end{array}\right\} \text{descending node}$$

Of course, both the positive values of $\Psi$ higher than +38;34 and all the negative values of $\Psi$ do not relate to possible eclipses; such values, in fact, show that no eclipse will take place. To deal with this, Neugebauer [1955] vol. 1, p. 57 imagined a function $\Psi^*$ such that $\Psi^* = \Psi$ when $\Psi$ is less than 17;24 fingers and $\Psi^*$ = 34;38 – $\Psi$ when $\Psi$ is greater than 17;24 fingers.

$\Psi$ was normally computed only for syzygies that were believed to be eclipse possibilities, separated by 5 or 6 months. Many Ephemerides contain instead of column $\Psi$ a column dubbed by Neugebauer $\Psi'$. This is a linear zigzag function providing values of $\Psi$ for all syzygies. For small values of E —i.e., less than 2,24 še = 2°—$\Psi' = \Psi$ at the ascending node and $\Psi' = -\Psi$ at the descending node. If E is greater than +2,24 še,

$$\Psi' = \frac{2E \mp 2{,}24}{6} \pm 17;24$$

at the ascending and descending node respectively. This function is derived from that for E, and therefore has changes in d at the jumping points of column B and is twice as rapidly changing ($\frac{E}{6}$) in the nodal zone as outside of it ($2 \cdot \frac{E}{6}$).

$M = +2,0$ fingers
$m = -2,0$ fingers
$d_1 = 42;5,14$
$d_2 = 39;35,14$

The values of $d_1$ and $d_2$ are a third of the values of $d_1$ and $d_2$ in the linear zigzag function for $E_c$.

The Procedure Texts that deal with Ψ and Ψ′ are all fairly obscure; they are section 4 of ACT 200 and sections 1 to 4 of ACT 204.

The excellence of this Babylonian theory of eclipses (columns B, E, and Ψ) is demonstrated in Aaboe-Henderson [1975]. They also published (pp. 220-221) a small fragment—BM 40754—that contains columns $C_1$, $E_1$, and $\Psi'_1$ for -131; it is parallel to BM 46042 (ACT 7a) which contains columns $T_1$, $\Phi_1$, $B_1$, $C_1$, $G_1$, $C'_1$, $J_1$, and $K_1$ for -131 to -129.

The relationship of column F, which gives the daily velocity of the Moon, to column Φ has already been commented on; they have the same periods and reach their maxima and minima at the same times. The parameters of F are:

$M = 15;56,54,22,30^{\circ/d}$
$m = 11;4,4,41,15^{\circ/d}$
$\mu = 13;30,29,31,52,30^{\circ/d}$
$d = 0;42^{\circ/d}$
$\Delta = 4;52,49,41,15^{\circ/d}$
$P = 13;56,39,6,...$ synodic months
$\Pi = 6247$ synodic months.

These parameters are used in MM 86.11.405 + VAT 209 (ACT 18) and are given in several Procedure Texts: in section 5 of ACT 200, in section 5 of ACT 204, and in CBS 1493 (ACT 208).

A variant version, with the same periods as above, was used in computing column F in BM 34581+34610 (ACT 92). This is an unusual, undated Ephemeris from Babylon with its columns in the order: T, Φ, F, B, D, E, G. The parameters are:

$M = 15;52,43,7,30^{\circ/d}$
$m = 10;59,53,26,15^{\circ/d}$
$\mu = 13;26,18,16,52,30^{\circ/d}$
$d = 0;42^{\circ/d}$
$\Delta = 4;52,49,41,15^{\circ}$

So M and m in this system equal M – 0;4,11,15 and m – 0;4,11,15 in the previous system; therefore, the difference between the two μ's is also 0;4,11,15. The maxima and minima occur about half an interval (i.e., half a synodic month) later than do the maxima and minima of Φ.

But most Ephemerides use an abbreviated linear zigzag function of F. Its parameters are:

$M = 15;57^{\circ/d}$
$m = 11;4^{\circ/d}$
$\mu = 13;30,30^{\circ/d}$
$d = 0;42^{\circ/d}$
$\Delta = 4;53^{\circ}$
$P = 13;57,8,\ldots$ synodic months
$\Pi = 293$ synodic months.

Clearly, periodic adjustments are necessary for those who use this system; that such adjustments took place is known from the fact that, uniquely among all the columns in System B, the entries in column F do not form a single unbroken sequence over the centuries of its use.

One Ephemeris to which we have already referred, ACT 81, has a column F* which gives the velocities of the Moon in degrees per day for successive tithis (sic!). Column F was presumably tabulated from such a column F*; see Neugebauer [1975] vol. 1, pp. 500-501. Its parameters, then, are the same as for F except that:

$$d^* = \frac{2\Delta + d}{30} = \frac{10;27,39,22,30}{30} = 0;20,55,18,45.$$

We have already discussed above (pp. 226-227) the relation of column G to column Φ and the derivation of G from two columns of Φ. It should suffice here to remind the reader that column G represents the excess measured in large hours of a synodic month over 29 days on the assumption that the solar velocity is a constant $30^{\circ/m}$, and to list its parameters:

$M = 5;4,57,2,13,20^{H}$
$m = 2;4,59,45,11,6,40^{H}$
$\mu = 3;34,58,23,42,13,20^{H}$
$d = 0;25,48,38,31,6,40^{H}$
$\Delta = 2;59,57,17,2,13,20^{H}$
$P = \frac{2\Delta}{d} = \frac{1,44,7}{7,28} = 13;56,39,6,\ldots$
$\Pi = 1,44,7 = 6247$ synodic months.

Since column G is computed on the assumption that the Sun has a constant velocity, $30^{\circ/m}$, a correction is needed for the months in which the solar velocity is $28;7,30^{\circ/m}$. That correction is $-0;57,3,45^{H}$ in months during which the Sun is always in its “slow arc”, zero in months during which it is always in its “fast arc”, and a negative number between these extremes when it straddles a jumping point.

We have also previously (p. 230) discussed column C′ which corrects the time of sunset by half the difference between the lengths of daylight in two successive months as recorded in column C. Then column K gives the algebraic sum of the entries in columns G, J, and C′, which combines the length of the synodic month beyond 29 days with the correction for the increase or decrease in the length of daylight. Then column M gives at conjunctions the date in month, day, and large hours *before* sunset, and at oppositions the date in month, day, and large hours *after* sunset—, i.e.,

$$M_1(n+1) = M_1(n) - K_1(n+1)$$
$$M_2(n+1) = M_2(n) + K_2(n+1)$$

Clearly, if the addition of K leads to a number greater than $6^H$, a day must be added; if the subtraction of K leads to a number less than 0, a day must be subtracted. The final date of the beginning of the month must be judged by taking into consideration column $P_1$, for which are computed the times between sunset and moonset on the evening following conjunction, or on the evening after that if $M_1$ is small. There were at least two tables intermediate between $M_1$ and $P_1$ accounting for the elongation of the Moon from the Sun at sunset of the day on which first visibility of the lunar crescent is expected and for the effect of the lunar latitude. $P_2$ represents the time between sunrise and opposition and between sunset and opposition, and $P_3$ the time at last visibility, between moonrise and sunrise. Some information concerning the computation of columns $P_1$, $P_2$, and $P_3$ is given in the Procedure Texts (sections 15 and 16 of ACT 200; BM 35399 (ACT 201); BM 35076 (ACT 201a); and BM 35125 (ACT 201aa)).

In brief summary, then, through columns Φ, B, C, E, and Ψ of System A one can predict lunar and solar eclipse possibilities, and through these columns plus F, G, J, C′, K, M, and P the evening on which the lunar crescent will first be visible. The whole is based on internally consistent and accurate period relations that lead to a set of continuous functions that extend without variation over the centuries of their use.

*4. 2a.2. System B of the Moon*

Whereas System A is a tightly-knit whole in which many columns are directly related to each other through sharing the same periods and in which each individual occurrence of a temporally limited column on one tablet can be extended forwards and backwards to link with all other occurrences of that column, so that one may imagine a super Ephemeris extending over many centuries of which each of the extant Ephemerides represents a horizontal slice, the Ephemerides of System B exhibit no such coherence. One characteristic that distinguishes a System B text, then, is its incompatibility not only with a System A text, but frequently also with other System B texts. Other distinguishing features are some characteristic parameters, the use of a linear zigzag function for the velocity of the Sun, presented in column A, instead of the step-function that is embedded in column B of System A, and the absence of some of System A's columns together with the presence of Columns not found in System A. To demonstrate this last point the full array of a complete System B lunar Ephemeris is given below.

T. Column of dates as in System A.
(Column Φ of System A is omitted).
A. Velocity of the Sun. Omitted in System A.
B. Longitude of the Moon as in System A.
C. Length of daylight as in System A.
D. Half length of the night. Omitted in System A.
E. Latitude of the Moon as in System A; no examples extant.
Ψ. Eclipse magnitude as in System A.
F. Lunar velocity as in System A.
G. Length of the month assuming constant solar motion as in System A.
H. Differences of column J. Omitted in System A.
J. Corrections to column G due to varying solar motion as in System A.
K. Length of the month as in System A.
L. Date of syzygy using midnight epoch. Omitted in System A.
M. Date of syzygy using evening or morning epoch; attested corresponding column in System A uses evening epoch.
N. Time between syzygy and sunset or sunrise; not attested for System A.
O. Elongation of Moon from Sun at first or last visibility; not attested for System A.
Q. Influence of the obliquity of the ecliptic on first or last visibility; not attested for System A.
R. Influence of the lunar latitude on first or last visibility; not attested for System A.
P. Duration of first or last visibility as in System A.

We have noted above that the earliest tablet containing System B material from Babylon is dated -257/6, from Uruk -207/6, while the earliest Ephemerides are dated -169/8 and -205/4 respectively. System B, then, certainly antedates Hipparchus, and its influence on Hipparchus, first recognized by Kugler [1900a] pp. 24-25, 40, 46, 50-53, and 114, is certain.

For column A, which gives the velocity of the Sun in degrees per synodic month, the unabbreviated parameters of the linear zigzag function are:

$M = 30;1,59°$
$m = 28;10,39,40°$
$\mu = 29;16,19,20°$
$\Delta = 1;51,19,20°$
$d = 0;18°$
$P = 12;22,8,53,20$ synodic months (an "anomalistic" year)
$\Pi = 2,46,59 = 10,019$ synodic months.

The derivation of such linear zigzag functions from empirical data is explained in Neugebauer [1968].

The abbreviated parameters are:

$M = 30;2°$
$m = 28;10,40°$
$\mu = 29;6,20°$
$\Delta = 1;51,20°$
$d = 0;18°$
$P = 12;22,13,20$ synodic months
$\Pi = 5,34 = 334$ synodic months.

ACT 185 for SE 124 = -187/6 (from Uruk), 186 (from Uruk), and 187 (from Uruk) are all tables of daily solar motion. The daily velocity is assumed to be a constant 0;59,9°. A fragmentary tablet, BM 37089, published by Aaboe [1964] may also be a table of daily positions of the Sun, based on a linear zigzag function of velocities whose $d = 0;0,1,43,42,13,20^{\circ/d}$; the maximum and minimum are not preserved. Neugebauer [1955] p. 194 was able to reconstruct from the single surviving line of section 8 of the Procedure Text ACT 200 a linear zigzag function of solar velocities in the twelve zodiacal signs. The parameters are:

$M = 1;2,44^{\circ/d}$ in Sagittarius
$m = 0;55,32^{\circ/d}$ in Gemini
$\mu = 0;59,8^{\circ/d}$ in Virgo and Pisces
$\Delta = 0;7,12°$
$d = 0;1,12°$
$P = 12$ synodic months.

Column B gives the longitudes of the Moon at the syzygies simply as the sum of the entries in Column A, though $B_2$ is set off by 180° from the computed values of the accumulated solar motion at the middle of each synodic month.

Column C gives the length of daylight based on a scheme somewhat different from that used in System A. Here, the Vernal Equinox occurs when the Sun's longitude is 8° of Hired Man. The values of the lengths of daylights are found by adding or subtracting a given quantity for each degree in each 30° measured from the solstices; these given quantities are in the ratio to each other of 1 : 2 : 3, and the ratio of the longest to the shortest day is 3 : 2 or 3,36$^{uš}$ : 2,24$^{uš}$.

| | | | | | |
|---|---|---|---|---|---|
| ♑ 8° | 2,24$^{uš}$ | +0;12$^{uš}$ | ♋ 8° | 3,36$^{uš}$ | −0;12$^{uš}$ |
| ♒ 8° | 2,30$^{uš}$ | +0;24$^{uš}$ | ♌ 8° | 3,30$^{uš}$ | −0;24$^{uš}$ |
| ♓ 8° | 2,42$^{uš}$ | +0;36$^{uš}$ | ♍ 8° | 3,18$^{uš}$ | −0;36$^{uš}$ |
| ♈ 8° | 3,0$^{uš}$ | +0;36$^{uš}$ | ♎ 8° | 3,0$^{uš}$ | −0;36$^{uš}$ |
| ♉ 8° | 3,18$^{uš}$ | +0;24$^{uš}$ | ♏ 8° | 2,42$^{uš}$ | −0;24$^{uš}$ |
| ♊ 8° | 3,30$^{uš}$ | +0;12$^{uš}$ | ♐ 8° | 2,30$^{uš}$ | −0;12$^{uš}$ |

Then the ἀναφοραί of each 30° arc on the ecliptic are, according to System B:

| | | |
|---|---|---|
| 1 | 21° | 12 |
| 2 | 24° | 11 |
| 3 | 27° | 10 |
| 4 | 33° | 9 |
| 5 | 36° | 8 |
| 6 | 39° | 7 |

This scheme, with 2d substituted for d between Cancer and Leo and between Sagittarius and Capricorn, was also well known to the Greeks.

Column D gives half the length of the night, which is simply 6,0$^{uš}$ diminished by the entry in column C and divided by 2. A column D′ is also attested, which is simply 6,0$^{uš}$ diminished by the entry in column C.

However, no extant Ephemeris of System B contains a column E, though one must have existed since there are examples of columns Ψ, Ψ′, and Ψ′′, where, as in System A, $\Psi = c \pm \frac{E}{6}$. Ψ′ is also dependent on E. In all cases, of course, since E is the latitude of the Moon at a syzygy, and the longitude of the Moon (and therefore its elongation from its node) depends on the variable solar velocity, in System B column Ψ′ is related to the linear zigzag function of column A; see Neugebauer [1975] vol. 1, p. 524.

Column Ψ′′ is another linear zigzag function with a discontinuity after the function passes 0, from positive to negative or from negative to positive. The parameters are:

M = 9;52,15°
m = –9;52,15°
μ = 0
Δ = 19;44,30°
d = 3;52,30° per synodic month
c = 1;30°

$$P = \frac{4(M+c)}{4(M+c)+d} = \frac{1,30,58}{1,38,43} = \frac{5458 \text{ syn. mos.}}{5923 \text{ drac. mos.}}$$

The period relation 5458 synodic months = 5923 draconitic months was known to Hipparchus.

The discontinuity is computed by replacing d with d–2c = 3;52,30°–3° = 0;52,30° when the function has crossed 0 and would be more than 2c if d were added to it.

The maximum eclipse occurs when E = 0 so that Ψ = c = 1;30° = 18 fingers (in System A c = 17;24 fingers). In BM 45694 (ACT 123) and VAT 1770 (ACT 125) Ψ′′ is measured in units of eclipse-magnitudes, so that the ratio of the parameters is 1 : 1;30°—i.e., the parameters in degrees are multiplied by 0;40 to give parameters in eclipse-magnitudes (e.g., M = 9;52,15° × 0;40 = 6;34,50 eclipse magnitudes).

There also exist examples of columns of the differences of column Ψ′; these differences form a linear zigzag function whose unabbreviated parameters are:

$M = 48;13,4,26,40^f$
$m = 44;46,55,33,20^f$
$\mu = 46;30^f$
$\Delta = 3;26,8,53,20^f$
$d = 0;33,20^f$
P = 12;22,8 synodic months (one year)

The abbreviated parameters are:

$M = 48;13,4,30^f$
$m = 44;46,55,30^f$
$\mu = 46;30^f$
$\Delta = 3;26,9^f$
$d = 0;33,20^f$
P = 12;22,8,24 synodic months.

From the column of the differences of Ψ′ are directly computed the values entered into column Ψ′ with the following restrictions:

$M = 1,58;27^f$
$m = -1,58;27^f$
$c = 18^f$ (greatest possible eclipse).

Then, when the function crosses 0 and the next entry would be greater than $2c = 36^f$, 2c is subtracted from the difference before it is added to or subtracted from the preceding entry.

Column F represents the daily lunar velocity at the syzygy in the form of a linear zigzag function whose parameters are:

$M = 15;16,5°$
$m = 11;5,5°$
$\mu = 13;10,35°$
$\Delta = 4;11°$
$d = 0;36°$
$P = \frac{4,11}{18} = 13;56,40$
$p = \frac{2\Delta}{2\Delta+d} = \frac{4,11}{4,29} = \frac{251}{269}$

The period relation, 251 synodic months = 269 anomalistic months, is attributed to Kidenas by an anonymous early third century commentator on Ptolemy (Jones [1990] p. 21), and was known to Hipparchus.

Again, there is evidence for abbreviated parameters:

$M = 15;16°$
$m = 11;5°$
$\mu = 13;10,30°$
$\Delta = 4;11°$
$d = 0;36°$

The relationship of column F to computed modern values of lunar velocity was exhaustively investigated by Riley [1994]. He found quite good agreement, though the minimum daily velocity is too low. He explained this by noting that, in order to keep the function in phase with the anomalistic month while avoiding any change in the slope of the linear zigzag function (such as the leveling off that occurs in the truncated form of column Φ), it was necessary to extend the function below "real" values for the velocity.

A group of auxiliary tablets—MLC 1880 (ACT 190), U 96 (ACT 191), A 3408 (ACT 192), U 122+142 (ACT 193), AO 6492 (ACT 194), BM 45818+45838+46192 (ACT 194a), BM 34623 (ACT 194b), and Rm 777 (ACT 196)—give daily lunar velocities for sequences of days (F*) or daily longitudes for such sequences (ΣF*) or both. F* forms a linear zigzag function with parameters:

$$M = 15;14,35^\circ$$
$$m = 11;6,35^\circ$$
$$\mu = 13;10,35^\circ$$
$$\Delta = 4;8^\circ$$
$$d = 0;18^\circ$$
$$p = \frac{248 \text{ days}}{9 \text{ anom. mos.}} = 27;33,20^{d/m}.$$

The period relation 248 days = 9 anomalistic months had a widespread influence on the astronomies of Eurasia, from Greece and Western Europe to India and Central Asia; see Jones [1983].

As well as tables of the daily velocities of the Moon, there exists a column F′ giving the velocity of the Moon for a large hour; the entries are, of course, derived from column $F_1$ (giving the velocity of the Moon on the day of conjunction) by division by 6. The result is a linear zigzag function with the following parameters:

$$M = 2;32,40,50^\circ$$
$$m = 1;50,50,50^\circ$$
$$\mu = 2;11,45,50^\circ$$
$$\Delta = 0;41,50^\circ$$
$$d = 0;6^\circ$$
$$P = 13;56,40 \text{ synodic months}$$
$$\Pi = 251 \text{ synodic months.}$$

Column G in System B, like its parallel in System A, gives the excess in large hours of the synodic month over 29 days on the assumption that solar velocity is constant. It again forms a linear zigzag function with the same period as column F. The parameters are:

$M = 4;29,27,5^H$
$m = 1;52,34,35^H$
$\mu = 3;11,0,50^H$ (= $0;31,50,8,20^d$)
$\Delta = 2;36,52,30^H$
$d = 0;22,30^H$
$P = \frac{4,11}{18} = 13;56,40$

An average synodic month, then, is $29;31,50,8,20^d$ long. This length of a synodic month was also known to Hipparchus. A papyrus fragment of a Greek adaptation of column G was published by Neugebauer [1988].

In order to account for the varying velocity of the Sun, the Babylonians constructed a column J that corrects column G; in order to compute column J one uses column H, which gives the differences of column J as a linear zigzag function with the following parameters:

$M = 0;21^H$
$m = 0^H$
$\mu = 0;10,30^H$
$\Delta = 0;21^H$
$d = 0;6,47,30^H$
$P = \frac{16,48}{2,43} = 6;11,2,34,36,4,...$

As in the case of the derivation of $\Psi'$ from its differences, the entries in column J are directly derived from those in column H with the proviso that one remain within specified limits:

$M = 0;32,28,6^H$
$m = -0;32,28,6^H$
$\Delta = 1;4,56,12^H$
$d = 0;10,30^H$ (this is $\mu$ of column H)
$P = 12;22,8$ synodic months.

Examples are also found in which abbreviated parameters are used:

$M = 0;32,28^H$
$m = -0;32,28^H$
$\Delta = 1;4,56^H$
$d = 0;10,30^H$
$P = 12;22,5,42,51,15,...$ synodic months.

Column K contains the sum of the entries in columns G and J, so that the synodic month = 29 days + K large hours.

Column K is used as a column of differences for column L, which records the time interval between the preceding midnight and the time of the syzygy, measured in large hours. When L is greater than $6^H$, $6^H$ is subtracted from it and 1 day is added to the date. In many Ephemerides the name of the month (identical to that in column T) is noted in column L.

Column M gives the time of the syzygy before or after sunrise or sunset; it is computed from columns D and L, where D is $\frac{6^H - C}{2}$ (the time from nidnight to the preceding sunset or the succeeding sunrise, or, when increased by C, the time from midnight to the preceding sunrise or the succeeding sunset).

Columns N to P lead one from the time of the syzygy to the time-intervals for the "Lunar Six", given in $P_1$, $P_2$, and $P_3$:

I From sunset to moonset in the evening before conjunction: $P_1$
II From moonset to sunrise in the morning before opposition: $P_2$
III From sunrise to moonset in the morning after opposition: $P_2$
IV From moonrise to sunset in the evening before opposition: $P_2$
V From sunset to moonrise in the evening after opposition: $P_2$
VI From moonrise to sunrise in the morning before conjunction: $P_3$

First, column $N_1$ gives the time between the time of the conjunction (in column L) and the sunset of the evening on which the first visibility is expected, where $N_1 = C + D - L$; but, if $N_1$ is less than 5 large hours, one must add a full day ($6^H$) or, sometimes, 2 full days ($12^H$), with the proviso that $N_1$ should be less than $14^H$. Column $N_2$ gives the time from opposition during daylight to the succeeding sunrise or from the opposition during night to the preceding sunrise. In the two cases, $N_2$ equals respectively $D - L + 6^H$ and $L - D$, to which, when appropriate, $6^H$ must be added sometimes.

Column O gives the elongation between the Moon and the Sun at the designated sunset or sunrise. This is computed by multiplying the time-interval of column N by the velocity of the Moon in a large hour (given in column F′) diminished by the velocity of the Sun in a large hour (0;10°); the result—$N \cdot (F' - 0;10°)$—is diminished by a small variable correction (between 0;30° and 3;40°). The purpose of this correction is uncertain; Schaumberger [1935] p. 388 suggested that it might be a correction for twilight, which is probably at least partially correct; Sidersky [1919] had previously suggested parallax, which is certainly not the sole element reflected in the correction, but may be part of it.

Column Q converts the elongation of column O into an oblique ascensional difference, and column R gives a correction for the effects on the "Lunar Six" of the Moon's latitude. Then the final result, P, for each of the "Lunar Six" is the sum of the appropriate entries in columns O, Q, and R.

### *4.2b. The Planets*

ACT vol. 2 contains a relatively small set of Ephemerides of the planets: for Mercury 10 (8 from Babylon, 2 from Uruk), for Venus 9 (7 from Babylon, 2 from Uruk), for Mars 8 (4 each from Babylon and Uruk), for Jupiter 40 (27 from Babylon, 13 from Uruk), and for Saturn 12 (4 from Babylon, 8 from Uruk). Their dates range from -307/6 (BM 36723+37234; ACT 300a; Mercury) to -8/7 (U 179; ACT 401; Venus), and they represent a wide variety of permutations of Systems A and B and, for Venus, purely *ad planetam* solutions. The largest block among the 79 planetary Ephemerides are the 40 pertaining to Jupiter—almost exactly half. These have been reduced to 39 by Huber [1957], who combined four tablets recognized in ACT as giving the day by day longitudes of Jupiter with four or five fragments of unidentified content to form three larger tablets of daily motion, all from Uruk. These are: U 152+176 (ACT 652) + U 149 (ACT 1015) + U 158 (ACT 1016) for -195/4; U 174 (ACT 650) + U 163 (ACT 651) + U 180(18) (ACT 653) for -194/3; and U 148 (ACT 1014) + perhaps U 180(29) (ACT 1021) + U 141 (ACT 1032) for -190/1; see Neugebauer [1975] vol. 1 pp. 413-418. The only other planetary Ephemerides which have come to light since the publication of ACT in 1955 are a tablet containing the last appearances of Mercury in the evening (Ω) for -423 to -401 computed according to System $A_3$ (it was published in Aaboe-Britton-Henderson-Neugebauer-Sachs [1991] pp. 34-43), a tablet listing successive first appearances of Mercury in the morning (Γ) computed according to System $A_1$ (it was published by Aaboe [1977]), and a tablet for Ω of Saturn from -311 to -290 (it was published by von Weiher [1993] p. 110, no. 170).

However, a large number of "templates" giving the longitudes, without dates, of the occurrences of the Greek-letter phenomena of a planet in one of its periods was published in Aaboe-Sachs [1966]. BM 36300+36753+37156 +37210+37336 gives Γ, Φ, Ψ, and Ω of Saturn according to System A; BM 36814 lists the same four phenomena for Saturn, also according to System A; BM 78080 gives the longitudes of an unidentified phenomenon of Saturn according to System A′; BM 36762+36779+36813+37070+37082+37125+ 37174 lists Γ, Φ, Θ, Ψ, and Ω of Jupiter according to System A; BM 32218 gives the longitudes of an unidentified phenomenon of Jupiter according to System A″; BM 36810+36947 and BM 36811, which are probably fragments of the same tablet, give the longitudes of Γ, Φ, Θ, Ψ, and Ω of Mars according to System A and, for the retrograde arc, Scheme S, as well as dates of the summer solstice; BM 36957 and BM 37236 are two separate tablets listing the longitudes of all five Greek-letter phenomena of Mars according to System A; and BM 37196 gives the longitudes of an unspecified phenomenon of Mars according to System A. Similar templates are found among the Greek papyri from Oxyrhynchus published by Jones [1999].

Volume 2 of ACT also contains 27 Procedure Texts—11 from Uruk and 16 from Babylon (there were no Procedure Texts for the Moon from Uruk). Ten of these are for Jupiter and one each for Jupiter with Saturn and Mars and with Venus; seven are for Mercury and two for Mercury with Saturn; four are for Mars; and one each for Saturn and Venus alone. In addition eleven Ephemerides have colophons that include Procedure Texts. An additional Procedure Text for Mars, BM 37024, from Babylon, was published by Aaboe [1987].

The basic structure of Babylonian planetary theory is the period relation that asserts that a particular Greek-letter phenomenon occurs $\Pi$ times in a period of Y years with the synodic arcs, $\Delta\lambda$, adding up to 360° Z times and the planet making R rotations. From this structure follow several important equations:

$P = \frac{\Pi}{Z}$ is the number of $\overline{\Delta\lambda}$'s that fit into 360°

$\overline{\Delta\lambda} = \frac{360°}{P} = \frac{360° \times Z}{\Pi}$

Further, it is demonstrable that

$Y = i\Pi + Z$

$R = k\Pi + Z$

| | | | |
|---|---|---|---|
| Then, | for Saturn | $i = 1, k = 0$ | $Y = \Pi + Z, R = Z$ |
| | for Jupiter | $i = 1, k = 0$ | $Y = \Pi + Z, R = Z$ |
| | for Mars | $i = 2, k = 1$ | $Y = 2\Pi + Z, R = \Pi + Z$ |
| | for Venus | $i = 1, k = 1$ | $Y = R = \Pi + Z$ |
| | for Mercury | $i = 0, k = 0$ | $Y = R = Z$ |

These values of i and k reflect the relative velocities of the planets and the Sun. The Sun makes one rotation, the planet none, between successive occurrences of a phenomenon of Saturn or Jupiter; the Sun makes two rotations, the planet one in the case of Mars; the Sun makes one rotation, the planet one in the case of Venus; and the Sun makes no revolutions, the planet also none in the case of Mercury.

We have described previously (II C 2.1) the texts that give planetary periods. More are found in Procedure Texts: BM 33801 (ACT 811) sections 2 (Saturn) and 3 (Mars); BM 34221+34299+35119+35206+35445+45702 (ACT 812) section 10 (Jupiter); BM 34081+34622+34846+45851+46135 (ACT 813) section 1 (Jupiter); and DT 183 (ACT 814) section 1 (Jupiter). Those periods that were employed in the ACT texts are based on combinations of Goal-Year and shorter Periods:

| | | |
|---|---|---|
| Saturn | $Y = 265$ yrs $= 59 \times 4 + 29$ | |
| Jupiter | $Y = 427$ yrs $= 71 \times 5 + 12 \times 6$ | |
| Mars | $Y = 284$ yrs $= 47 \times 4 + 32 \times 3$ $(= 79 \times 3 + 47)$ | |
| Venus | $Y = 1151$ yrs $= 8 \times 144 - 1$ | |
| Mercury | $Y = 848$ yrs $= 46 \times 18 + 20$ | for $\Gamma$ |
| | $Y = 388$ yrs $= 46 \times 8 + 20$ | for $\Sigma$ |
| | $Y = 480$ yrs $= 46 \times 10 + 20$ | for $\Xi$ |
| | $Y = 217$ yrs $= 46 \times 4 + 20 + 13$ | for $\Omega$ |

These combinations are designed so that the inaccuracy of one period is cancelled by the inaccuracy in the opposite direction of the other period. ACT periods were known to the Greeks and, through them, in India (see Pingree [1998] pp. 135-136).

With the knowledge of crude periods for the rotations of the planets, these values of Y, and the formulae given above, it is simple to compute Π and Z.

| | | | |
|---|---|---|---|
| Saturn | 1 rot. ≈ 29;30 yrs | R = Z = 9 ≈ $\frac{265}{29;30}$ | Π = 265 - 9 = 256 |
| Jupiter | 1 rot. ≈ 11;50 yrs | R = Z = 36 ≈ $\frac{427}{11;50}$ | Π = 427 - 36 = 391 |
| Mars | 1 rot. ≈ 1;52 yrs | R = 151 ≈ $\frac{284}{1;52}$ | Z = 151 - Π |
| | 284 = Π + 151 | Π = 133 | Z = 18 |
| Venus | 1 rot. = 1 yr | R = 1151 | |
| | Π = 144 × 5 = 720 (from Π = 5 if Y = 8) | | Z = 431 |
| Mercury | 1 rot. = 1 yr | R = 848 = Z | Π = 2673 (from Π = 145 if Y = 46) |
| | | R = 388 = Z | Π = 1223 |
| | | R = 480 = Z | Π = 1513 |
| | | R = 217 = Z | Π = 684 |

Now that we have Π and Z, we can compute P in each case as well as $\overline{\Delta\lambda}$.

| | | |
|---|---|---|
| Saturn | P = $\frac{256}{9}$ = 28;26,40 | $\overline{\Delta\lambda}$ = 12;39,22,30° |
| Jupiter | P = $\frac{391}{36}$ = 10;51,40 | $\overline{\Delta\lambda}$ = 33;8,44,52,...° |
| Mars | P = $\frac{133}{18}$ = 7;23,20 | $\overline{\Delta\lambda}$ = 48;43,18,29,...° |
| Venus | P = $\frac{720}{431}$ = 1;40,13,... | $\overline{\Delta\lambda}$ = 3,35;30° |
| Mercury | P = $\frac{2673}{848}$ = 3;9,7,38,... | $\overline{\Delta\lambda}$ = 1,54;12,31,30,...° |
| | P = $\frac{1223}{388}$ = 3;9,7,25,... | $\overline{\Delta\lambda}$ = 1,54;12,39,26,...° |
| | P = $\frac{1513}{480}$ = 3;9,7,30,... | $\overline{\Delta\lambda}$ = 1,54;12,36,38,...° |
| | P = $\frac{684}{217}$ = 3;9,7,32,... | $\overline{\Delta\lambda}$ = 1,54;12,37,53,...° |

Finally, if we located all of the successive longitudes of Π phenomena in order on the 360° of the ecliptic, the intervals between these steps, $\overline{\delta}$, would be $\frac{360°}{\Pi} = \frac{\overline{\Delta\lambda}}{Z}$. This means that each $\overline{\Delta\lambda}$ is exactly Z × $\overline{\delta}$.

| | |
|---|---|
| Saturn | $\overline{\delta}$ = 1;24,22,30° |
| Jupiter | $\overline{\delta}$ = 0;55,14,34,...° |
| Mars | $\overline{\delta}$ = 2;42,24,21,...° |
| Venus | $\overline{\delta}$ = 0;30° |
| Mercury | $\overline{\delta}$ = 0;8,4,50,...° |
| | $\overline{\delta}$ = 0;17,39,41,...° |
| | $\overline{\delta}$ = 0;14,16,34,...° |
| | $\overline{\delta}$ = 0;31,34,44,...° |

It was also realized by the Babylonians that, in System A and approximately in System B, $\overline{\Delta t}$, the mean interval of time between successive occurrences of a phenomenon, equals $\overline{\Delta\lambda}$ + C, where

$C = \frac{Z}{\Pi} \times \varepsilon = \frac{\overline{\Delta\lambda}}{6,0} \times 11;4^{\tau} = \overline{\Delta\lambda} \times 0;1,50,40$ (plus 1 yr = 12;22,8 synodic months (or 11;4$^{\tau}$) for Saturn, Jupiter, and Venus, or plus 24;44,16 synodic months (or 22;8$^{\tau}$) for Mars).

| | |
|---|---|
| Saturn | C = 11;27,20,37,30 |
| Jupiter | C = 12;5,8,8,5,14,40 |
| Mars | C = 23;37,51,52,45,29,20 |
| Venus | C = 17;41,28,40 |
| Mercury | C = 3;30,39,6,6 |
| | C = 3;30,39,20,43,57,20 |
| | C = 3;30,39,15,34,5,20 |
| | C = 3;30,39,17,52,25,20 |

Some of these parameters are found in Procedure Texts:

Saturn.

Y = 265, Π = 256, R = 9 (= 54,0°). AO 6477 (ACT 801) section 6; A 3418 (ACT 802) section 4; and BM 34765 (ACT 819c) reverse.

Jupiter.

Π = 391, $\overline{\Delta t}$ = 45;14$^{\tau}$. ACT 812 section 1.

$\overline{\Delta\lambda}$ = 33;8,45°

C = 33;8,45° × 0;1,50,40 + 11;4 = 12;5,8,8,20. ACT 812 section 2; ACT 813 section 13.

Y = 427, Π = 391, R = 36 (= 3,36,0°), $\overline{\Delta\lambda}$ = 33;8,45°. ACT 813 section 21.

Mars.

Y = 284. ACT 811 section 3.

C = 28;37,52. BM 34676 (ACT 811a) section 3 (cf. sections 4, 5, and 6)

Y = 284, Π = 133, R = 151 (= 15,6,0°), $\overline{\Delta\lambda}$ = 48;43,18,30. ACT 811a section 11.

Venus.

$\overline{\Delta\lambda}$ = 9,35;30°. ACT 812 sections 28 and 29.

Mercury.

Y = 46, Π = 144 (a mistake for 145. This mistake leads the scribe to compute that $\overline{\Delta\lambda}$ = 1,55 and that P = 3;7,49,33,54,46). U 136 (ACT 800).

C = 3;30,39,4,20. AO 6477 (ACT 801) section 2.

However, the mean values, $\overline{\Delta\lambda}$ and $\overline{\Delta t}$, do not correspond to reality. In order to deal with this fact one must introduce a method of accounting for the variability of Δλ and Δt; this was done by varying the planet's velocity in different sections of the ecliptic while keeping the velocity of the Sun unchanged; then the phase would occur whenever the "true" planet was at an established elongation from the "mean" Sun. Two mathematical devices were available to accomplish this: the step-function (System A) and the linear zigzag function (System B). Both were applied to the superior planets, System A alone to Mercury and to Venus. The details are as follows.

*Saturn*

System A. There are no Ephemerides belonging to System A, but two Procedure Texts—ACT 801 sections 3-5 and ACT 802 sections 1-3—and especially the "template" tablets of Saturn (Aaboe-Sachs [1966] 1-7) do belong. From the first two "template" tablets (BM 36300+36753+37156+37210+37336 (Text A) and BM 36814 (Text B)) we can recover System A in which $\Pi = 256$, $Z = 9$, $Y = \Pi + Z = 265$, and $P = \Pi/Z = 28;26,40$. The 360° of the ecliptic are divided into two arcs, $\alpha_1$ and $\alpha_2$, in each of which a $\Delta\lambda$, $w_1$ and $w_2$, is applied.

| | | | |
|---|---|---|---|
| $\alpha_1$ | ♌ 10° → ♓ 0° | 3,20° | $w_1 = 11;43,7,30°$ |
| $\alpha_2$ | ♓ 0° → ♌ 10° | 2,40° | $w_2 = 14;3,45°$ |

$\frac{\alpha_1}{\alpha_2} = \frac{3}{2}$ and $\frac{w_1}{w_2} = \frac{5}{6}$. Furthermore, $\delta_1 = \frac{w_1}{Z} = 1;18,7,30°$ and $\delta_2 = \frac{w_2}{Z} = 1;33,45°$. And $\frac{\alpha_1}{w_1} + \frac{\alpha_2}{w_2} = \frac{256}{9} = \frac{\Pi}{Z} = 28;26,40 = P$.

These parameters are related to the subdivision of the synodic arc in the slow and the fast arcs as presented in the "template" tablets:

| | slow arc ($\alpha_1$) | fast arc ($\alpha_2$) |
|---|---|---|
| Γ→Φ | 7;30° | 9° |
| Φ→Ψ | –6;40° | –8° |
| Ψ→Ω | 7;33,7,30° | 9;3,45° |
| Ω→Γ | 3;20° | 4° |
| | 11;43,7,30° | 14;3,45° |

Similar subdivisions of the synodic arcs of all the planets are found in Sanskrit texts as well as, it seems likely, Egyptian tables (see Pingree [1998] p. 136).

A corrupt parallel to this is found in the Procedure Texts mentioned above. These texts present the same values of $\alpha_1$ and $\alpha_2$. Since P is fixed, the values of $w_1$ and $w_2$ are also the same; and the ratio of motions in the slow arc to those in the fast arc is 5 : 6. What we find is:

| | slow arc | fast arc |
|---|---|---|
| Ω→Γ | $0;5^{\circ/d}$ | $0;6^{\circ/d}$ |
| Γ→Γ′ | $0;5^{\circ/d} \times 30^d = 2;30°$ | $0;6^{\circ/d} \times 30^d = 3°$ |
| Γ′→Φ | $0;3,20^{\circ/d} \times 90^d = 5°$ | $0;4^{\circ/d} \times 90^d = 6°$ |
| Φ→Θ | $-0;4,[13,]40^{\circ/d} \times [52;]30^d =$ –3;41,57,30° | $-0;5,4,24^{\circ/d} \times 52;30^d =$ –4;26,21° |
| Θ→Ψ | $-0;3,20^{\circ/d} \times 60^d = -3;20°$ | $-0;4^{\circ/d} \times 60^d = -4°$ |
| Ψ→Ω′ | $0;3,35,30^{\circ/d} \times 90^d = 5;23,15°$ | $0;4,18,40^{\circ/d} \times 90^d = 6;28°$ |
| Ω′→Ω | $0;5^{\circ/d} \times 30^d = 2;30°$ | $0;6^{\circ/d} \times 30^d = 6°$ |

These numbers yield 8;21,17,30° + Ω → Γ in the slow arc, 13;1,39° + Ω → Γ in the fast arc. There is no choice for the days in Ω → Γ that will bring both of these motions to their respective w's. Therefore, the system is corrupt. Moreover, the texts state that Φ→Ψ in the slow arc is –7;33,7,30° (against the attested –7;1,57,30°) and in the fast arc –9;3,45° (against the attested –8;26,21°); Aaboe-Sachs take 7;33,7,30° and 9;3,45° to refer to Ψ →Ω as in the correct System A scheme, even though the scheme in the Procedure Texts leads to 7;53,20° and 9;28° respectively.

From the third "template" tablet, BM 78080 (Text C), Aaboe-Sachs have reconstructed a System A′ with the following parameters:

| | | | |
|---|---|---|---|
| $\alpha_1$ | ♌ 20° → ♒ 10° | 2;50° | $w_1$ = 11;43,7,30° |
| $\alpha_2$ | ♒ 10° → ♌ 20° | 3,10° | $w_2$ = 14;3,45° |

$\frac{\alpha_1}{w_1} + \frac{\alpha_2}{w_2}$ = 28;1,4 = P = $\frac{\Pi}{Z}$ = $\frac{6304}{225}$

$\delta_1$ = 0;3,7,30° $\delta_2$ = 0;3,45°.

$\overline{\Delta\lambda}$, then, is $\frac{6,0}{28;1,4}$ = 12;50,56,20,...°, which is considerably higher than it should be.

System B. The second mathematical device is a linear zigzag function, System B, to which all of the extant Ephemerides of Saturn belong. In these Ephemerides column I gives the year-number in the Seleucid era with an indication of when a year contains an intercalary month XII or month VI; column II gives the entries in a linear zigzag function for the times of a phenomenon; column III lists the month-names and the numbers of the tithis and their fractions; column IV gives the entries in a linear zigzag function for the longitudes of a phenomenon; and column V lists the zodiacal signs and the numbers of the degrees and their fractions. For each additional phenomenon four additional columns are required, though often the columns for the linear zigzag function are omitted. There are Ephemerides for Γ, Θ, Ψ, and Ω singly as well as for Γ, Φ, Θ, for Γ, Φ, and for Ψ, Ω in combination. The parameters for the longitudes are:

M = 14;4,42,30°
m = 11;14,2,30°
μ = 12;39,22,30°
Δ = 2;50,40°
d = 0;12°
P = 28;26,40.

See ACT 801 section 8 and 802 section 6.

The parameters for the dates are:

M = 25;32,3,7,$30^{\tau}$
m = 22;41,23,7,$30^{\tau}$
μ = 24;6,43,7,$30^{\tau}$
Δ = 2;50,$40^{\tau}$
d = 0;$12^{\tau}$

$P = 28;26,40.$

Since P and Δ are the same in the two functions,

$C = \mu^{\tau} - \mu^{\circ} = 11;27,20,37,30.$

The maximum Δλ occurs at about Taurus 10°.

The entries in most of the Ephemerides are truncated at one, two, or three places of the sexagesimal fraction. Other Ephemerides use an approximation for the linear zigzag function for time:

$M = 25;32,5^{\tau}$
$m = 22;41,25^{\tau}$
$\mu = 24;6,45^{\tau}$
$\Delta = 2;50,40^{\tau}$
$d = 0;12^{\tau}$
$P = 28;26,40.$

See ACT 801 section 7 and 802 section 5.

*Jupiter*

System A. The parameters are:

| | | |
|---|---|---|
| $\alpha_1$ ♐ 0° → ♊ 25° | 3,25° | $w_1 = 36°$ |
| $\alpha_2$ ♊ 25° → ♐ 0° | 2,35° | $w_2 = 30°$ |

$\frac{w_1}{w_2} = \frac{6}{5} \cdot \frac{\alpha_1}{w_1} + \frac{\alpha_2}{w_2} = \frac{391}{36} = \frac{\Pi}{Z} = 10;51,40 = P.$

$\delta_1 = 1°$, $\delta_2 = 0;50°$. In the Ephemerides $C = 12;5,10$.

These parameters yield the following scheme for the subdivisions of the synodic arcs (ACT 813 section 2 and 814 section 2):

| | slow arc | fast arc |
|---|---|---|
| Γ→Φ | 16;15° | 19;30° |
| Φ→Θ | −4° | −4;48° |
| Θ→Ψ | −6° | −7;12° |
| Ψ→Ω | 17;45° | 21;18° |
| Ω→Γ | 6° | 7;12° |
| | 30° | 36° |

For the retrograde arc in the slow arc see BM 35241 (ACT 819b) section 2. A similar scheme lies behind the first "template" for Jupiter, BM 36762+36779+36813+37070+37082+37125 and 37174 (Text D), except that for Φ→Θ and for Θ→Ψ the slow arc has −5°, the fast arc −6°. See also the third "template", BM 37474 (Text F).

In the Procedure Texts are several variant schemes that retain the $\alpha_1$ and $\alpha_2$ of System A; see ACT 813 section 10 (velocities; broken) and BM 35943 (ACT 813a) (velocities; broken).

In ACT 813 sections 23 and 30 we have:

| | slow arc | fast arc |
|---|---|---|
| $\Omega \to \Gamma$ | [6];15° | $29^{\tau}$ |
| $\Gamma \to \Phi$ | 16;15° | $4^{m}\ 4^{\tau}$ |
| $\Phi \to \Theta$ | −5;10° | $58^{\tau}$ |
| $\Theta \to \Psi$ | −3;15° | $2^{m}\ 4^{\tau}$ |
| $\Psi \to \Omega$ | 15° | $4^{m}\ 10^{\tau}$ |
| | 29;5°(!) | $13^{m}\ 15^{\tau}$ (= $6,45^{\tau}$) |

Another variant, with the boundaries of $\alpha_1$ and $\alpha_2$ slightly shifted, is found in ACT 813 section 11. See also, for the fast arc, ACT 813 section 24.

System A′ is based on a division of the 360° of the ecliptic into four arcs:

| | | | |
|---|---|---|---|
| $\alpha_1$ | ♋ 9° → ♏ 9° | 2,0° | $w_1 = 30°$ |
| $\alpha_2$ | ♏ 9° → ♑ 2° | 53° | $w_2 = 33;45°$ |
| $\alpha_3$ | ♑ 2° → ♉ 17° | 2,15° | $w_3 = 36°$ |
| $\alpha_4$ | ♉ 17° → ♋ 9° | 52° | $w_4 = 33;45°$ |

$\frac{w_1}{w_2} = \frac{8}{9}$, $\frac{w_2}{w_3} = \frac{15}{16}$. $\frac{\alpha_1}{w_1} + \frac{\alpha_2}{w_2} + \frac{\alpha_3}{w_3} + \frac{\alpha_4}{w_4} = \frac{391}{36} = \frac{\Pi}{Z} = 10;51,40 = P$.

$\delta_1 = 0;50°$, $\delta_2 = 0;56,15° = \delta_4$, and $\delta_3 = 1°$. In the Ephemerides C = 12;5,10.

The bounds of the four α's and their respective w's are given in U 180(10) (ACT 805) section 2; see also ACT 812 section 3; and 813 section 25; and 814 section 2. BM 33869 (ACT 810) gives this same information in sections 1 and 2, and continues in sections 3-6 with the subdivisions of the slow arc ($\alpha_1$), the intermediate arcs ($\alpha_2$ and $\alpha_4$), and the fast arc ($\alpha_3$).

| | slow arc | intermediate arc |
|---|---|---|
| $\Omega \to \Gamma$ | $0;12,30^{\circ/d}$ (× $30^{d}$ = 6;15°) | $0;14,3,45^{\circ/d}$ (× $30^{d}$ = 7;1,52,30°) |
| $\Gamma \to \Gamma'$ | $0;12,30^{\circ/d}$ × $30^{d}$ (= 6;15°) | $0;14,3,45^{\circ/d}$ × $30^{d}$ (= 7;1,52,30°) |
| $\Gamma' \to \Phi$ | $0;6,40^{\circ/d}$ × $90^{d}$ (= 10°) | $0;7,30^{\circ/d}$ × $90^{d}$ (= 11;15°) |
| $\Phi \to \Psi$ | $-0;4,10^{\circ/d}$ × $120^{d}$ (= −8;20°) | $-0;4,41,15^{\circ/d}$ ×$120^{d}$ (= -9;22,30°) |
| $\Psi \to \Omega'$ | $0;6,23,20^{\circ/d}$ × $90^{d}$ (= 9;35°) | $0;7,11,15^{\circ/d}$×$90^{d}$ (=10;46,52,30°) |
| $\Omega' \to \Omega$ | $0;12,30^{\circ/d}$ × $30^{d}$ (= 6;15°) | $0;14,3,45^{\circ/d}$ × $30^{d}$ (= 7;1,52,30°) |
| | 30° | 33;45° |

For the slow arc see also ACT 812 section 4; 813 section 9; and 818 sections 1 and 2.

The results for the intermediate arc are, of course, simply those for the slow arc multiplied by $\frac{w_2}{w_1} = \frac{33;45}{30} = \frac{9}{8} = 1;7,30$. The entries for the fast arc should be those for the intermediate arc multiplied by $\frac{w_3}{w_2} = \frac{16}{15} = 1;4$. Instead the scribe has multiplied the entries for the intermediate arc by $\frac{w_3}{w_1} = \frac{6}{5} = 1;12$.

| | fast arc (text) | fast arc (correct) |
|---|---|---|
| Ω→Γ | 0;16,52,30°/d (× 30d = 8;26,15°) | 0;15°/d (× 30d = 7;30°) |
| Γ→Γ′ | 0;16,52,30°/d × 30d (= 8;26,15°) | 0;15°/d × 30d (= 7;30°) |
| Γ′→Φ | 0;9°/d × 90d (= 13;30°) | 0;8°/d × 90d (= 12°) |
| Φ→Ψ | −0;5,37,30°/d × 120d (= −11;15°) | −0;5°/d × 120d (= −10°) |
| Ψ→Ω′ | 0;8,37,30°/d × 90d (= 12;55,15°) | 0;7,40°/d × 90d (= 11;30°) |
| Ω′→Ω | 0;16,52,30°/d × 30d (= 8;26,15°) | 0;15°/d × 30d (= 7;30°) |
| | 40;30° | 36° |

These subdivisions show that, from conjunction, the Sun travels about 15d to Γ; from Γ about 120d to Φ; from Φ about 120d to Ψ; from Ψ about 120d to Ω; and from Ω about 15d to conjunction, for a total of 390 days. If we counted 1° of solar motion in each day, this would be all right for the slow arc where $\Delta\lambda = 30°$, but impossible for the intermediate and fast arcs. For System A′ see also ACT 813 sections 15-18.

ACT 813 section 9 gives a variant subdivision of the slow arc.

| | |
|---|---|
| Ω→Γ | 0;12,30°/d (× 30d) = 6;15° |
| Γ→Γ′ | 0;12,30°/d × 30d = 6;15° |
| Γ′→Φ | 0;6,15°/d × 90d = 9;22,30° |
| Φ→Ψ | −0;5°/d × 120d = −10° |
| Ψ→Ω′ | 0;7,55°/d × 90d = 11;52,30° |
| Ω′→Ω | 0;12,30°/d × 30d = 6;15° |
| | 30° |

In section 31 of the same Procedure Text is a corrupt subdivision of the fast arc.

In the Procedure Texts and in the second "template" of Jupiter, BM 32218 (Text E), there is evidence for a System A′′. In this the parameters are:

| | | | |
|---|---|---|---|
| $\alpha_1$ | ♋ 5;45° → ♏ 5;45° | 2,0° | $w_1 = 30°$ |
| $\alpha_2$ | ♏ 5;45° → ♑ 2° | 56;15° | $w_2 = 33;45°$ |
| $\alpha_3$ | ♑ 2° → ♉ 17° | 2,15° | $w_3 = 36°$ |
| $\alpha_4$ | ♉ 17° → ♋ 5;45° | 48;45° | $w_4 = 33;45°$ |

Since $\alpha_2+\alpha_4$ in System A′′ = $\alpha_2 + \alpha_4$ in System A′, $\frac{\alpha_1}{w_1}+\frac{\alpha_2}{w_2}+\frac{\alpha_3}{w_3}+\frac{\alpha_4}{w_4}=\frac{391}{36}$ again. Also the δ's are the same. The Procedure Text presenting this System A′′ is ACT 813 section 7.

A System A′′′ is recorded in ACT 813 section 8 and BM 36801 (ACT 813b) section 3.

| | | | |
|---|---|---|---|
| $\alpha_1$ | ♋ 9° → ♏ 9° | 2,0° | $w_1 = 30°$ |
| $\alpha_2$ | ♏ 9° → ♐ 27° | 48° | $w_2 = 33;45°$ |
| $\alpha_3$ | ♐ 27° → ♉ 17° | 2,20° | $w_3 = 36°$ |
| $\alpha_4$ | ♉ 17° → ♋ 9° | 52° | $w_4 = 33;45°$ |

Here we have the aberrant $\frac{\alpha_1}{w_1} + \frac{\alpha_2}{w_2} + \frac{\alpha_3}{w_3} + \frac{\alpha_4}{w_4} = \frac{293}{27} = 10;51,6,40 = P$.

A curious six-fold division of the 360° of the ecliptic is found in ACT 811 section 1.

| | | | |
|---|---|---|---|
| $\alpha_1$ | ♋ 0° → ♎ 30° | 2,0° | $w_1 = 30°$ |
| $\alpha_2$ | ♏ 0° → ♏ 30° | 30° | $w_2 = 32°$ |
| $\alpha_3$ | ♐ 0° → ♐ 30° | 30° | $w_3 = 34°$ |
| $\alpha_4$ | ♑ 0° → ♈ 30° | 2,0° | $w_4 = 36°$ |
| $\alpha_5$ | ♉ 0° → ♉ 30° | 30° | $w_5 = 34°$ |
| $\alpha_6$ | ♊ 0° → ♊ 30° | 30° | $w_6 = 32°$ |

$\frac{w_1}{w_2} = \frac{15}{16}; \frac{w_2}{w_3} = \frac{16}{17}; \frac{w_3}{w_4} = \frac{17}{18}. \frac{\alpha_1}{w_1} + \frac{\alpha_2}{w_2} + \frac{\alpha_3}{w_3} + \frac{\alpha_4}{w_4} + \frac{\alpha_5}{w_5} + \frac{\alpha_6}{w_6} \approx \frac{395}{36} = 10;58,20 = P$.

Finally, ACT 813 section 29 and BM 34757 (ACT 817) section 1 give a linear zigzag function for the motion of Jupiter from Ψ to Ω in each zodiacal sign. The parameters are:

$M = 6;20°$ in Capricorn
$m = 6;10°$ in Cancer
$\mu = 6;15°$ in Aries and Libra
$\Delta = 0;10°$
$d = 0;1,40°$

ACT 817 section 1 also presents an incomplete list of the periods of invisibility of Jupiter (Ω→Γ) in different zodiacal signs.

Of System A one Ephemeris contains Γ and Φ; one Φ, Θ, and Ψ; two Φ and Θ; one Φ alone; one Θ alone; two Ψ and Ω; one Ψ alone; and one Ω alone. Of System A′ one Ephemeris contains Γ, Φ, Θ, and Ψ; one Γ and Φ; one Γ alone; one Φ and Θ; one Θ, Ψ, and Ω; one Θ and Ψ; one Θ alone; and two Ψ and Ω.

System B. The parameters of System B for longitudes are:

$M = 38;2°$
$m = 28;15,30°$
$\mu = 33;8,45°$
$\Delta = 9;46,30°$
$d = 1;48°$
$P = \frac{2\Delta}{d} = \frac{391}{36} = 10;51,40.$

See ACT 805 section 1.

The parameters for the dates are:

$M = 50;7,15^{\tau}$
$m = 40;20,45^{\tau}$
$\mu = 45;14^{\tau}$
$\Delta = 9;46,30^{\tau}$
$d = 1;48^{\tau}$
$P = 10;51,40$

See ACT 812 section 1 and 813 section 12.

Since the period and Δ are the same in the two linear zigzag functions,

$$C = \mu^{\tau} - \mu^{\circ} = 45;14 - 33;8,45 = 12;5,15.$$

See ACT 812 section 2. The maximum $\Delta\lambda$ occurs in about Pisces 15°.

One Ephemeris—A 3426 (ACT 640) from Uruk—and one Procedure Text—ACT 813 sections 21-22 from Babylon—attest to the existence of a System B′ of Jupiter. Its parameters for longitudes are:

$$M = 37;58,20^{\circ}$$
$$m = 28;19,10^{\circ}$$
$$\mu = 33;8,45^{\circ}$$
$$\Delta = 9;39,10^{\circ}$$
$$d = 1;46,40^{\circ}$$
$$P = \frac{2\Delta}{d} = \frac{695}{64} = 10;51,33,45$$

The parameters for the dates are:

$$M = 50;3,31^{\tau}$$
$$m = 40;24,15^{\tau}$$
$$\mu = 45;13,33^{\tau}$$
$$\Delta = 9;39,16^{\tau}$$
$$d = 1;46,40^{\tau}$$
$$P = \frac{8689}{800} = 10;51,40,30$$

Since the periods and the $\Delta$'s are different, the differences between times and longitudes are not a constant.

Of System B Ephemerides one is for $\Gamma$ and $\Phi$, one for $\Gamma$ alone, one for $\Phi$ alone, one for $\Theta$,$\Psi$, and $\Omega$, one for $\Theta$ and $\Psi$, one for $\Theta$ alone, three for $\Psi$ and $\Omega$, one for $\Psi$ alone, and one for $\Omega$ alone. The Ephemeris belonging to System B′ has columns for $\Gamma$ and $\Psi$.

*Mars*

System A. The parameters are:

| | | | |
|---|---|---|---|
| $\alpha_1$ | ♉0° → ♊30° | 1,0° | $w_1 = 45°$ |
| $\alpha_2$ | ♋0° → ♌30° | 1,0° | $w_2 = 30°$ |
| $\alpha_3$ | ♍0° → ♎30° | 1,0° | $w_3 = 40°$ |
| $\alpha_4$ | ♏0° → ♐30° | 1,0° | $w_4 = 1,0°$ |
| $\alpha_5$ | ♑0° → ♒30° | 1,0° | $w_5 = 1,30°$ |
| $\alpha_6$ | ♓0° → ♈30° | 1,0° | $w_6 = 1,7;30°$ |

$\frac{w_1}{w_2} = \frac{3}{2}$. $\frac{w_2}{w_3} = \frac{3}{4}$. $\frac{w_3}{w_4} = \frac{2}{3}$. $\frac{w_4}{w_5} = \frac{2}{3}$. $\frac{w_5}{w_6} = \frac{4}{3}$. $\frac{w_6}{w_1} = \frac{3}{2}$. Furthermore,

$\delta_1 = 2;30°$, $\delta_2 = 1;40°$; $\delta_3 = 2;13,20°$; $\delta_4 = 3;20°$; $\delta_5 = 5°$; and $\delta_6 = 3;45°$.

$$\frac{\alpha_1}{w_1} + \frac{\alpha_2}{w_2} + \frac{\alpha_3}{w_3} + \frac{\alpha_4}{w_4} + \frac{\alpha_5}{w_5} + \frac{\alpha_6}{w_6} = \frac{133}{18} = 7;23,20 = P.$$

The value used for C is 23;37,52; see ACT 811a section 3.

This System is used to compute $\Gamma$, $\Phi$, and $\Omega$. A subdivision of the synodic arc and time based on these three phenomena is found in ACT 811a sections 4-9.

| | | |
|---|---|---|
| $\Omega \to \Gamma$ | 1,29;19,23,55° | $2,44;45,6,46,46,40^{\tau}$ |
| $\Gamma \to \Phi$ | 2,42;24,21,40° | $4,59;32,55,57,46,40^{\tau}$ |
| $\Phi \to \Omega$ | 2,36;59,32,56,40° | $4,49;33,50,5,51,6,40^{\tau}$ |
| | 6,48;43,18,31,40° | $12,33;51,52,50,24,26,40^{\tau}$ |

See Neugebauer [1975] vol. 1, pp. 408-410, and Aaboe [1987] p. 7.

Section 10 of the same tablet contains another subdivision scheme preceded by a linear zigzag function for the velocity of Mars in the arc after Γ in each zodiacal sign. The parameters of this zigzag function are:

$M = 0;50^{\circ/\tau}$ in Capricorn
$m = 0;30^{\circ/\tau}$ in Cancer
$\mu = 0;40^{\circ/\tau}$ in Aries and Libra
$\Delta = 0;20^{\circ}$
$d = 0;3,20^{\circ}$.

The subdivision scheme uses the mean velocity, $0;40^{\circ/\tau}$.

| | |
|---|---|
| $\Gamma \to \Gamma'$ | $0;40^{\circ/\tau} \times 190^{\tau}$ (= 2,6;40°) |
| $\Gamma' \to \Gamma''$ | $0;36^{\circ/\tau} \times 30^{\tau}$ (= 18°) |
| $\Gamma'' \to \Gamma'''$ | $0;24^{\circ/\tau} \times 30^{\tau}$ (= 12°) |
| $\Gamma''' \to \Phi$ | $0;12^{\circ/\tau} \times 30^{\tau}$ (= 6°) |
| | $4,40^{\tau}$ 2,42;40° |
| $\Phi \to \Psi$ | $-0;12^{\circ/\tau}$ |
| $\Psi \to \Psi'$ | $0;12^{\circ/\tau} \times 30^{\tau}$ (= 6°) |
| $\Psi' \to \Psi''$ | $0;24^{\circ/\tau} \times 30^{\tau}$ (= 12°) |
| $\Psi'' \to \Psi'''$ | $0;36^{\circ/\tau} \times 30^{\tau}$ (= 18°) |
| $\Psi''' \to \Omega$ | $0;40^{\circ/\tau} \times 233^{\tau}$ (= 2,35;20°) |
| | $5,23^{\tau}$ 3,11;20° |

For Ω→Γ one uses the same linear zigzag function to determine the velocity as was used for Γ→Γ′; the time is not given. Obviously also for Γ→Γ′ and the other subdivisions one must use the velocities appropriate to the zodiacal sign with the proportion 40 : 36 : 24 : 12 or 10 : 9 : 6 : 3.

For the retrograde arc, Φ→Ψ, there are four schemes, called by Neugebauer R, S, T, and U, for the stretch Φ→Θ and one, an extension of S, for Θ→Ψ. In all cases the longitude of Φ is known, and the amount to be subtracted from the longitude of Φ to arrive at the longitude of Θ depends on the arc, α, in which Φ occurs.

According to Scheme R one uses a linear zigzag function of which the parameters are:

$M = -7;12^{\circ}$ in $\alpha_2$
$m = -6^{\circ}$ in $\alpha_5$
$\mu = -6;36^{\circ}$
$d = -0;24^{\circ}$

Then there result –6° in $\alpha_5$; –6;24° in $\alpha_6$ and $\alpha_4$; –6;48° in $\alpha_1$ and $\alpha_3$; and –7;12° in $\alpha_2$. See U 150 (ACT 803) reverse.

According to Scheme S the quantities of Scheme R apply to the first zodiacal sign in each α; in the second signs the quantities in the first are diminished or increased by the number of degrees that the longitude of Φ occupies within the second sign multiplied by 0;0,48 = 0;24 : 30. See ACT 803 obverse.

According to Scheme T one uses a different linear zigzag function whose parameters are:

M = –7;30° at $\alpha_2$
m = –6° at $\alpha_5$
μ = –6;45°
Δ = 1;30°
d = –0;30°
P = 6

There result –6° in $\alpha_5$; –6;30° in $\alpha_6$ and $\alpha_4$; –7° in $\alpha_1$ and $\alpha_3$; and –7;30° in $\alpha_2$. See ACT 811a section 1.

Finally, Scheme U depends on a linear zigzag function derived from Scheme T in which the quantity to be subtracted changes with each degree of longitudinal change for Φ. Its parameters are:

M = –7;30° at Leo 0°
m = –6° at Aquarius 0°
μ = –6;45° at Taurus 0° and Scorpius 0°
Δ = 1;30°
d = –0;0,30°
P = 6,0.

See U 97+180(26) (ACT 804).

The rule for the retrograde arc from Θ to Ψ according to Scheme S is simply that it is 3/2 times that for Φ→Θ. The same rule probably applies in Schemes R, T, and U. For an explanation of why the retrograde arc of Mars is treated differently than those of Saturn and Jupiter see Aaboe [1987] pp. 9-13.

System A with Scheme S is used in the "templates" of Mars, namely BM 36810+36947 (Text G), BM 36811 (Text H), BM 36957 (Text I), and BM 37236 (Text J). See also BM 37196 (Text K).

Of the few published Ephemerides of System A of Mars one contains Γ and Ω; one Γ and Φ; one Φ and Θ; one Φ alone; one Θ and Ψ; and one, apparently, Ψ alone.

System B. One tablet, U 106 (ACT 510), contains an Ephemeris of Mars computed according to System B, as was recognized by P. Huber (see Aaboe [1958] p. 216). Its parameters are:

$M = 1,20;7,28,30°$
$m = 17;19,8,30°$
$\mu = 48;43,18,30°$
$\Delta = 1,2;48,20°$
$d = 17°$
$P = \frac{2,5;36,40}{17} = 7;23,20 = \frac{133}{18}$.

*Venus*

Strictly speaking, there are no Ephemerides of Venus that belong to either System A or System B. There is, however, a "template" belonging to System A—BM 32599 (ACT 1050) from Babylon—recognized as such by N. T. Hamilton (see Hamilton-Aaboe [1998]). Rather, the Babylonians constructed tables based on either $\overline{\Delta\lambda}$ or $\overline{\Delta t}$ (System $A_0$ in ACT) or on $\Delta\lambda$'s that vary with the arc of the 360° of the ecliptic in which they begin; this last variety uses either the relation $5\Delta\lambda = 18,0° - 2;30°$ (System $A_1$) or $5\Delta\lambda = 18,0° - 2;40°$ (System $A_2$). Since the extant material is minute many questions remain concerning how these "Systems" worked in practice.

The basic parameter behind System $A_1$, $5\Delta\lambda = 18,0° - 2;30°$, means that $\overline{\Delta\lambda} = \frac{17;57,30°}{5} = 3,35;30°$. From this P = 1;40,13,...; and, since for Venus Y = Π + Z, Y = 1151. This is equivalent to saying that in almost 8 years Venus passes through 5 synodic arcs and 8 rotations minus 2;30°. Therefore, in 144 × almost 8 years Venus passes through 144 × 5 synodic arcs and 144 × 8 rotations minus 144 × 2;30°. Since 144 × 2;30° = 360°, the almost 1152 years and the 1152 rotations must both be reduced by 1 to 1151.

We have seen above that correctly for Venus $\overline{C} = 6,0 + 17;41,28,40$; for this the Babylonians substituted 6,17;40. Then $\overline{\Delta t} = \overline{\Delta\lambda} + C = 9,53;10^{\tau}$, and $5\,\overline{\Delta t} = 49,25;50^{\tau} = 99^{m} - 4;10^{\tau}$. The total number of tithis we wish to subtract from 1152 years is $6,11;4^{\tau}$; but $144 \times 4;10^{\tau} = 10,0^{\tau}$ or $3,48;56^{\tau}$ or $7;37,52^{m}$ too much. This is because $99^{m} = 12;22,8^{m/y} \times 8;0,14,13,...^{y}$; $0;0,14,13^{y} \times 6,11;4^{\tau/y} = 1;27,55,19,52^{\tau}$; and $1;27,55,19,52^{\tau} \times 2,24 = 3,31;0,47,40,48^{\tau}$. Therefore, we must subtract $6,11;4^{\tau} + 3,31;1^{\tau} = 9,42;5^{\tau}$ from 1152 years. Complete agreement would be reached if the subtrahend were $\approx -4;2,32^{\tau}$; if the Babylonians had used C = 6,17;41,28,40, their subtrahend would have been $-4;2,36,40^{\tau}$.

There is one Ephemeris from Uruk, A 3415 (ACT 400), for Ξ which uses System $A_0$, i.e., in which $\Delta\lambda = \overline{\Delta\lambda} = 3,35;30°$ and $\Delta\tau = \overline{\Delta\tau} = 9,53;10^{\tau}$. A second Ephemeris from Uruk, U 179 (ACT 401), for Σ uses a modified System $A_0$ in which $\overline{\Delta t} = 99^{m} - 4;5^{d}$ ($4;5^{d} \approx 4;10^{\tau}$).

Two, or perhaps three, Ephemerides from Babylon attest to the use of System $A_1$; they are BM 34128+34222 (ACT 410) for Ξ, Ψ, Ω, Γ, and Φ and BM 35853 (ACT 412) for Ξ, Ψ, and Ω together with, perhaps, BM 34593 (ACT 411), for Ω, Γ, and Φ. To these must be added the columns for Γ and Σ in BM 35495+40102+46176 (ACT 420) with the Procedure Text it contains (ACT 821b). From these pieces Neugebauer [1955] vol. 2, p. 301 has reconstructed the following schemes, each consisting of five occurrences of a phenomenon. To this we have added a column for $c = \Delta\tau - \Delta\lambda$.

| | Zodiacal signs | $\Delta\lambda$ | $\Delta\tau$ | c |
|---|---|---|---|---|
| Ξ | ♊ | 3,44;10° | 10,0$^{\tau}$ | 6,15;50 |
| | ♌ | 3,30;30° | 9,23$^{\tau}$ | 5,52;30 |
| | ♎ | 3,32° | 9,45$^{\tau}$ | 6,13 |
| | ♑ | 3,36;20° | 10,11$^{\tau}$ | 6,34;40 |
| | ♓ | 3,34;30° | 10,7$^{\tau}$ | 6,32;30 |
| | | 17,57;30° | 49,26$^{\tau}$ | |
| Ψ | ♉ | 3,37;30° | 10,1$^{\tau}$ | 6,23;30 |
| | ♋ | 3,38;30° | 9,59$^{\tau}$ | 6,20;30 |
| | ♍, ♎ | 3,29;30° | 9,46$^{\tau}$ | 6,16;30 |
| | ♐ | 3,28;30° | 9,49$^{\tau}$ | 6,20;30 |
| | ♒ | 3,43;30° | 9,51$^{\tau}$ | 6,17;30 |
| | | 17,57;30° | 49,26$^{\tau}$ | |
| Γ | ♈, ♉, [♊] | 3,39;50° | 9,48$^{\tau}$ | 6,8;10 |
| | ♋, ♌ | 3,41;10° | 9,52$^{\tau}$ | 6,10;50 |
| | ♍, ♎, [♏] | 3,30;40° | 9,59$^{\tau}$ | 6,28;20 |
| | ♐, [♑] | 3,25;30° | 9,55$^{\tau}$ | 6,29;30 |
| | ♒, ♓ | 3,40;20° | 9,52$^{\tau}$ | 6,11;40 |
| | | 17,57;30° | 49,26$^{\tau}$ | |
| Σ | ♓, ♈ | 3,45;20° | 10,14$^{\tau}$ | 6,28;40 |
| | ♉, ♊, ♋ | 3,29;40° | 9,45$^{\tau}$ | 6,15;20 |
| | ♌, ♍ | 3,28;50° | 9,31$^{\tau}$ | 6,2;10 |
| | ♎, ♏, ♐ | 3,38;30° | 9,57$^{\tau}$ | 6,18;30 |
| | ♑, ♒ | 3,35;10° | 9,59$^{\tau}$ | 6,23;50 |
| | | 17,57;30° | 49,26$^{\tau}$ | |

All that is clear from this is that the sums of the five arcs and five periods must equal 17,57;30° and 49,26$^{\tau}$ respectively. But there is no evident relation between $\Delta\lambda$'s and $\Delta\tau$'s.

The columns of ACT 420 with the Procedure Text it preserves (ACT 821b) for Ψ, Ω, and Φ, and BM 42799+45777 (ACT 421a) for Ξ, Ψ, Ω, and Γ attest to a System $A_2$, as does also perhaps BM 35118 (ACT 421) for Ξ and Ψ. From these pieces Neugebauer [1955] vol. 2, p. 302, has reconstructed the following scheme; again we have added the values of c.

| | Zodiacal signs | Δλ | Δτ | c |
|---|---|---|---|---|
| Ψ | [♈], ♉, [♊] | 3,37;30° | 10,1$^{\tau}$ | 6,23;30 |
| | ♋, ♌ | 3,38;30° | 9,59$^{\tau}$ | 6,20;30 |
| | [♍], ♎, ♏ | 3,29;20° | 9,46$^{\tau}$ | 6,16;40 |
| | ♐, ♑ | 3,28;30° | 9,49$^{\tau}$ | 6,20;30 |
| | ♒, ♓ | 3,43;30° | 9,51$^{\tau}$ | 6,7;30 |
| | | 17,57;20° | 49,26$^{\tau}$ | |
| Ω | ♈, ♉, ♊ | 3,36° | 9,58$^{\tau}$ | 6,22 |
| | ♋, ♌ | 3,38° | 10,4$^{\tau}$ | 6,26 |
| | ♍, ♎, ♏ | 3,32° | 10,1$^{\tau}$ | 6,29 |
| | ♐, ♑ | 3,29;30° | 9,40$^{\tau}$ | 6,10;30 |
| | ♒, ♓ | 3,41;50° | 9,43$^{\tau}$ | 6,1;10 |
| | | 17,57;20° | 49,26$^{\tau}$ | |
| Φ | ♈, ♉ | 3,40;30° | 9,49$^{\tau}$ | 6,8;30 |
| | ♊, ♋, ♌ | 3,42;50° | 9,31$^{\tau}$ | 6,8;10 |
| | ♍, ♎ | 3,30;10° | 9,57$^{\tau}$ | 6,26;50 |
| | ♏, ♐, [♑] | 3,24;40° | 10,14$^{\tau}$ | 6,49;20 |
| | ♒, ♓ | 3,30;10° | 9,55$^{\tau}$ | 6,15;50 |
| | | 17,57;20° | 49,26$^{\tau}$ | |

It is unclear whether BM 55546 obverse (ACT 430) for Φ and Σ belongs to System $A_1$ or System $A_2$.

A scheme for subdividing Venus' synodic arc and period for each zodiacal sign is found in a Procedure Text, ACT 812 sections 11-23. This is quite fragmentary.

| Zodiacal sign | Γ→[Ξ] | | [Ξ]→Γ |
|---|---|---|---|
| ♋ | [5,2]7° | [5,2]6$^{\tau}$ | [4,36]$^{\tau}$ |
| ♌ | 5,28;30° | | 4,34$^{\tau}$ |
| ♍ | 5,30° | | 4,32$^{\tau}$ |
| ♎ | 5,31;30° | 5,23;30$^{\tau}$ | 4,30$^{\tau}$ |
| ♏ | 5 ........° | 5,2[....$^{\tau}$] | 4,30$^{\tau}$ |
| ♐ | [5,29°] | | |
| ♑ | 5,27° | | |
| ♒ | 5,28° | | |
| ♓ | 5,30° | | 4,30$^{\tau}$ |

| | | | |
|---|---|---|---|
| ♈ | 5,[31;30°] | 5,23;30$^{\tau}$ | 4,30$^{\tau}$ |
| ♉ | [5, .......°] | 5 [......$^{\tau}$] | 4,32$^{\tau}$ |
| ♊ | [5,]29° | | 4,34$^{\tau}$ |

| Zodiacal signs | Γ→Σ | Σ→Ξ | | Ξ→Ψ | | Ψ→Ω | | Ω→Γ | Γ→Φ |
|---|---|---|---|---|---|---|---|---|---|
| ♋ | [4,13]$^{\tau}$ | 56$^{\tau}$ | | 4,12$^{\tau}$ | | 17$^{\tau}$ | | 15$^{\tau}$ | |
| ♌ | 4,17$^{\tau}$ | 56$^{\tau}$ | | 4,7$^{\tau}$ | -6° | 17$^{\tau}$ | | 15$^{\tau}$ | |
| ♍ | 4,21$^{\tau}$ | 56$^{\tau}$ | | 4,2$^{\tau}$ | -6° | 17$^{\tau}$ | | | |
| ♎ | 4,25$^{\tau}$ | 56$^{\tau}$ | 4,14° | [3,58]$^{\tau}$ | | [19]$^{\tau}$ | | | |
| ♏ | 4,29$^{\tau}$ | 56$^{\tau}$ | | [3,58]$^{\tau}$ | | [21]$^{\tau}$ | | | |
| ♐ | 4,3[3]$^{\tau}$ | 1,0$^{\tau}$ | | 3,58$^{\tau}$ | | 22;30$^{\tau}$ | -2° | 4;30$^{\tau}$ | -10° |
| ♑ | [4,33]$^{\tau}$ | 1,2$^{\tau}$ | | 3,58$^{\tau}$ | | 23$^{\tau}$ | -1° | 1$^{\tau}$ | -9° |
| ♒ | 4,29$^{\tau}$ | 1,2$^{\tau}$ | 4,21° | 4,1$^{\tau}$ | | 23$^{\tau}$ | -1° | 1$^{\tau}$ | -9° |
| ♓ | 4,25$^{\tau}$ | 1,1$^{\tau}$ | | 4,3$^{\tau}$ | | 22$^{\tau}$ | -1° | 2$^{\tau}$ | |
| ♈ | 4,22$^{\tau}$ | 1,0$^{\tau}$ | | 4,6$^{\tau}$ | | [21]$^{\tau}$ | | | |
| ♉ | 4,19$^{\tau}$ | 1,0$^{\tau}$ | | 4,8$^{\tau}$ | | [20]$^{\tau}$ | | | |
| ♊ | 4,16$^{\tau}$ | 58$^{\tau}$ | | 4,10$^{\tau}$ | | 19$^{\tau}$ | | | |

See also sections 24-27 of the same text.

*Mercury*

System A. All the extant Ephemerides of Mercury belong to a System A. The earliest attested, System $A_3$, is for Ω; the next earliest, System $A_2$, is for Σ and Ω with "pushes" to Ξ and Γ; and the most common variety, System $A_1$, is for Γ and Ξ with "pushes" to Σ and Ω.

System $A_3$ is known to us through two later copies of originals of unknown date. The obverse of BM 36651+36719+37032+37053 was published in Aaboe-Britton-Henderson-Neugebauer-Sachs [1991] pp. 34-43; it gives a list of Mercury's first appearances in the evening (Ξ) from 41 Artaxerxes I (-423/2) to 2 Artaxerxes II (-401/0). Since in columns II and III there are still preserved numbers 1-17 indicating the first seventeen of the nineteen years of Darius II and, in column III, the numbers 1 and 2 referring to the first two years of the reign of Artaxerxes II, the original from which the extant tablet was copied was itself copied after March of -402. The other tablet is BM 36321 (ACT 816), which describes System $A_3$, and which was copied from a broken original by the well-known scribe Bēl-apla-iddin, descendant of Mušēzib.

There are two Models employed in System $A_3$. In Model I, which produces the longitudes of Ω in each of three columns where each entry is three synodic arcs later than its predecessor, the w's are negative and the zodiacal signs are counted in the reverse order. The parameters are:

| | | | |
|---|---|---|---|
| $\alpha_1$ | ♈ 30° → ♌ 30° | 4,0° | $w_1 = -16;52,30°$ |
| $\alpha_2$ | ♌ 30° → ♋ 20;37,30° | 39;22,30° | $w_2 = -16°$ |
| $\alpha_3$ | ♋ 20;37,30° → ♈ 30° | 1,20;37,30° | $w_3 = -20°$ |

$\delta_1 = 0;0,52,44,3,45°$, $\delta_2 = 0;0,50°$, and $\delta_3 = 0;1,2,30°$.

$P = \frac{23863}{1152} = 20;42,51,52,30$.

The values of the w's in this model are the measure of the amount by which three successive occurrences of Ω in the arcs α fall short of 360°.

In Model II, which produces the longitudes of the phenomenon in each line, the parameters are:

| | | | |
|---|---|---|---|
| $\alpha_1$ | ♌ 30° → ♈ 30° | 4,0° | $w_1 = 1,50;56,15°$ |
| $\alpha_2$ | ♈ 30° → ♋ 20° | 1,20° | $w_2 = 2,11;28,53,20°$ |
| $\alpha_3$ | ♋ 20° → ♌ 30° | 40° | $w_3 = 1,45;11,6,40°$ |

$\delta_1 = 0;18,45°$, $\delta_2 = 0;17,46,40°$, and $\delta_3 = 0;22,13,20°$.

$P = \frac{\Pi}{Z} = \frac{1119}{355} = 3;9,7,36,20,\ldots$

Since $\delta = \frac{w}{Z}$, and each sequence of three occurrences contains 3Z intervals, we multiply $3 \times Z$ and find 1065—i.e., $\Pi - 54$. This means that $54\delta$ is the measure of the amount by which three occurrences are less than 6,0°; and, using the δ's of Model II,

$$-54 \times \delta_1 = -16;52,30°$$
$$-54 \times \delta_2 = -16°$$
$$-54 \times \delta_3 = -20°.$$

The results are the w's of Model I. But the two Models do not fit together because of the difference in the boundary of $\alpha_2$ which produces a different period for each. The parameters of System $A_3$ are confirmed by ACT 816. This Procedure Text also gives confused accounts of a 20-year period for Ω and of a one- and a 20-year period for Ξ; see Neugebauer [1955] vol. 1, pp. 470-471. Section 3 refers to Mercury's stations.

System $A_2$ is represented by two early Ephemerides from Babylon: BM 36723+37234 (ACT 300a) covers the years -307/6 to -289/8, giving the dates and longitudes of Σ, Ξ, Ω, and Γ, and BM 36922+37115 (ACT 300) covers the years -301/0 to -293/2, with only the longitude of Ξ, the dates and longitudes of Ω, and the dates of Γ being represented.

In System $A_2$ the dates and longitudes of the last visibilities, Σ and Ω, are computed independently, and then the dates and longitudes of Ξ are derived from those for Σ, those of Γ from those of Ω. For both Σ and Ω the 360° of the ecliptic are divided into four arcs. For Σ the parameters are:

| | | | |
|---|---|---|---|
| $\alpha_1$ | ♋ 0° → ♍ 30° | 1,30° | $w_1 = 1,47;46,40°$ |
| $\alpha_2$ | ♎ 0° → ♑ 6° | 1,36° | $w_2 = 2,9;20°$ |
| $\alpha_3$ | ♑ 6° → ♈ 5° | 1,29° | $w_3 = 1,37°$ |
| $\alpha_4$ | ♈ 5° → ♊ 30° | 1,25° | $w_4 = 2,9;20°$ |

$\frac{w_1}{w_2} = \frac{5}{6}$, $\frac{w_2}{w_3} = \frac{4}{3}$, $\frac{w_3}{w_4} = \frac{3}{4}$; $\delta_1 = 0;16,40°$, $\delta_2 = \delta_4 = 0;20°$, $\delta_3 = 0;15°$.

$P = \frac{\Pi}{Z} = \frac{1223}{388} = 3;9,7,25,\ldots$

The derivations of the dates of Σ and of the longitudes and dates of Ξ remain obscure; see Neugebauer [1955] vol. 2, pp. 296 and 298.

The step function for Ω has the following parameters:

| | | | |
|---|---|---|---|
| $\alpha_1$ | ♋ 0° → ♐ 30° | 3,0° | $w_1 = 1,48;30°$ |
| $\alpha_2$ | ♑ 0° → ♒ 30° | 1,0° | $w_2 = 2,0;33,20°$ |
| $\alpha_3$ | ♓ 0° → ♈ 30° | 1,0° | $w_3 = 1,48;30°$ |
| $\alpha_4$ | ♉ 0° → ♊ 30° | 1,0° | $w_4 = 2,15;37,30°$ |

$\frac{w_1}{w_2} = \frac{9}{10},\ \frac{w_2}{w_3} = \frac{10}{9},\ \frac{w_3}{w_4} = \frac{4}{5};\ \delta_1 = \delta_3 = 0;30°,\ \delta_2 = 0;33,20°,\ \delta_4 = 0;37,30°.$

$P = \frac{\Pi}{Z} = \frac{684}{217} = 3;9,7,32,....$

The computation of the dates of Ω remains obscure; see Neugebauer [1955] vol. 2, p. 298. But the derivation of the dates and times of Γ is understood.

The longitudinal "pushes" from Ω to Γ depend on the longitude of Ω in the following fashion:

| beginning of arc | change per degree | difference at each degree | end of arc | | |
|---|---|---|---|---|---|
| longitude of Ω | | | longitude of Ω | longitude of Γ | Γ–Ω |
| ♈ 0° | +0;24° | | ♈ 15° | ♈ 21° | +6° |
| ♈ 15° | –0;4° | | ♉ 15° | ♉ 19° | +4° |
| ♉ 15° | –0;12° | | ♋ 15° | ♋ 9° | –6° |
| ♋ 15° | | –8° | ♎ 15° | ♎ 7° | –8° |
| ♎ 15° | –0;8° | | ♏ 15° | ♏ 3° | –12° |
| ♏ 15° | | –12° | ♓ 15° | ♓ 3° | –12° |
| ♓ 15° | +0;48° | | ♈ 0° | ♈ 0° | 0° |

A similar scheme produces the dates of Γ:

| beginning of arc | difference at beginning | change per degree | end of arc | |
|---|---|---|---|---|
| longitude of Ω | | | longitude of Ω | time of Γ |
| ♈ 15° | $+38^\tau$ | $-0;6^\tau$ | ♉ 15° | $+35^\tau$ |
| ♉ 15° | $+35^\tau$ | $-0;13^\tau$ | ♋ 15° | $+22^\tau$ |
| ♋ 15° | $+22^\tau$ | $0^\tau$ | ♎ 15° | $+22^\tau$ |
| ♎ 15° | $+22^\tau$ | $-0;16^\tau$ | ♏ 15° | $+14^\tau$ |
| ♏ 15° | $+14^\tau$ | $-0;3,12^\tau$ | ♑ 30° | $+10^\tau$ |
| ♒ 0° | $+10^\tau$ | $+0;12^\tau$ | ♓ 15° | $+19^\tau$ |
| ♓ 15° | $+19^\tau$ | $+0;38^\tau$ | ♈ 15° | $+38^\tau$ |

The 360° of the ecliptic are divided into three arcs for the step-functions of System $A_1$. The longitudes and times of Σ and of Ω are derived from those of Γ and Ξ.

The parameters of the step-function for Γ are:

| | | | |
|---|---|---|---|
| $\alpha_1$ | ♌ 1° → ♑ 16° | 2,45° | $w_1$ = 1,46° |
| $\alpha_2$ | ♑ 16° → ♉ 30° | 2,14° | $w_2$ = 2,21;20° |
| $\alpha_3$ | ♊ 0° → ♌ 1° | 1,1° | $w_3$ = 1,34;13,20° |

$\frac{w_1}{w_2} = \frac{3}{4}$, $\frac{w_2}{w_3} = \frac{3}{2}$, $\frac{w_3}{w_1} = \frac{9}{10}$; $\delta_1$ = 0;7,30°, $\delta_2$ = 0;10°, $\delta_3$ = 0;6,40°.

P = $\frac{\Pi}{Z} = \frac{2673}{848}$ = 3;9,7,38,....

The values of the α's and w's are also given in the Procedure Text ACT 801, section 1, where also it is stated that C = 3;30,39,4,20$^{\tau}$; in practice the value of C is truncated to 3;30,39$^{\tau}$.

The parameters of the step-function for Ξ are:

| | | | |
|---|---|---|---|
| $\alpha_1$ | ♋ 6° → ♎ 26° | 1,50° | $w_1$ = 2,40° |
| $\alpha_2$ | ♎ 26° → ♓ 10° | 2,14° | $w_2$ = 1,46;40° |
| $\alpha_3$ | ♓ 10° → ♋ 6° | 1,56° | $w_3$ = 1,36° |

$\frac{w_1}{w_2} = \frac{3}{2}$, $\frac{w_2}{w_3} = \frac{10}{9}$, $\frac{w_3}{w_1} = \frac{3}{5}$; $\delta_1$ = 0;20°, $\delta_2$ = 0;13,20°, $\delta_3$ = 0;12°.

P = $\frac{\Pi}{Z} = \frac{1513}{480}$ = 3;9,7,30.

The values of the α's and the w's are given in ACT 801, section 2. For computing the dates of Ξ, C is still 3;30,39$^{\tau}$.

The arcs and times for Γ→Σ and for Ξ→Ω are computed by means of schemes similar to those used for Ω→Γ in System $A_2$. For the arcs for Γ→Σ:

| beginning of arc | difference at beginning | change per degree | end of arc | |
|---|---|---|---|---|
| longitude of Γ | | | longitude of Γ | longitude of Σ |
| ♈ 15° | +12° | +0;4° | ♉ 15° | +14° (♉ 29°) |
| ♉ 15° | +14° | +0;8° | ♊ 15° | +18° (♋ 3°) |
| ♊ 15° | +18° | +0;8° | ♋ 15° | +22° (♌ 7°) |
| ♋ 15° | +22° | +0;8° | ♌ 15° | +26° (♍ 11°) |
| ♌ 15° | +26° | +0;8$^{c}$ | ♍ 15° | +30° (♎ 15°) |
| ♍ 15° | +30° | +0;8° | ♎ 15° | +34° (♏ 19°) |
| ♎ 15° | +34° | +0;20° | ♏ 15° | +44° (♐ 29°) |
| ♏ 15° | +44° | 0° | ♐ 15° | +44° (♑ 29°) |
| ♐ 15° | +44° | −0;4° | ♑ 15° | +42° (♒ 27°) |
| ♑ 15° | +42° | −0;16° | ♒ 15° | +34° (♓ 19°) |
| ♒ 15° | +34° | −0;20° | ♓ 15° | +24° (♈ 9°) |
| ♓ 15° | +24° | −0;24° | ♈ 15° | +12° (♈ 27°) |

For the times for Γ→Σ:

| beginning of arc | difference at beginning | change per degree | end of arc | |
|---|---|---|---|---|
| longitude of Γ | | | longitude of Γ | time of Σ |
| ♈ 15° | 14$^{\tau}$ | +0;4$^{\tau}$ | ♉ 15° | +16$^{\tau}$ |
| ♉ 15° | 16$^{\tau}$ | +0;6$^{\tau}$ | ♊ 15° | +19$^{\tau}$ |
| ♊ 15° | 19$^{\tau}$ | +0;10$^{\tau}$ | ♋ 15° | +24$^{\tau}$ |
| ♋ 15° | 24$^{\tau}$ | +0;6$^{\tau}$ | ♌ 15° | +27$^{\tau}$ |
| ♌ 15° | 27$^{\tau}$ | +0;6$^{\tau}$ | ♍ 15° | +30$^{\tau}$ |
| ♍ 15° | 30$^{\tau}$ | +0;12$^{\tau}$ | ♎ 15° | +36$^{\tau}$ |
| ♎ 15° | 36$^{\tau}$ | +0;20$^{\tau}$ | ♏ 15° | +46$^{\tau}$ |
| ♏ 15° | 46$^{\tau}$ | 0$^{\tau}$ | ♐ 15° | +46$^{\tau}$ |
| ♐ 15° | 46$^{\tau}$ | –0;4$^{\tau}$ | ♑ 15° | +44$^{\tau}$ |
| ♑ 15° | 44$^{\tau}$ | –0;20$^{\tau}$ | ♒ 15° | +34$^{\tau}$ |
| ♒ 15° | 34$^{\tau}$ | –0;20$^{\tau}$ | ♓ 15° | +24$^{\tau}$ |
| ♓ 15° | 24$^{\tau}$ | –0;20$^{\tau}$ | ♈ 15° | +14$^{\tau}$ |

For the arcs for Ξ→Ω:

| beginning of arc | difference at beginning | change per degree | end of arc | |
|---|---|---|---|---|
| longitude of Ξ | | | longitude of Ξ | longitude of Ω |
| ♈ 15° | +36° | +0;12° | ♉ 15° | +42° (♊ 27°) |
| ♉ 15° | +42° | +0;8° | ♊ 15° | +46° (♌ 1°) |
| ♊ 15° | +46° | –0;8° | ♋ 15° | +42° (♌ 27°) |
| ♋ 15° | +42° | –0;12° | ♌ 15° | +36° (♍ 21°) |
| ♌ 15° | +36° | –0;20° | ♍ 15° | +26° (♎ 11°) |
| ♍ 15° | +26° | –0;16° | ♎ 15° | +14° (♎ 29°) |
| ♎ 15° | +14° | 0° | ♏ 15° | +14° (♏ 29°) |
| ♏ 15° | +14° | +0;4° | ♐ 15° | +16° (♑ 1°) |
| ♐ 15° | +16° | +0;8° | ♑ 15° | +20° (♒ 5°) |
| ♑ 15° | +20° | +0;4° | ♒ 15° | +22° (♓ 7°) |
| ♒ 15° | +22° | 0° | ♓ 15° | +22° (♈ 7°) |
| ♓ 15° | +22° | +0;28° | ♈ 15° | +36° (♉ 21°) |

For the times for Ξ→Ω:

| beginning of arc longitude of Ξ | difference at beginning | change per degree | end of arc longitude of Ξ | time of Ω |
|---|---|---|---|---|
| ♈ 15° | $36^{\tau}$ | $+0;12^{\tau}$ | ♉ 15° | $42^{\tau}$ |
| ♉ 15° | $42^{\tau}$ | $+0;12^{\tau}$ | ♊ 15° | $48^{\tau}$ |
| ♊ 15° | $48^{\tau}$ | $-0;8^{\tau}$ | ♋ 15° | $44^{\tau}$ |
| ♋ 15° | $44^{\tau}$ | $-0;12^{\tau}$ | ♌ 15° | $38^{\tau}$ |
| ♌ 15° | $38^{\tau}$ | $-0;36^{\tau}$ | ♍ 15° | $20^{\tau}$ |
| ♍ 15° | $20^{\tau}$ | $-0;10^{\tau}$ | ♎ 15° | $15^{\tau}$ |
| ♎ 15° | $15^{\tau}$ | $0^{\tau}$ | ♏ 15° | $15^{\tau}$ |
| ♏ 15° | $15^{\tau}$ | $+0;2^{\tau}$ | ♐ 15° | $16^{\tau}$ |
| ♐ 15° | $16^{\tau}$ | $+0;12^{\tau}$ | ♑ 15° | $22^{\tau}$ |
| ♑ 15° | $22^{\tau}$ | $+0;4^{\tau}$ | ♒ 15° | $24^{\tau}$ |
| ♒ 15° | $24^{\tau}$ | $0^{\tau}$ | ♓ 15° | $24^{\tau}$ |
| ♓ 15° | $24^{\tau}$ | $+0;24^{\tau}$ | ♈ 15° | $36^{\tau}$ |

Babylonian equivalents of these tables for arcs and times and for Σ→Γ and Ω→Ξ are found in A 3409 (ACT 800a), U 169 (ACT 800b), U180(27) (ACT 800c), U 151+164 (ACT 800d), and Warka X 40 (ACT 800e).

There remains one important tablet, A 3425 (ACT 310), that records the days (*not* tithis) of each month, the differences in longitude, and the longitudes of Mercury for one year, probably -189/8. In the margin beside the third column are written the names of the zodiacal signs on the day that the planet enters each and the Greek-letter phenomena Γ, Σ, Ξ, and Ω, though Ω and Γ are noted where the direction of the planet changes—i.e., at Ψ and Φ. The differences in the second column are constant for long stretches, and then for somewhat shorter stretches change daily, often with constant second differences. The table, then, is computed, but not by any of the three known Systems of Mercury. For further discussion see Neugebauer [1975] vol. 1, pp. 418-420.

The earliest dated Ephemeris belongs to System $A_2$ of Mercury; BM 36723+37234 (ACT 300a) covers the years from -307/6 to -289/8. However, the Ephemeris of System $A_3$ mentioned above, though copied after -400, contains the longitudes of Ξ from -423 till -402. Even though we have no dated Ephemerides for Saturn before -311/0 (W 22755/3; von Weiher [1993] no. 170), for Mars before -222/1 (U 101; ACT 500), or for Jupiter before -198/7 (AO 6476+U 104, ACT 600, and U 102, ACT 606), it is likely that the ability to compute Ephemerides for all of them existed by the middle of the third century B.C. if not by the beginning of the Seleucid Era in -311. But van der Waerden [1968] claims that the "templates" for Mars work best if they began in -498 or -419 and the first "template" for Saturn if it began in -539, -510, -451, -392, or -333. He then argues from the similarities of the "templates" for the two planets that their respective initial dates were -498 and -510. Pointing to the facts that there exist observational records from the sixth century B.C. and that he had argued that System A of the Moon must be dated to between -620 and -440, van der Waerden concludes that the Systems A of both the Moon and the planets were invented during the reign of the Achaemenid Darius I (-520 to -484). These arguments are repeated and the astronomer who invented these Systems A was identified with Nabû-rimannu in van der Waerden [1974], pp. 281-283. It is, however, extremely difficult to understand why we have no extant Ephemerides dated securely in the fifth and pre-Seleucid fourth centuries B.C., but rather a handful of texts providing planetary longitudes that do *not* rely on the strict mathematics of System A, but on subdivisions of the synodic arcs (see II C 2.3 above). The existence of these texts seems to preclude the use of Systems A much before the Seleucid Era. It is argued by Ferrari d'Occhieppo [1978] that the most likely date for the discovery of Venus' 1151-year period was between -359 and -327; the assumptions that he needs to make in order to propose this theory make us inclined to disbelieve it.

A fundamental problem for the mathematical planetary astronomy of the Babylonians is that of the method(s) by which they derived the rules and the parameters of their Systems from the observational data available to them. The phenomena recorded in the Diaries and tabulated in the Ephemerides—first and last visibilities, first and second stations, and acronychal rising—all depend primarily on the elongation of the planet from the Sun. Van der Waerden [1957] showed, among other things, that the Babylonian planetary theories were generally based on the principle that each Greek-letter phenomenon of each planet occurs when the planet is at a fixed elongation from the mean Sun. This means that, while column B in both of the Systems of the Moon assumes variable solar velocity but constant lunar velocity, the time and longitude columns of the planetary Ephemerides assume constant solar velocity and variable planetary velocity. Therefore, the intervals in time and in longitude between successive occurrences of Greek-letter phenomena vary solely because of the variation of the planet's velocity associated with particular arcs of the ecliptic. Van der Waerden also showed that for System A of Mars—and, by extension, the Systems A of the other planets—the division of the ecliptic into a number of arcs in which there are different intervals between successive occurrences of the same Greek-letter phenomena is equivalent to dividing each of these arcs into an integer number of steps of a fixed length for each arc, so that the sum of these steps is the total number of the occurrences of the phenomenon within the ACT period. Thus, for Mars there are six arcs, each of 60°.

| | Arc | Steps | Length of step |
|---|---|---|---|
| I | ♉ 0° → ♊ 30° | 24 | 2;30° |
| II | ♋ 0° → ♌ 30° | 36 | 1;40° |
| III | ♍ 0° → ♎ 30° | 27 | 2;13,20° |
| IV | ♏ 0° → ♐ 30° | 18 | 3;20° |
| V | ♑ 0° → ♒ 30° | 12 | 5° |
| VI | ♓ 0° → ♈ 30° | 16 | 3;45° |
| | | 133 | |

As there are 133 steps or occurrences of a phenomenon in the period of 284 years during which Mars passes 151 times through 360° of longitude, between any two successive occurrences of a phenomenon it travels 360° plus 18 steps.

The problem was approached again by Aaboe [1958], who, while noting that in System A the difference in time-intervals ($\Delta t$) equals the difference in longitude-intervals ($\Delta\lambda$) plus a constant, C, determined that $\Delta\lambda$ is a function of $\lambda$ alone. In System B, on the other hand, the zigzag functions for $\Delta\tau$ and $\Delta \lambda$ are slightly out of phase with each other, though the difference between their mean values equals C, and $\Delta\lambda$ is a function of the number of the occurrence of the phenomenon within the period. In the end, given that the observational reports give only the times of occurrence of first and last visibilities and acronychal rising, but the distance from a Normal Star in the case of first and last stations, Aaboe concluded that it might be that the schemes for $\Delta\lambda$ and the relation $\Delta t = \Delta\lambda + C$ arose from the study of observations of stations, while schemes for $\Delta t$ were fundamental for the other Greek-letter phenomena.

A few years later appeared Aaboe [1965], which extended van der Waerden's analysis of the steps in System A of Mars to the Sun and to the planets (with the exception of Venus). The only cases where an arc is not divided into an integer number of steps are arcs 2 and 4 in Systems A′ and A‴ of Jupiter and both arcs of System A of Saturn; and it is only arcs 2 and 4 of System A″ of Jupiter that do not contain an integer number of degrees. Aaboe next notes that, in each scheme of synodic arcs (w) or of steps, the ratios of these arcs or steps to each other is normally that of two regular and consecutive integers (where a regular integer is one that contains no prime factors other than 2, 3, and 5, so that its reciprocal has a finite sexagesimal expansion). From this Aaboe suggests that the schemes for longitudes in System A, like those in the "templates", are "constructed ... by means of techniques which all belong to elementary arithmetic," and that they are "completely independent of dates."

Schmidt [1969] addresses the problem of providing an explanation of the fact that, in System A, the ratio of the sum of the arcs ($\alpha$) to the sum of the synodic arcs (w) is equal to the ratio of the number of occurrences of the phenomenon in a period (Π) to the number of times the sum of the mean synodic arcs ($\overline{\Delta\lambda}$) equals 360° (Z)—i.e.,

$$\frac{\alpha_1}{w_1}+\frac{\alpha_2}{w_2}+\ldots+\frac{\alpha_n}{w_n}=\frac{\Pi}{Z}=P.$$

In other words, the left hand quantity is the time measured in mean synodic periods that it takes the phenomenon to "travel" 360°. In the course of this exposition, Schmidt states, without discussion, that the times in a System A planetary Ephemeris are derived from the longitudes by the relation $\Delta t = \Delta\lambda + C$. He suggests that

$$\frac{\alpha_1}{w_1}+\frac{\alpha_2}{w_2}+\ldots+\frac{\alpha_n}{w_n}=\frac{\Pi}{Z}$$

is derived from a hypothetical Ephemeris of $\overline{\Delta\lambda}$ and $\overline{\Delta t}$ produced from observed values of Π and Z; this is changed so that the observed different values of Δλ, i.e. $w_1$, $w_2$, etc., are associated with the appropriate arcs of the ecliptic, i.e. $\alpha_1$, $\alpha_2$, etc. From the resulting values of Δλ are computed the appropriate values of Δt.

Aaboe [1980] tackled the problem of the derivation of System A (he used Mars as an example) from observations such as those recorded in the Diaries, which normally give only the date and the zodiacal sign for the Greek-letter phenomena. He imagines an a priori division of the 360° of the ecliptic into six equal arcs (α) of two zodiacal signs each, and then the distribution into these arcs of observed occurrences of a phenomenon. One starts by knowing the parameters of the period, Π = 133, Z = 18, and Y = 2Π + Z = 284 years. This tells one that there are 133 steps and that each Δλ contains 18 steps, as shown by van der Waerden. From the tallying over a relatively short period of time of all the occurrences of a chosen phenomenon in each of the six arcs it is immediately clear that the arc consisting of Cancer and Leo contains the largest number of occurrences, that containing Capricorn and Aquarius the least. The number of steps in each arc must be determined by applying to the crude data the rules previously discussed by Aaboe and others: the sum of all the steps is 133 and the number of steps in each arc must be regular. Then the crude results from tallying the steps over a period of 79 years, say, would be corrected as follows:

| Arc | "Observed" steps 79 yrs. | Calculated steps 284 yrs. | ACT steps | Actual steps 284 yrs. |
|---|---|---|---|---|
| ♉, ♊ | 7 | 25 6/37 | 24 | 25 |
| ♋, ♌ | 9 | 32 13/37 | 36 | 32 |
| ♍, ♎ | 8 | 28 28/37 | 27 | 28 |
| ♏, ♐ | 5 | 17 36/37 | 18 | 18 |
| ♑, ♒ | 3 | 10 29/37 | 12 | 14 |
| ♓, ♈ | 5 | 17 36/37 | 16 | 16 |
| | 37 | 133 | 133 | 133 |

The ACT roundings are intended to preserve nice ratios:
$\frac{24}{36}=\frac{2}{3}, \frac{36}{27}=\frac{4}{3}, \frac{27}{18}=\frac{3}{2}, \frac{18}{12}=\frac{3}{2}, \frac{12}{16}=\frac{3}{4}, \frac{16}{24}=\frac{2}{3}$.
The small deviations in the number of steps in each arc are of minor import; the continued accuracy of the system depends on the retention of the period, $P=\frac{\Pi}{Z}$.

As Aaboe remarks, to find $C = \overline{\Delta t} - \overline{\Delta\lambda}$ we need to know only that 1 year = 12;22,8 synodic months in addition to Π and Z. For $\overline{\Delta\lambda} = \frac{Z \times 360^\circ}{\Pi}$ and, for Mars, $\overline{\Delta t} = Y \times 12;22,8 \text{ months} \times 30\tau = \frac{2\Pi + Z}{\Pi} \times 12;22,8 \text{ months} \times 30\tau$. Therefore,

$$C = \overline{\Delta t} - \overline{\Delta\lambda} = 24 \text{ mos.} + (2+\tfrac{Z}{\Pi}) \times 11;4^{\tau} = 24 \text{ mos.} + 23;37,51,52,\ldots^{\tau}.$$

The Ephemerides of System A use C = 23;37,52. The only problem remaining is that of determining an initial longitude and time.

Recently Swerdlow [1998] offered a different theory of how the Babylonians derived the ACT theories of the planets from the observational material available to them. He notes (p. 51) that the positions of the occurrences of first and last visibilities given in the Diaries are not precise, but are simply by zodiacal sign (at least after ca. -400) with the only modification being an occasional indication that the occurrence was at the beginning or end of the zodiacal sign, whereas (pp. 42-44) there existed an (as yet uninvestigated) method of using the time-degrees between planet-rise and sunrise or between sunset and planet-set to correct the observed date to an estimated "ideal" date. On the other hand, the dates of stations are difficult to determine, but the distance of the stationary planet from a Normal Star is recorded. But he denies that these distances could provide "longitudes" precise enough for use. From these and other considerations he concludes that the basis of the Babylonian planetary systems were the observed or estimated times of the occurrences, and that the longitudes were computed by applying the rule that there is a constant difference between the entries for time and for longitude in the Ephemerides. Much of the book is devoted to a detailed demonstration of how the Babylonians could have developed their planetary theories from a careful examination of the lengths of synodic times in relation to zodiacal signs to generate first approximations that preserved the basic period relations, followed by adjustments designed to facilitate computational practices much like the adjustments proposed by Aaboe in his theory.

Swerdlow (pp. 183-186) objects to Aaboe's theory on the grounds that is is based on only Mars, where the arcs in System A are each exactly two zodiacal signs in length, whereas the arcs in the Systems A of the other planets are of varying length. This objection can be met, however, by hypothesizing that the distribution of occurrences was not by zodiacal signs, but by estimated longitudes. He also questions how System B could have been arrived at by Aaboe's method; the answer was provided by Neugebauer [1968].

In the course of his argument Swerdlow makes several assumptions that we believe to be erroneous. He insists that the purpose of the Diaries was connected with omens, and that the Ephemerides were designed to predict ominous events. But only first visibilities and first invisibilities among the Greek-letter phenomena were ever treated as ominous in Babylonian texts, and then only as they occurred in different months or near certain constellations; the "precision" of the Ephemerides is irrelevant to omens. Acronychal rising was not an omen; and the arc of retrogression was ominous, but not the stations themselves.

He also questions the ability of the Babylonian astronomers to observe longitudes, and even insists (p. 190) that they "were motivated above all by avoiding the requirement of precise measurement of longitude for the obvious reason that they knew perfectly well that they could do no such thing." This assumption ignores the existence of the velocity schemes of the planets that can be dated to the late fifth and fourth centuries B.C., the observed (in the Diaries) and computed (in the Almanacs) entries of the planets into the zodiacal signs, the accuracy of which is assured by Huber [1958], and texts such as A 3405, a third century B.C. tablet from Uruk, that gives the degrees of longitudes of the planets at their phases and of the Moon at its eclipses. Swerdlow's claim that passings-by of the Normal Stars and measurements of the distances of planetary stations from Normal Stars can not yield useful longitudes is controverted by Huber's results and by the fact that they are at least as accurate as estimated times. Indeed, if longitudes could not be observed with some degree of accuracy, we fail to see why they should be included in the Ephemerides at all.

These observations lead us to conclude that we are not yet in a position to decide among Aaboe's hypothesis, Swerdlow's hypothesis, and some mixed hypothesis. While it is true that the observations of planetary phenomena in the Diaries include "longitudes" only for the stations of the superior planets (there is an observation of $\Phi$ of Venus "2 fingers in front of $\beta$ Tauri" on 29 IX -136), it is also true that the recording of many planetary phenomena was skipped by the scribes of the Diaries. The Diaries must have been supplemented from other sources (including, probably, computation), or there may even have been another set of records of observations that we have not yet found. No one has yet demonstrated that it is possible to derive the Systems A of any planet from just the information in the Diaries.

In answer to Aaboe's question about initial longitudes at "epoch" dates, Swerdlow (pp. 141sqq.) suggests that the Babylonians used the principle that van der Waerden [1957] "discovered" for Mars, that the planet at each phase is distant from the mean Sun by a constant, $\eta$, particular to that phase. This principle might be checked by comparing dates and longitudes of phases in the Ephemerides with the longitudes of the mean Sun computed, e.g., with ACT 185-187, which use 0;59,9° as the mean daily velocity of the Sun. It seems to us improbable that one will find dates within the probable time-frame of the beginning of the Ephemerides (in the first half of the third century B.C.) when the ACT longitudes of the planets' phases are $\eta°$ distant from the ACT mean Sun.

If a Babylonian astronomer compared ACT predictions with observations (including those recorded in the Diaries), as many must have, he would find many disagreements. This did not discourage either the computers of Ephemerides or the compilers of Diaries from continuing their work. In the third century B.C., when the Ephemerides begin, Babylonian astronomers also began to compute, but *not* with ACT methods, the Normal Star Almanacs and the Almanacs and to compile from observational reports often identical to those in the Diaries and from computations the Goal-Year Texts. All five projects—Diaries, Ephemerides, Normal Star Almanacs, Almanacs, and Goal-Year Texts—happily coexisted, as far as we can tell, for the remainder of the life of cuneiform culture, until the third quarter of the first century A.D. We know that, at least for a time during the 2nd century B.C., the Esagil in Babylon supported astronomical observers—presumably for compiling the Diaries, and also for producing Ephemerides and Almanacs[57]; who sponsored the other projects is unclear. Also unanswered are the questions of why this astronomical activity was regarded as important and of what role each of the projects played in it. The question of who the sponsors were may be answered by new archival material; some light at least might be thrown on the other questions by imaginative investigations into the designs of the five projects and their interrelations.

[57] The contracts issued by the Esagil have been treated by G. McEwan, Priest and Temple in Hellenistic Babylonia, pp. 17 and 19; see also van der Spek [1985] pp. 548-555. The astronomers ("Scribes of (the collection) Enūma Anu Enlil") are to make observations (*ša naṣār inaṣṣarū*; this probably refers to the Diaries which are labeled *naṣār ša ginê*) and to deliver *tersētu u meš-ḫi*[meš]. Tersītu occurs, probably as a title, in the colophons of Ephemerides; *meš-ḫi* is the beginning of the title of Almanacs or Normal Star Almanacs; the terminology varies between Babylon and Uruk.

## APPENDIX

### Catalogue of Constellations and Star-names with Tentative Identifications

In this book, we use to a large extent literal translations of ancient star names because many identifications are uncertain. The following is a list of these names; where no modern star name is given, we think that an identification is not yet possible.

| | | |
|---|---|---|
| Abundant One | | β Comae Berenices ? |
| Anunītu | | Eastern fish and part of line of Pisces |
| Bright star of Anunītu | | |
| Fin of Anunītu | | |
| Bright star in ribbon of Fish *(nūnu)* | | η Piscium |
| Arrow | | Canis Maior, Canis Minor, and parts of Puppis and Pyxis |
| Constellation of Arrow | | ρ Puppis |
| Elbow of Arrow | | α Pyxis |
| Left foot of Arrow | | π Puppis |
| Bark | | ε Sagittarii |
| Barley-stalk | See Furrow | |
| Bison | | |
| Head of Bison | | |
| Left foot of Bison | | |
| Left hand of Bison | | |
| Middle of Bison | | |
| Right foot of Bison | | |
| Right hand of Bison | | |
| Bow | | δ, ε, σ, ω Canis Minoris and κ Puppis |
| Bristle | See Stars | |
| Bull of Heaven | | Taurus |
| Bright star of Bull of Heaven | | |
| Jaw of the Bull | | α Tauri and the Hyades |
| Chariot | | Northern part of Taurus |
| Northern reins of Chariot | | β Tauri |
| Southern reins of Chariot | | ζ Tauri |
| Circle | | Corona Borealis |
| Clothing | | |

| | | |
|---|---|---|
| Crab | | Cancer |
| Front star of Crab to North | | η Cancri |
| Front star of Crab to South | | ϑ Cancri |
| Rear star of Crab to North | | γ Cancri |
| Rear star of Crab to South | | δ Cancri |
| Crook | | Auriga |
| Hand of Crook | | ϑ and υ Aurigae |
| Dead Man | | Delphinus ? |
| Deleter | | β Andromedae |
| Dignity | See Circle | |
| Dog | | Southern part of Hercules |
| Mouth of Dog | | |
| Tongue of Dog | | |
| Doublets | | β, γ Herculis |
| Eagle | | Aquila |
| Bright star of Eagle | | α Aquilae |
| Enmešarra | See Chariot | |
| Bright star in front of Enmešarra | | |
| EN.TE.NA.BAR.ḪUM | See Ḫabaṣirānu | |
| Eridu | | Parts of Puppis and Vela |
| Hands of Eridu | | γ Velorum |
| Erragal | | ζ Lyrae |
| Eru, Erua | See Frond of Eru | |
| Ewe | | Northeastern part of Boötes ? |
| Field | | α, β, γ Pegasi and α Andromedae |
| Fish | | Piscis Austrinus |
| Middle of Fish | | |
| Fox | | 80-86 Ursae Maioris ? |
| Frond of Eru | | γ Comae Berenices |
| Furrow | | Virgo |
| Base of Furrow | | |
| Bright star of Furrow | | α Virginis |
| Single star in front of Furrow | | γ Virginis |
| Goat | See She-goat | |
| Goat-Fish | | Capricorn |
| Front star of Goat-Fish | | γ Capricorni |
| Horn of Goat-Fish | | β Capricorni |
| Middle of Goat-Fish | | |
| Rear star of Goat-Fish | | δ Capricorni |

| | | |
|---|---|---|
| Great One | | Aquarius |
| Front basket of Great One | | φ or χ Aquarii |
| Rear basket of Great One | | λ Piscium |
| Great Twins | See Twins | |
| GU.LA | See Great One | |
| Ḫabaṣirānu | | Most of Centaurus |
| Head of Ḫabaṣirānu | | δ Centauri |
| Left foot of Ḫabaṣirānu | | |
| Ḫaniš | | ε or ν Centauri ? |
| Harness | | η, υ Boötis. See ŠU.PA |
| Second star of Harness | | ε Boötis |
| Stars following Harness | | ε, ξ, ο, π Boötis |
| Ḫarriru | See Rainbow | |
| Harrow | | Eastern part of Vela |
| Bite of Harrow | | μ Velorum |
| Heir of the Sublime Temple | | α Ursae Minoris ? |
| Hired Man | | Aries |
| Front star of head of Hired Man | | β Arietis |
| Rear star of head of Hired Man | | α Arietis |
| Middle of Hired Man ? | | |
| Hitched Yoke | | α Draconis ? |
| Horse | | α, β, γ, δ Cassiopeiae |
| Head of Horse | | |
| Jaw of the Bull | See Bull of Heaven | |
| Kidney | | |
| King | | α Leonis |
| Lady of Life | See She-goat | |
| Lamma | | α Lyrae |
| Latarak | | $\pi^4$ Orionis ? |
| Lion | | Leo |
| Breast of Lion | | α, γ, ζ, η Leonis. See King |
| Small star 4 cubits behind King | | ρ Leonis |
| Head of Lion | | ε, μ Leonis |
| Right front foot of Lion | | π Leonis |
| Foot in middle of Lion | | ρ Leonis |
| Rear foot of Lion | | β Virginis |
| Rump or Thigh of Lion | | δ, ϑ Leonis |
| Tail of Lion | | β Leonis |
| Dusky stars in tail of Lion | | 5 or 21 Leonis |

| | | |
|---|---|---|
| Lisi | | $\alpha$ Scorpii |
| Little Twins | See Twins | |
| Lulal | | $\pi^3$ Orionis ? |
| Lulim | See Stag | |
| | | |
| Mad Dog | | Lupus |
| Eye of Mad Dog | | $\gamma$ Lupi |
| Left hand of Mad Dog | | |
| Right hand of Mad Dog | | |
| Left foot of Mad Dog | | |
| Rear foot of Mad Dog | | $\zeta^2$ Scorpii |
| Middle of Mad Dog | | $\eta$ Lupi |
| | | |
| naṣrapu | | b, c Persei |
| NIN.MAḪ | | Part of Vela |
| Hand of NIN.MAḪ | | $\varphi$ Velorum |
| Nin-SAR | | $\varepsilon$ Lyrae |
| Numušda | | $\eta$ or $\kappa$ Centauri |
| NUN.KI | See Eridu | |
| | | |
| Old Man | | Perseus |
| Breast of Old Man | | |
| Bright star of Old Man | | $\alpha$ or $\eta$ Persei |
| Dusky stars of Old Man | | h, $\chi$ Persei |
| Right foot of Old Man | | |
| Shoulder of Old Man | | |
| | | |
| Pabilsag | | Sagittarius and part of Ophiuchus |
| 4 stars in East of Pabilsag | | 13-16 Sagittarii |
| 4 stars in West of Pabilsag | | $\nu^1$, $\nu^2$, $\xi^1$, $\xi^2$ Sagittarii |
| Left hand of Pabilsag on bow | | $\delta$ Sagittarii |
| Right hand of Pabilsag on arrow | | $\varphi$ Sagittarii |
| Bright star on tip of Pabilsag's arrow | | $\vartheta$ Ophiuchi |
| Sting of Pabilsag | | |
| Pabilsag above Bark | | $\iota$ Sagittarii |
| Panther | | Cygnus, Lacerta, and parts of Cassiopeia and Cepheus |
| Breast of the Panther | | $\alpha$ Cygni |
| Crown of the Panther | | |
| Foot of the Panther | | |
| Basis of the left foot of the Panther | | |
| Right foot of the Panther | | |
| Toe of the right foot of the Panther | | |

| | |
|---|---|
| Heel of the Panther | λ Andromedae |
| Knee of the Panther | α, β Lacertae |
| Shoulder of the Panther | β or γ Cygni |
| Pig | Head and first coil of Draco ? |
| Breast of the Pig | |
| Pelvis of the Pig | |
| Plow | α, β Trianguli and γ Andromedae |
| | |
| Rainbow | 18, 31, 32 Andromedae ? |
| Raven | Corvus and Crater |
| Middle of the Raven | γ Corvi |
| Tail of the Raven | α Crateris |
| Rooster | Lepus |
| Middle of the Rooster | α Leporis |
| | |
| Šargaz | υ Scorpii |
| Šarur | λ Scorpii |
| Scales | Libra |
| Front pan of the Scales (or Southern part of the Scales) | α Librae |
| Middle of the Scales (or Northern part of the Scales) | β Librae |
| Middle of the Scales | γ Librae |
| Front (star) of the Scales | |
| Scorpion | Scorpius |
| Breast of the Scorpion | α Scorpii. See Lisi |
| Head of the Scorpion | |
| Middle star of the head of the Scorpion | δ Scorpii |
| Upper star of the head of the Scorpion | β Scorpii |
| Horn of the Scorpion | γ Scorpii |
| Right horn of the Scorpion | |
| Base of the upright of the Scorpion | |
| Upraised tail of thc Scorpion | $\mu^1$ Scorpii |
| Sting of the Scorpion See Šarur and Šargaz | |
| She-goat | Lyra |
| Crook of the She-goat | α Lyrae. See Lamma |
| Left hand of the She-goat | |
| Foot of the She-goat | |
| Knee of the She-goat | μ Herculis |
| Single star | μ or ϑ Herculis |
| Sitting gods (of Ekur) | μ Virginis |

| | |
|---|---|
| Snake | Hydra |
| Head of the Snake | |
| Middle of the Snake | ϑ, κ Hydrae |
| Southern Yoke | |
| Stag | Eastern part of Andromeda |
| Hip of the Stag | |
| Horn of the Stag | |
| PA.SI of the Stag | |
| Standing gods (of Ekur) | ζ, η Herculis ? |
| Stars | Pleiades (η+ Tauri) |
| Šullat | μ Centauri ? |
| ŠU.PA | Boötes |
| Swallow | ε, ζ, ϑ Pegasi, α Equulei, and the western fish of Pisces |
| Head of the Swallow | |
| Tails of the Swallow | λ Piscium |
| Triplets | α, δ Herculis |
| True Shepherd of Anu | Orion |
| Crown of the True Shepherd of Anu | |
| Right hand of the True Shepherd of Anu | ξ Orionis |
| Rear heel of the True Shepherd of Anu | κ Orionis |
| Twins | Gemini |
| Great Twins | α, β Geminorum |
| Front Twin star | α Geminorum |
| Rear Twin star | β Geminorum |
| Front star of Twins' feet | η Geminorum |
| Rear star of Twins' feet | μ Geminorum |
| Hands of front Great Twin | ϑ Geminorum |
| Little Twins | ζ, λ Geminorum |
| Front star of Little Twins | |
| Rear (star) of Little Twins | |
| Twins' star near the True Shepherd of Anu | γ Geminorum |
| Wagon | Ursa Maior |
| Wagon of Heaven | Ursa Minor |
| Wildcat | |
| Head of the Wildcat | |
| Wolf | α Trianguli. See Plow |
| Yoke | α Boötis. See ŠU.PA |

| | |
|---|---|
| Zababa | Parts of Ophiuchus and Serpens |
| Crown of Zababa | |
| Eye of Zababa | η Ophiuchi |
| Shoulder of Zababa | |
| Left shoulder of Zababa | |
| Right shoulder of Zababa | |
| Middle of Zababa | ν Ophiuchi |
| Shin of Zababa | η Serpentis |
| Foot of Zababa | λ Aquilae |
| Left foot of Zababa | |
| Right foot of Zababa | |

# BIBLIOGRAPHY


Aaboe [1958] - Aaboe, A., "On Babylonian Planetary Theories," Centaurus 5, 204-277

Aaboe [1964] - Aaboe, A., "A Seleucid Table of Daily Solar(?) Positions," JCS 18, 31-34

Aaboe [1965] - Aaboe, A., "On Period Relations in Babylonian Astronomy," Centaurus 10, 213-231

Aaboe [1966] - Aaboe, A., "On a Babylonian Scheme for Solar Motion of the System A Variety," Centaurus 11, 302-303

Aaboe [1968] - Aaboe, A., Some Lunar Auxiliary Tables and Related Texts from the Late Babylonian Period, KDVSMM 36/12, København

Aaboe [1969] - Aaboe, A., A Computed List of New Moons for 319 B.C. to 316 B.C. from Babylon: B.M. 40094, KDVSMM 37/3, København

Aaboe [1971] - Aaboe, A., Lunar and Solar Velocities and the Length of Lunation Intervals in Babylonian Astronomy, KDVSMM 38/6, København

Aaboe [1972] - Aaboe, A., "Remarks on the Theoretical Treatment of Eclipses in Antiquity," JHA 3, 105-118

Aaboe [1977] - Aaboe, A., "A Computed Cuneiform Text for Mercury from Babylon: B.M. 48147," Ποίσματα, Festschrift Willy Hartner, Wiesbaden, 1-8

Aaboe [1987] - Aaboe, A., "A Late Babylonian Procedure Text for Mars, and Some Remarks on Retrograde Arcs," From Deferent to Equant, Festschrift E. Kennedy, New York, 1-14

Aaboe [1991] - Aaboe, A., "Babylonian Mathematics, Astrology, and Astronomy," Cambridge Ancient History III/2, 276-292

Aaboe-Britton-Henderson-Neugebauer-Sachs [1991] - Aaboe, A., Britton, J., Henderson, J., Neugebauer, O., and Sachs, A., Saros Cycle Dates and Related Babylonian Astronomical Texts, Philadelphia

Aaboe-Henderson [1975] - Aaboe, A. and Henderson, J., "The Babylonian Theory of Lunar Latitude and Eclipses According to System A," AIHS 25, 181-222

Aaboe-Huber [1977] - Aaboe, A. and Huber, P. J., "A Text Concerning Subdivision of the Synodic Motion of Venus from Babylon: BM 37151," Essays on the Ancient Near East in Memory of Jacob Joel Finkelstein, Hamden, CT, 1-4

Aaboe-Sachs [1966] - Aaboe, A. and Sachs, A. J., "Some Dateless Computed Lists of Longitudes of Characteristic Planetary Phenomena From the Late Babylonian Period," JCS 20, 1-33

Aaboe-Sachs [1969] - Aaboe, A. and Sachs, A. J., "Two Lunar Texts of the Achaemenid Period from Babylon," Centaurus 14, 1-22

Arnaud [1987] - Arnaud, D., Recherches au pays d'Aštata: Emar VI/4, Paris

Arnaud [1996] - Arnaud, D., "L'edition ougaritaine de la série astrologique «Éclipses du dieu-soleil»," Semitica 45, 7-18

Ashfaque [1974-1986] - Ashfaque, S. M., "Constellations in the Harappan Seals," Pakistan Archaeology 10-22, 135-167

Bauer [1936] – Bauer, Th., "Eine Sammlung von Himmelsvorzeichen," ZA 43, 308-314

Beaulieu [1993] - Beaulieu, P.-A., "The Impact of Month-lengths on the Neo-Babylonian Cultic Calendars," ZA 83, 66-87

Beaulieu-Britton [1994] - Beaulieu, P.-A. and Britton, J. P., "Rituals for an Eclipse Possibility in the 8th Year of Cyrus," JCS 46, 73-86

Becker [1992] - Becker, U., "Babylonische Sternbilder und Sternnamen, Teil I," Wissenschaft und Fortschritt 42, 235-237

Bernsen [1969] - Bernsen, L., "On the Construction of Column B in System A of the Astronomical Cuneiform Texts," Centaurus 14, 23-28

Bezold [1911] - Bezold, C., Astronomie, Himmelsschau und Astrallehre bei den Babyloniern, SHAW, Phil.-hist. Kl. 1911/2

Bezold-Kopff-Boll [1913] - Bezold, C., Kopff, A., and Boll, F., Zenit- und Aequatorialgestirne am babylonischen Fixsternhimmel, SHAW, Phil.-hist. Kl., 1913/11

Bidez-Cumont [1938] - Bidez, J. and Cumont, F., Les mages hellénisés, 2 vols., Paris

Boll [1911] - Boll, F., "Zur babylonischen Planetenordnung," ZA 25, 372-377

Boll [1913] - Boll, F., "Neues zur babylonischen Planetenordnung," ZA 28, 340-351

Boll [1916] - Boll, F., "Antike Beobachtungen farbiger Sterne," Abh. d. Bayr. Akad. d. Wiss., phil.-hist. Kl., 30/1

Borger [1956] - Borger, R., Die Inschriften Asarhaddons, Königs von Assyrien, Archiv für Orientforschung, Beiheft 9, Graz

Borger [1973] - Borger, R., "Der astrologische Text LB 1321," Festschrift F. M. Th. de Liagre Böhl, Leiden, 38-43

Bottéro [1950] - Bottéro, J., "Autres textes de Qatna," RA 44, 105-112

Brack-Bernsen [1980] - Brack-Bernsen, L., "Some Investigations on the Ephemerides of the Babylonian Moon Texts, System A," Centaurus 24, 36-50

Brack-Bernsen [1990] - Brack-Bernsen, L., "On the Babylonian Lunar Theory: A Construction of Column Φ from Horizontal Observations," Centaurus 33, 39-56

Brack-Bernsen [1993] - Brack-Bernsen, L., "Babylonische Mondtexte: Beobachtung und Theorie," Die Rolle der Astronomie in den Kulturen Mesopotamiens, Graz, 331-358

Brack-Bernsen [1994] - Brack-Bernsen, L., "Konsistenz zwischen Kolonne phi und babylonischen Aufzeichnungen der "Lunar Four"," in Ad Radices. Festband zum fünfzigjährigen Bestehen des Instituts für Geschichte der Naturwissenschaften der Johann Wolfgang Goethe Universität, Stuttgart, 45-64

Brack-Bernsen [1997] - Brack-Bernsen, L., Zur Entstehung der babylonischen Mondtheorie, Stuttgart

Brack-Bernsen-Schmidt [1994] - Brack-Bernsen, L. and Schmidt, O., "On the Foundations of the Babylonian Column Φ: Astronomical Significance of Partial Sums of the Lunar Four," Centaurus 37, 183-209

Bremner [1993] - Bremner, R. W., "The Shadow Length Table in Mul-Apin," Die Rolle der Astronomie in den Kulturen Mesopotamiens, Graz, 367-382

Brinkman-Kennedy [1983] - Brinkman, J. A. and Kennedy, D. A., "Documentary Evidence For the Economic Base of Early Neo-Babylonian Society," JCS 35, 1-90

Britton [1989] - Britton, J. P., "An Early Function for Eclipse Magnitudes in Babylonian Astronomy," Centaurus 32, 1-52

Britton [1993] - Britton, J., "Scientific Astronomy in Pre-Seleucid Babylon," Die Rolle der Astronomie in den Kulturen Mesopotamiens, Graz, 61-76

Britton-Walker [1991] - Britton, J. and Walker, C. B. F., "A 4th Century Babylonian Model for Venus: B.M. 33552," Centaurus 34, 97-118

Britton-Walker [1996] - Britton, J. and Walker, C., "Astronomy and Astrology in Mesopotamia," Astronomy before the telescope, ed. C. Walker, London, 42-67

Çağırgan [1984] - Çağırgan, G., "Three more duplicates to Astrolabe B," Belleten 48, 399-416

Cassianus Bassus, Geoponica: Geoponica sive Cassiani Bassi Scholastici De re rustica eclogae, ed. H. Beckh, Leipzig 1895

Chadwick [1991] - Chadwick, R., "Celestial Episodes and Celestial Objects in Ancient Mesopotamia," Bull. Can. Soc. Mesop. Studies 22, 43-50

Cohen [1990] - Cohen, M. E., "$^{iti}$úd-duru$_5$: The Reading for the Nippur Month $^{iti}$ZIZ$_2$.A," NABU 1990 no. 134

Cornelius [1942] - Cornelius, F., "Berossos und die altorientalische Chronologie," Klio 35, 1-16

Dietrich [1996] - Dietrich, M., "Altbabylonische Omina zur Sonnenfinsternis," WZKM 86, 99-106

Dietrich-Loretz [1990] – Dietrich, M. - Loretz, O., Mantik in Ugarit, Münster

Dindorf [1855] - Dindorf, G., Scholia Graeca in Homeri Odysseam, 2 vols., Oxford

Dossin [1927] - Dossin, G., Autres textes sumériens et accadiens, Paris

Donbaz-Koch [1995] - Donbaz, V. and Koch, J., "Ein Astrolab der dritten Generation: NV. 10," JCS 47, 63-84

Durand [1988] - Durand, J.-M., Archives épistolaires de Mari I/1 = Archives royales de Mari 26, Paris

Epping [1881] - Epping, J., "Zur Entzifferung der astronomischen Tafeln der Chaldäer," Stimmen aus Maria Laach 21, 1881, 277-292

Epping [1889a] - Epping, J., with J. N. Strassmaier, Astronomisches aus Babylon, Ergänzungsheft zu den Stimmen aus Maria Laach 44, Freiburg im Breisgau

Epping [1889b] - Epping, J., "Aus einem Briefe des Herrn Professor J. Epping an J. N. Strassmaier," ZA 4, 76-82

Epping [1890a] - Epping, J., "Die babylonische Berechnung des Neumondes," Stimmen aus Maria Laach 39, 225-240

Epping [1890b] - Epping, J., "Sachliche Erklärung des Tablets No. 400 der Cambyses-Inschriften," ZA 5, 281-288

Epping-Strassmaier [1890/1891] - Epping, J. and Strassmaier, J. N., "Neue babylonische Planeten-Tafeln," ZA 5, 341-366; and ZA 6, 89-102 and 217-244

Epping-Strassmaier [1892] - Epping, J. and Strassmaier, J. N., "Babylonische Mondbeobachtungen aus den Jahren 38 und 79 der Seleuciden-Aera," ZA 7, 220-254

Epping-Strassmaier [1893] - Epping, J. and Strassmaier, J. N., "Der Saros-Canon der Babylonier," ZA 8, 149-178

Farber [1993] - Farber, W., "Zur Orthographie von EAE 22: Neue Lesungen und Versuch einer Deutung," Die Rolle der Astronomie in den Kulturen Mesopotamiens, Graz, 247-257

Fatoohi-Stephenson-al-Dargazelli [1999] - Fatoohi, L. J., Stephenson, F. R., and al-Dargazelli, S. S., "The Babylonian First Visibility of the Lunar Crescent: Data and Criterion," JHA 30, 51-72

Ferrari d'Occhieppo [1978] - Ferrari d'Occhieppo, K., "Wann wurde die 1151jährige Venus-Periode entdeckt?," SÖAW, math.-nat. Kl., Abt. II, 186, 441-447

Fotheringham [1928] - Fotheringham, J. K., "The Indebtedness of Greek to Chaldaean Astronomy," The Observatory 51, 301-315

Fotheringham [1932] - Fotheringham, J. K., in The Observatory 703, Dec. 1932, 338-340

Foxvog [1993] - Foxvog, D. A., "Astral Dumuzi," The Tablet and the Scroll (Festschrift W. W. Hallo), Bethesda, 103-108

Freydank [1992] - Freydank, H., Beiträge zur mittelassyrischen Chronologie und Geschichte, Berlin

Gadd [1967] - Gadd, C. J., "Omens expressed in numbers," JCS 21, 52-63

Gehlken [1991] - Gehlken, E., "Der längste Tag in Babylon (MUL.APIN und die Wasseruhr)," NABU 1991 no. 95

Geller [1990a] - Geller, M. J., "Astronomical Diaries and Corrections of Diodorus," BSOAS 53, 1-7

Geller [1990b] - Geller, M. J., "Astronomy and Authorship," BSOAS 53, 209-213

Gera-Horowitz [1997] - Gera, D. and Horowitz, W., "Antiochus IV in Life and Death: Evidence from the Babylonian Astronomical Diaries," JAOS 117, 240-252

Gleßmer [1996] - Gleßmer, U., "Horizontal Measuring in the Babylonian Astronomical Compendium MUL.APIN and in the Astronomical Book of 1 En," Henoch 18, 259-282

Goldstein [1980] - Goldstein, B. R., "Babylonian Solar Theory Reconsidered," AIHS 30, 189-191

Goldstine [1973] - Goldstine, H. H., New and Full Moons 1001 B.C. to A.D. 1651, Philadelphia

Gössmann [1950] - Gössmann, F., Planetarium Babylonicum, Rom

Grasshoff [1998] - Grasshoff, G., "Normal Star Observations in Late Babylonian Astronomical Diaries," Ancient Astronomy and Celestial Divination, Cambridge, Mass.

Grayson [1975] - Grayson, A. K., Assyrian and Babylonian Chronicles, Locust Valley, NY

Güterbock [1988] – Güterbock, H. G., "Bilingual Moon Omens from Boğazköy," A Scientific Humanist: Studies in Memory of Abraham Sachs; 161-173

Hamilton-Aaboe [1998] - Hamilton, N. T. and Aaboe, A., "A Babylonian Venus Text Computed According to System A: ACT No. 1050," AHES 53, 215-221

Heimpel [1982] - Heimpel, W., "The Sun at Night and the Doors of Heaven in Babylonian Texts", JCS 38, 127-151

Hommel [1909] - Hommel, F., "Die babylonisch-assyrischen Planetenlisten," Hilprecht Anniversary Volume, Leipzig, 170-188

Horowitz [1989] - Horowitz, W., "The Akkadian Name for Ursa Minor: $^{mul}$mar.gíd.da.an.na = eriqqi šamê/šamāmi," ZA 79, 242-244

Horowitz [1989/1990] - Horowitz, W., "Two MUL.APIN Fragments," AfO 36/37, 116-117

Horowitz [1990] - Horowitz, W., "More Writings for Ursa Major with Determinative giš," NABU 1990 no. 4

Horowitz [1993] - Horowitz, W., "The Reverse of the Neo-Assyrian Planisphere CT 33,11," Die Rolle der Astronomie in den Kulturen Mesopotamiens, Graz, 149-159

Horowitz [1994] - Horowitz, W., "Two New Ziqpu-star Texts and Stellar Circles," JCS 46, 89-98

Horowitz [1998] - Horowitz, W., Mesopotamian Cosmic Geography, Winona Lake, Indiana

Høyrup [1997/1998] - Høyrup, J., "A Note on Water-Clocks and on the Authority of Texts," AfO 44/45, 192-194

Huber [1957] - Huber, P., "Zur täglichen Bewegung des Jupiter nach babylonischen Texten," ZA 52, 265-303

Huber [1958] - Huber, P., "Ueber den Nullpunkt der babylonischen Ekliptik," Centaurus 5, 192-208

Huber [1982] - Huber, P. J., Astronomical Dating of Babylon I and Ur III, Malibu

Huber [1987a] – Huber, P. J., "Dating by Lunar Eclipse Omens with Speculations on the Birth of Omen Astrology," From Ancient Omens to Statistical Mechanics (Festschrift Asger Aaboe), Copenhagen, 3-13

Huber [1987b] - Huber, P. J., "Astronomical Evidence for the Long and against the Middle and Short Chronologies," High, Middle or Low?, ed. P. Åström, part 1, Gothenburg, 5-17

Hunger [1969] - Hunger, H., "Kryptographische astrologische Omina," AOAT 1, 133-145

Hunger [1972] - Hunger, H., "Neues von Nabû-zuqup-kēna," ZA 62, 99-101

Hunger [1976a] - Hunger, H., Spätbabylonische Texte aus Uruk, Teil I, Berlin

Hunger [1976b] - Hunger, H., "Astrologische Wettervorhersagen," ZA 66, 234-260

Hunger [1982] - Hunger, H., "Zwei Tafeln des astronomischen Textes MUL.APIN im Vorderasiatischen Museum zu Berlin," Forschungen und Berichte 22, 127-135

Hunger [1988] - Hunger, H., "A 3456: eine Sammlung von Merkurbeobachtungen," A Scientific Humanist: Studies in Memory of Abraham Sachs, Philadelphia, 201-223

Hunger [1991] - Hunger, H., "Schematische Berechnungen der Sonnenwenden," Baghdader Mitteilungen 22, 513-519

Hunger [1992] - Hunger, H., Astrological Reports to Assyrian Kings, Helsinki

Hunger [1996] - Hunger, H., "Ein astrologisches Zahlenschema," WZKM 86, 191-196

Hunger [1998] - Hunger, H., "Zur Lesung sumerischer Zahlwörter," AOAT 253, 179-183

Hunger [1999] - Hunger, H., "Saturnbeobachtungen aus der Zeit Nebukadnezars II.," forthcoming

Hunger-Dvorak [1981] - Hunger, H. and Dvorak, R., Ephemeriden von Sonne, Mond und hellen Planeten von -1000 bis -601, Wien

Hunger-Pingree [1989] - Hunger, H. and Pingree, D., MUL.APIN. An Astronomical Compendium in Cuneiform, AfO Beiheft 24, Horn

Jacobsen [1946] - Jacobsen, Th., "Sumerian Mythology. A Review Article", JNES 5, 128-152

Jacobsen [1990] - Jacobsen, Th., "Enuma Elisj: De Schepping van der Wereld," De Schepping van der Wereld, ed. D. van der Plas, B. Becking, and D. Meijer, Muiderberg, no page nos.

Jeyes [1980] - Jeyes, U., "The Act of Extispicy in Ancient Mesopotamia: An Outline," Assyriological Miscellanies 1, 13-32

Jeyes [1991/1992] - Jeyes, U., "Divination as a Science in Ancient Mesopotamia," JEOL 32, 23-41

Jones [1983] - Jones, A., "The Development and Transmission of 248-day Schemes for Lunar Motion in Ancient Astronomy," AHES 29, 1-36

Jones [1990] - Jones, A., "Ptolemy's First Commentator," TAPS 80/7, Philadelphia

Jones [1993] - Jones, A., "Evidence for Babylonian arithmetical schemes in Greek astronomy," Die Rolle der Astronomie in den Kulturen Mesopotamiens, Graz, 77-94

Jones [1999] - Jones, A., Astronomical Papyri from Oxyrhynchus, Philadelphia

de Jong-van Soldt [1987/1988] - de Jong, T. and van Soldt, W. H., "Redating an Early Solar Eclipse Record (KTU 1.78)," JEOL 30, 65-77

Kennedy [1986] - Kennedy, D. A., "Documentary Evidence For the Economic Base of Early Neo-Babylonian Society, Part II" JCS 38, 172-244

King [1913] - King, L. W., "A Neo-Babylonian Astronomical Treatise in the British Museum and its Bearing on the Age of Babylonian Astronomy," PSBA 43, 41-46

Koch [1989] - Koch, J., Neue Untersuchungen zur Topographie des babylonischen Fixsternhimmels, Wiesbaden

Koch [1991] - Koch, J., "Der Mardukstern Nēberu," WO 22, 48-72

Koch [1991/1992a] - Koch, J., "Zu einigen astronomischen 'Diaries'," AfO 38/39, 101-109

Koch [1991/1992b] - Koch, J., "Irrungen und Wirrungen einer Rezension," AfO 38/39, 125-130

Koch [1992] - Koch, J., "Der Sternenkatalog BM 78161," WO 23, 39-67

Koch [1993] - Koch, J., "Das Sternbild $^{\text{mul}}$maš-tab-ba-tur-tur," Die Rolle der Astronomie in den Kulturen Mesopotamiens, Graz, 185-198

Koch [1995] - Koch, J., "Der Dalbanna-Sternenkatalog," WO 26, 43-85

Koch [1995/1996] - Koch, J., "MUL.APIN II i 68-71," AfO 42/43, 155-162

Koch [1996] - Koch, J., "AO 6478, MUL.APIN und das 364 Tage-Jahr," NABU 1996 No. 111

Koch [1999] - Koch, J., "Die Planeten-Hypsomata in einem babylonischen Sternenkatalog," JNES 58, 19-31

Koch-Westenholz [1990] - Koch-Westenholz, U., "Eine neue Interpretation der Kudurru-Symbole," AHES 41/2, 93-114

Koch-Westenholz [1993] - Koch-Westenholz, U., "Mesopotamian Astrology at Hattusas," Die Rolle der Astronomie in den Kulturen Mesopotamiens, Graz, 231-246

Koch-Westenholz [1995] - Koch-Westenholz, U., Mesopotamian Astrology, Copenhagen

Kramer [1960] - Kramer, S. N., Two Elegies on a Pushkin Museum Tablet, Moscow

Kramer [1961] - Kramer, S. N., Sumerian Mythology, New York

Kudlek-Mickler [1971] - Kudlek, M. and Mickler, E. H., Solar and Lunar Eclipses of the Ancient Near East from 3000 B.C. to 0 with Maps, Kevelaer/Neukirchen-Vluyn

Kugler [1900a] - Kugler, F. X., Die babylonische Mondrechnung, Freiburg im Breisgau

Kugler [1900b] - Kugler, F. X., "Zur Erklärung der babylonischen Mondtafeln, I. Mond- und Sonnenfinsternisse," ZA 15, 178-209

Kugler [1900c] - Kugler, F. X., "Astronomische Masse der Chaldäer," ZA 15, 383-392

Kugler [1902] - Kugler, F. X., "Astronomische und meteorologische Finsternisse (Eine assyriologisch-kosmologische Untersuchung)," ZDMG 56, 60-70

Kugler [1903] - Kugler, F. X., "Eine rätselvolle astronomische Keilinschrift (Strm. Kambys. 400)," ZA 17, 203-238

Kugler [1907] - Kugler, F. X., Sternkunde und Sterndienst in Babel I. Babylonische Planetenkunde, Münster in Westfalen

Kugler [1909] - Kugler, F. X., "Darlegungen und Thesen über altbabylonische Chronologie," ZA 22, 63-78

Kugler [1909/1910] - Kugler, F. X., Sternkunde und Sterndienst in Babel II. Babylonische Zeitordnung und ältere Himmelskunde, Teil I, Münster in Westfalen

Kugler [1911a] - Kugler, F. X., "Chronologisches und Soziales aus der Zeit Lugalanda's und Urukagina's," ZA 25, 275-280

Kugler [1911b] - Kugler, F. X., "Some New Lights on Babyloniam Astronomy," ZA 25, 304-320

Kugler [1912] - Kugler, F. X., Sternkunde und Sterndienst in Babel II, Teil II, Heft 1, Münster in Westfalen

Kugler [1913] - Kugler, F. X., Sternkunde und Sterndienst in Babel. Ergänzungen zum ersten und zweiten Buch, Teil I, Münster in Westfalen

Kugler [1914a] - Kugler, F. X., Sternkunde und Sterndienst in Babel. Erg. Teil II, Münster in Westfalen

Kugler [1914b] - Kugler, F. X., "Distances entre étoiles fixes d'après une tablette de l'époque des Séleucides. Deuxième partie," RA 11, 1-21

Kugler [1924] - Kugler, F. X., Sternkunde und Sterndienst in Babel II, Teil II, Heft 2, Münster in Westfalen

Kugler-Schaumberger [1933a] - Kugler, F. X. and Schaumberger, J., "Drei planetarische Hilfstafeln," Analecta Orientalia 6, 3 12

Kugler-Schaumberger [1933b] - Kugler, F. X. and Schaumberger, J., "Drei babylonische Planetentafeln der Seleukidenzeit," Orientalia NS 2, 97-116

Kümmel [1967] - Kümmel, H. M., Ersatzrituale für den hethitischen König, Wiesbaden

Labat [1965] - Labat, R., Un calendrier babylonien des travaux, des signes et des mois, Paris

Lacheman [1937] - Lacheman, E. R., "An omen text from Nuzi," RA 34, 1-8

Lambert [1989] - Lambert, W. G., "The Month Names of Old Babylonian Sippar," NABU 1989 no. 90

Lanfranchi-Parpola [1990] - Lanfranchi, G. B. and Parpola, S., The Correspondence of Sargon II, Part II, Helsinki

Langdon [1923] - Langdon, S., The Weld-Blundell Collection II, Oxford editions of cuneiform texts 2, Oxford

Langdon [1935] - Langdon, S., Babylonian Menologies and the Semitic Calendars, London

Largement [1957] - Largement, R., "Contribution à l'étude des astres errants dans l'astrologie chaldéenne," ZA 52, 235-264

Larsen [1987] - Larsen, M. T., "The Mesopotamian Lukewarm Mind," Language, Literature, and History (Festschrift Erica Reiner), 203-225

Lasserre [1966] - Lasserre, F., Die Fragmente des Eudoxos von Knidos, Berlin

Leibovici [1956] – Leibovici, M., "Un texte astrologique akkadien de Boghazköi," RA 50, 11-21

Lenormant [1873] - Lenormant, F., Choix de textes cunéiformes inédits ou incomplétement publiés, Paris

Lieberman [1990] - Lieberman, S., "Canonical and Official Cuneiform Texts: Towards an Understanding of Assurbanipal's Personal Tablet Collection," Lingering over Words (Festschrift W. L. Moran), pp. 305-336

Livingstone [1986] - Livingstone, A., Mystical and Mythological Explanatory Works of Assyrian and Babylonian Scholars, Oxford

Mahler [1891] - Mahler, E., "Der babylonische Schaltcyclus," ZA 6, 457-464

Mahler [1894] - Mahler, E., "Der Schaltcyclus der Babylonier," ZA 9, 42-61

Mahler [1898] - Mahler, E., "Der Schaltcyklus der Babylonier," ZDMG 52, 227-246

Mahler [1909] - Mahler, E., "Der Kalender der Babylonier," Hilprecht Anniversary Volume, Leipzig, 1-13

Maneveau [1992] - Maneveau, B., "Astronomie: contribution de l'Assyriologie à la couleur de Sirius," NABU 1992 no. 7

Martiny [1932] - Martiny, G., Die Kultrichtung in Mesopotamien, Berlin

Maul [1994] - Maul, S. M., Zukunftsbewältigung, Mainz

De Meis-Hunger [1998] - De Meis, S. and Hunger, H., Astronomical Dating of Assyrian and Babylonian Reports, Roma

De Meis-Meeus [1991] - De Meis, S. and Meeus, J., "A propos d'occultation de planètes par la Lune dans des textes babyloniens," L'astronomie, Juin 1991, 1-3

Meissner [1924] - Meissner, B., "Zur neubabylonischen Schaltungspraxis," ZA 35, 42-43

Miller [1988] - Miller, R. A., "Pleiades Perceived. MUL.MUL to Subaru," JAOS 108, 1-25

Moesgaard [1980] - Moesgaard, K. P., "The Full Moon Serpent. A Foundation Stone of Ancient Astronomy?," Centaurus 24, 51-96

Neugebauer [1929] - Neugebauer, O., "Zur Frage der astronomischen Fixierung der babylonischen Chronologie," OLZ 32, 913-921

Neugebauer [1936a] - Neugebauer, O., "Über eine Untersuchungsmethode astronomischer Keilschrifttexte," ZDMG 90, 121-134. Reprinted in Neugebauer [1983], 217-230

Neugebauer [1936b] - Neugebauer, O., "Jahreszeiten und Tageslängen in der babylonischen Astronomie," Osiris 2, 517-550

Neugebauer [1937a] - Neugebauer, O., "Untersuchungen zur antiken Astronomie I," QS B4, 29-33

Neugebauer [1937b] - Neugebauer, O., "Untersuchungen zur antiken Astronomie II. Datierung und Rekonstruktion von Texten des Systems II der Mondtheorie," QS B4, 34-91

Neugebauer [1937c] - Neugebauer, O., "Untersuchungen zur antiken Astronomie III. Die babylonische Theorie der Breitenbewegung des Mondes," QS B4, 193-347

Neugebauer [1938] - Neugebauer, O., "Untersuchungen zur antiken Astronomie V. Der Halleysche "Saros" und andere Ergänzungen zu UAA III," QS B4, 407-411
Neugebauer [1939] - Neugebauer, O., "Chronologie und babylonischer Kalender," OLZ 42, 403-414
Neugebauer [1941a] - Neugebauer, O., "The Chronology of the Hammurabi Age," JAOS 61, 58-61
Neugebauer [1941b] - Neugebauer, O., "On a Special Use of the Sign "Zero" in Cuneiform Astronomical Texts," JAOS 61, 213-215
Neugebauer [1941c] - Neugebauer, O., "Some Fundamental Concepts in Ancient Astronomy," Studies in the History of Science, Philadelphia, 13-29, reprinted in Neugebauer [1983], 5-21
Neugebauer [1942] - Neugebauer, O., "On Some Astronomical Papyri and Related Problems in Ancient Geography," TAPS, NS 32,2, 251-263
Neugebauer [1945] - Neugebauer, O., "Studies in Ancient Astronomy VII. Magnitudes of Lunar Eclipses in Babylonian Mathematical Astronomy," Isis 36, 10-15. Reprinted in Neugebauer [1983], 232-237
Neugebauer [1947a] - Neugebauer, O., "A Table of Solstices from Uruk," JCS 1, 143-148
Neugebauer [1947b] - Neugebauer, O., "Studies in Ancient Astronomy VIII. The Water Clock in Babylonian Astronomy," Isis 37, 37-43. Reprinted in Neugebauer [1983], 239-245
Neugebauer [1948a] - Neugebauer, O., "Solstices and Equinoxes in Babylonian Astronomy during the Seleucid Period," JCS 2, 209-222
Neugebauer [1948b] - Neugebauer, O., "Arithmetical Methods for the Dating of Babylonian Astronomical Texts," Studies and Essays Presented to R. Courant, New York, 265-275
Neugebauer [1950] - Neugebauer, O., "The Alleged Babylonian Discovery of the Precession of the Equinoxes," JAOS 70, 1-8, reprinted in Neugebauer [1983], 247-254
Neugebauer [1951] - Neugebauer, O., "The Babylonian Method for the Computation of the Last Visibilities of Mercury," PAPS 95, 110-116
Neugebauer [1953] - Neugebauer, O., "The Rising-times in Babylonian Astronomy," JCS 7, 100-102
Neugebauer [1954] - Neugebauer, O., "Babylonian Planetary Theory," PAPS 98, 60-89
Neugebauer [1955] - Neugebauer, O., Astronomical Cuneiform Texts, 3 vols., London
Neugebauer [1956] - Neugebauer, O., "Astronomical Commentary,", in The Code of Maimonides. Book Three, Treatise Eight. Sanctification of the New Moon, translated by Solomon Gandz, New Haven, 113-149
Neugebauer [1957] - Neugebauer, O., "Saros" and Lunar Velocity in Babylonian Astronomy, KDVSMM 31/4, København
Neugebauer [1968] - Neugebauer, O., "The Origin of "System B" of Babylonian Astronomy," Centaurus 12, 209-214
Neugebauer [1975] - Neugebauer, O., A History of Ancient Mathematical Astronomy, 3 vols., Berlin-Heidelberg-New York
Neugebauer [1983] - Neugebauer, O., Astronomy and History. Selected Essays, New York
Neugebauer [1988] - Neugebauer, O., "A Babylonian Lunar Ephemeris from Roman Egypt," A Scientific Humanist: Studies in Memory of Abraham Sachs, Philadelphia, 301-304
Neugebauer-Parker [1960/1969] - Neugebauer, O., and Parker, R. A., Egyptian Astronomical Texts, 3 vols. in 4 parts, Providence

Neugebauer-Sachs [1967] - Neugebauer, O., and Sachs, A., "Some Atypical Astronomical Cuneiform Texts, I," JCS 21, 183-217

Neugebauer-Sachs [1969] - Neugebauer, O., and Sachs, A., "Some Atypical Astronomical Cuneiform Texts, II," JCS 22, 92-113

Neugebauer [1934] - Neugebauer, P. V., review of Martiny [1932], Vierteljahrsschrift der Astronomischen Gesellschaft 69, 68-78

Neugebauer-Schott [1934] - Neugebauer, P. V. and Schott, A., reviews of Martiny [1932], ZA 42, 198-204 (Neugebauer) and 204-217 (Schott)

Neugebauer-Weidner [1915] - Neugebauer, P. V. and Weidner, E. F., "Ein astronomischer Beobachtungstext aus dem 37. Jahre Nebukadnezars II. (-567/66)," Berichte über die Verhandlungen d. Königl. Sächs. Ges. d. Wiss., Phil.-hist. Kl. 67, 2. Heft

Neugebauer-Weidner [1931/1932] - Neugebauer, P. V. and Weidner, E. F., "Die Himmelsrichtungen bei den Babyloniern," AfO 7, 269-271

Neumann [1991/1992] - Neumann, H., "Anmerkungen zu Johannes Koch, Neue Untersuchungen zur Topographie des babylonischen Fixsternhimmels," AfO 38/39, 110-124 + 131

Oelsner-Horowitz [1998] - Oelsner, J. and Horowitz, W., "The 30-Star-Catalogue HS 1897 and The Late Parallel BM 55502," AfO 44/45, 176-185

Oppenheim [1969] - Oppenheim, A. L., "Divination and Celestial Observation in the last Assyrian Empire," Centaurus 14, 97-135

Oppenheim [1974] – Oppenheim, A. L., "A Babylonian Diviner's Manual," JNES 33, 197-220

Oppenheim [1977] - Oppenheim, A. L., Ancient Mesopotamia (2nd ed.), Chicago

Oppert [1889] - Oppert, J., "L'éclipse lunaire de l'an 232 de l'ère des Arsacides (23 mars 24 a. J.-C.)," ZA 4, 174-185

Oppert [1891] - Oppert, J., "Un texte babylonien astronomique et sa traduction grecque d'après Claude Ptolémée," ZA 6, 103-123

Pannekoek [1916] - Pannekoek, A., "Calculation of Dates in the Babylonian Tables of Planets," Proc. KAW Amsterdam 19, 1916, 684-703

Pannekoek [1917] - Pannekoek, A., "The Origin of the Saros," Proc. KAW Amsterdam 20, 943-955

Pannekoek [1941] - Pannekoek, A., "Some Remarks on the Moon's Diameter and the Eclipse Tables in Babylonian Astronomy," Eudemus 1, 9-22

Pannekoek [1948] - Pannekoek, A., Planetary Theories, 1948 (reprinted from Popular Astronomy 55, 1947, and 56, 1948)

Pannekoek [1951] - Pannekoek, A., "Periodicities in Lunar Eclipses," Proc. KAW, Ser. B, 54, 13-24

Papke [1978] - Papke, W., Die Keilschriftserie MUL.APIN. Dokument wissenschaftlicher Astronomie im 3. Jahrtausend, Allenstein

Papke [1993] - Papke, W., "Die Sterne von Babylon. Die geheime Botschaft des Gilgamesch - nach 4000 Jahren entschlüsselt (1989)," WO 24, 213-222

Parker [1959] - Parker, R. A., A Vienna Demotic Papyrus on Eclipse- and Lunar Omina, Providence

Parker-Dubberstein [1956] - Parker, R. A. and Dubberstein, W. H., Babylonian Chronology 626 B. C. - A. D. 75, Providence

Parpola [1983] - Parpola, S., Letters from Assyrian Scholars to the Kings Esarhaddon and Assurbanipal, part II, Kevelaer/Neukirchen-Vluyn

Parpola [1993] - Parpola, S., Letters from Assyrian and Babylonian Scholars, Helsinki

Peters [1992] - Peters, C. A., "The Mesopotamian Astrologers' Universe: Celestial and Terrestrial," Bull. Can. Soc. Mesop. St. 23, 33-44

Pinches [1900] - Pinches, T. G., review of R. Brown, Researches into the Origin of the Primitive Constellations of the Greeks, Phoenicians, and Babylonians, vol. 2, London 1900, JRAS 1900, 571-577

Pingree [1963] - Pingree, D., "Astronomy and Astrology in India and Iran," Isis 54, 229-246
Pingree [1973] - Pingree, D., "The Mesopotamian Origin of Early Indian Mathematical Astronomy," JHA 4, 1-12
Pingree [1982] - Pingree, D., "Mesopotamian Astronomy and Astral Omens in Other Civilizations," Berliner Beiträge zum Vorderen Orient 1, 613-631
Pingree [1987a] - Pingree, D., "Venus Omens in India and Babylon," Language, Literature, and History (Festschrift Erica Reiner), New Haven, 293-315
Pingree [1987b] - Pingree, D., "Babylonian Planetary Theory in Sanskrit Omen Texts," From Ancient Omens to Statistical Mechanics (Festschrift Asger Aaboe), Copenhagen, 91-99
Pingree [1989] - Pingree, D., "MUL.APIN and Vedic Astronomy," DUMU-$E_2$-DUB-BA-A. Studies in Honor of Åke W. Sjöberg, Philadelphia, 439-445
Pingree [1993] - Pingree, D., "Venus Phenomena in Enūma Anu Enlil," Die Rolle der Astronomie in den Kulturen Mesopotamiens, Graz, 259-273
Pingree [1997] - Pingree, D., From Astral Omens to Astrology. From Babylon to Bīkāner, Rome
Pingree [1998] - Pingree, D., "Legacies in Astronomy and Celestial Omens," The Legacy of Mesopotamia, Oxford, 125-137
Pingree-Morrissey [1989] - Pingree, D. and Morrissey, P., "On the Identification of the Yogatārās of the Indian Nakṣatras," JHA 20, 99-119
Pingree-Reiner [1974/1977] - Pingree, D., and Reiner, E., "A Neo-Babylonian Report on Seasonal Hours," AfO 25, 50-55
Pingree-Reiner [1975] - Pingree, D. and Reiner, E., "Observational Texts Concerning the Planet Mercury," RA 69, 175-180
Pingree-Walker [1988] - Pingree, D., and Walker, C. B. F., "A Babylonian Star-Catalogue: BM 78161," A Scientific Humanist: Studies in Memory of Abraham Sachs, Philadelphia, 313-322
Pliny, Natural History: C. Plini Secundi Naturalis Historia, ed. C. Mayhoff, vol. 1, Leipzig 1933, and vol. 3, Leipzig 1892
Al-Rawi-George [1991/1992] - Al-Rawi, F. N. H. and George, A. R., "Enūma Anu Enlil XIV and other Early Astronomical Tables," AfO 38/39, 52-73
Reiner [1993] - Reiner, E., "Two Babylonian Precursors of Astrology," NABU 1993 no. 26
Reiner [1995] - Reiner, E., Astral Magic in Babylonia, Philadelphia
Reiner-Pingree [1975] - Reiner, E. and Pingree, D., The Venus Tablet of Ammiṣaduqa, BPO 1, Malibu
Reiner-Pingree [1981] - Reiner, E. and Pingree, D., Enūma Anu Enlil, Tablets 50-51, BPO 2, Malibu
Reiner-Pingree [1998] - Reiner, E. and Pingree, D., Babylonian Planetary Omens, Part Three, BPO 3, Groningen
Riley [1994] - Riley, L., "The Lunar Velocity Function in System B First-crescent Ephemerides," Centaurus 37, 1-51
Rochberg-Halton [1983] - Rochberg-Halton, F., "Stellar Distances in Early Babylonian Astronomy: A New Perspective on the Hilprecht Text (HS 229)," JNES 42, 209-217
Rochberg-Halton [1984a] - Rochberg-Halton, F., "New Evidence for the History of Astrology," JNES 43, 115-140
Rochberg-Halton [1984b] - Rochberg-Halton, F., "Canonicity in Cuneiform Texts," JCS 36, 127-144
Rochberg-Halton [1987a] - Rochberg-Halton, F., "The Assumed 29th *aḫû* Tablet of *Enūma Anu Enlil*," Language, Literature, and History (Festschrift Erica Reiner), 327-350

Rochberg-Halton [1987b] - Rochberg-Halton, F., "TCL 6, 13: Mixed Traditions in Late Babylonian Astrology," ZA 77, 207-228
Rochberg-Halton [1988a] - Rochberg-Halton, F., Aspects of Babylonian Celestial Divination: the Lunar Eclipse Tablets of Enūma Anu Enlil, AfO Beiheft 22, Horn
Rochberg-Halton [1988b] - Rochberg-Halton, F., "Elements of the Babylonian Contribution to Hellenistic Astrology," JAOS 108, 51-62
Rochberg-Halton [1988c] – Rochberg-Halton, F., "Benefic and Malefic Planets in Babylonian Astrology," A Scientific Humanist: Studies in Memory of Abraham Sachs, Philadelphia, 323-328
Rochberg-Halton [1989] - Rochberg-Halton, F., "Babylonian Horoscopes and their Sources," Orientalia 58, 102-123
Rochberg-Halton [1991a] - Rochberg-Halton, F., "Between Observation and Theory in Babylonian Astronomical Texts," JNES 50, 107-120
Rochberg-Halton [1991b] - Rochberg-Halton, F., "The Babylonian Astronomical Diaries," JAOS 111, 323-332
Rochberg-Halton [1993] - Rochberg-Halton, F., "The cultural locus of astronomy in Late Babylonia," Die Rolle der Astronomie in den Kulturen Mesopotamiens, Graz, 31-45
Rochberg [1998] - Rochberg, F., Babylonian Horoscopes, TAPS 88/1, Philadelphia
Roughton-Canzoneri [1992] - Roughton, N. A. and Canzoneri, G. L., "Babylonian Normal Stars in Sagittarius," JHA 23, 193-200
Sachs [1948] - Sachs, A. J., "A Classification of the Babylonian Astronomical Tablets of the Seleucid Period," JCS 2, 271-290
Sachs [1952a] - Sachs, A. J., "Babylonian Horoscopes," JCS 6, 49-74
Sachs [1952b] - Sachs, A. J., "A Late Babylonian Star Catalog," JCS 6, 146-150
Sachs [1952c] - Sachs, A. J., "Sirius Dates in Babylonian Astronomical Texts of the Seleucid Period," JCS 6, 105-114
Sachs [1955] - Sachs, A. J., with the co-operation of J. Schaumberger, Late Babylonian Astronomical and Related Texts copied by T. G. Pinches and J. N. Strassmaier, Providence
Sachs [1974] - Sachs, A. J., "Babylonian Observational Astronomy," Philosophical Transactions of the Royal Society 276, 43-50
Sachs [1976] - Sachs, A. J., "The Latest Datable Cuneiform Tablets," Kramer Anniversary Volume, Kevelaer/Neukirchen-Vluyn, 379-398
Sachs-Hunger [1988, 1989, 1996] - Sachs, A. J. and Hunger, H., Astronomical Diaries and Related Texts from Babylonia, vols. I -III, Vienna
Sachs-Neugebauer [1956] - Sachs, A. and Neugebauer, O., "A Procedure Text Concerning Solar and Lunar Motion: BM 36712," JCS 10, 131-136
Sachs-Walker [1984] - Sachs, A. J. and Walker, C. B. F., "Kepler's View of the Star of Bethlehem and the Babylonian Almanac for 7/6 B.C.," Iraq 46, 43-55
Sayce [1887] - Sayce, A. H., "Miscellaneous Notes 18. An Assyrian Augural Staff," ZA 2, 335-337
Schaumberger [1925] - Schaumberger, J., "Textus cuneiformis de stella Magorum?," Biblica 6, 444-449
Schaumberger [1926] - Schaumberger, J., "Iterum: Textus cuneiformis de stella Magorum?," Biblica 7, 294-301
Schaumberger [1935a] - Schaumberger, J., 3. Ergänzungsheft zum ersten und zweiten Buch (of Kugler's SSB), Münster in Westfalen
Schaumberger [1935b] - Schaumberger, J., "Der jüngste datierbare Keilschrifttext," Analecta Orientalia 12, 279-287
Schaumberger [1943] - Schaumberger, J., "Ein neues Keilschriftfragment über den angeblichen Stern der Weisen," Biblica 24, 162-169
Schaumberger [1952] - Schaumberger, J., "Die ziqpu-Gestirne nach neuen Keilschrifttexten," ZA 50, 214-229

Schaumberger [1954-56] - Schaumberger, J., "Astronomische Untersuchung der "historischen" Mondfinsternisse in Enûma Anu Enlil," AfO 17, 89-92
Schaumberger [1955] - Schaumberger, J., "Anaphora und Aufgangskalender in neuen Ziqpu-Texten," ZA 51, 237-251
Schaumberger-Schott [1938] - Schaumberger, J. and Schott, A., "Die Konjunktion von Mars und Saturn im Frühjahr 669 v. Chr. nach Thompson, Reports Nr. 88 und anderen Texten," ZA 44, 271-289
Scheil [1917a] - Scheil, V., "Déchiffrement d'un document anzanite relatif aux présages," RA 14, 29-59
Scheil [1917b] - Scheil, V., "Un fragment susien du livre Enuma Anu (ilu) Ellil," RA 14, 139-142
Schiaparelli [1906/1907] - Schiaparelli, G., "Venusbeobachtungen und Berechnungen der Babylonier," Weltall 6, Heft 27 und 7, Heft 2
Schiaparelli [1908a] - Schiaparelli, G., "Die Oppositionen des Mars nach babylonischen Beobachtungen," Weltall 9, Heft 1
Schiaparelli [1908b] - Schiaparelli, G., "I primordi dell'astronomia presso i Babilonesi," Scientia 3, 32- and 4, ; reprinted in Schiaparelli [1925] 41-123
Schiaparelli [1925] - Schiaparelli, G., Scritti sulla storia della astronomia antica, parte prima, tomo primo, Bologna
Schmidt [1969] - Schmidt, O., "A Mean Value Principle in Babylonian Planetary Theory," Centaurus 14, 267-286
Schnabel [1923] - Schnabel, P., Berossos und die babylonisch-hellenistische Literatur, Leipzig 1923, reprinted Hildesheim 1968
Schnabel [1924a] - Schnabel, P., "Neue babylonische Planetentafeln," ZA 35, 99-112
Schnabel [1924b] - Schnabel, P., "Die Sarosperiode der Finsternisse schon in der Sargonidenzeit bekannt," ZA 35, 297-318
Schnabel [1925a] - Schnabel, P., "Der jüngste datierbare Keilschrifttext," ZA 36, 66-70
Schnabel [1925b] - Schnabel, P., "Zur astronomischen Fixierung der altbabylonischen Chronologie mittels der Venustafeln der Ammizaduqa-Zeit," ZA 36, 109-122
Schnabel [1927] - Schnabel, P., "Kidenas, Hipparch und die Entdeckung der Präzession," ZA 37, 1-60
Schoch [1925] - Schoch, C., Ammizaduga, Berlin
Schott [1934] - Schott, A., "Das Werden der babylonisch-assyrischen Positions-Astronomie und einige seiner Bedingungen," ZDMG 88, 302-337
Schott [1936] - Schott, A., "Marduk und sein Stern," ZA 43, 124-145
Schott-Schaumberger [1942] - Schott, A., and Schaumberger, J., "Vier Briefe Mâr-Ištars an Asarhaddon über Himmelserscheinungen der Jahre -670/668," ZA 47, 89-129
Seidenberg [1983] - Seidenberg, A., "The Separation of Sky and Earth at Creation (III)," Folklore 94, 192-200
Sidersky [1916] - Sidersky, D., Étude sur la chronologie assyro-babylonienne, Paris
Sidersky [1919] - Sidersky, D., "Le calcul chaldéen des néoménies," RA 16, 21-36
Sidersky [1940] - Sidersky, D., "Nouvelle étude sur la chronologie de la dynastie ḫammurapienne," RA 37, 45-54
Šileiko [1927] – Šileiko, V., "Mondlaufprognosen aus der Zeit der ersten babylonischen Dynastie," Comptes-Rendus de l'Académie des Sciences de l'URSS, 125-128
Sivin [1969] - Sivin, N., Cosmos and Computation in Early Chinese Astronomy, Leiden
Slotsky [1993] - Slotsky, A. L., "The Uruk Solstice Scheme Revisited," Die Rolle der Astronomie in den Kulturen Mesopotamiens, Graz, 359-365
Slotsky [1997] - Slotsky, A. L., The Bourse of Babylon, Bethesda
Smith [1940] - Smith, S., Alalakh and Chronology, London

Smith [1969] - Smith, S., "Babylonian Time Reckoning," Iraq 31, 74-81
van Soldt [1989] - van Soldt, W. H., "De babylonische astronomie, het begin van een wetenschap," Phoenix 35/2, 39-56
van Soldt [1995] - van Soldt, W. H., Solar Omens of Enūma Anu Enlil: Tablets 23 (24) - 29 (30), Istanbul
van Soldt-de Jong [1989] - van Soldt, W. H. and de Jong, T., "The Earliest Known Solar Eclipse Record Redated," Nature 338, 238-240
van der Spek [1985] - van der Spek, R. J., "The Babylonian Temple during the Macedonian and Parthian Domination," BiOr 42, 542-562
van der Spek [1993] - van der Spek, R. J., "The Astronomical Diaries as a Source for Achaemenid and Seleucid History," BiOr 50, 91-101
Starr [1983] - Starr, I., The Rituals of the Diviner, Malibu
Steele [1997] - Steele, J. M., "Solar Eclipse Times Predicted by the Babylonians," JHA 28, 133-139
Steele-Stephenson [1997] - Steele, J. M. and Stephenson, F. R., "Lunar Eclipse Times Predicted by the Babylonians," JHA 28, 119-131
Stephenson-Walker [1985] - Stephenson, F. R. and Walker, C. B. F. (eds.), Halley's Comet in History, London
Stol [1988] - Stol, M., "Nisan 26-28 in KAR 178," NABU 1988 no. 47
Strassmaier [1888] - Strassmaier, J. N., "Arsaciden-Inschriften," ZA 3, 129-158
Strassmaier [1890] - Strassmaier, J. N., Inschriften von Cambyses, König von Babylon, Leipzig
Strassmaier [1892] - Strassmaier, J. N., "Einige chronologische Daten aus astronomischen Rechnungen," ZA 7, 197-204
Strassmaier [1895] - Strassmaier, J. N., "Der Saros-Canon Sp. II, 71," ZA 10, 64-69
Swerdlow [1998] - Swerdlow, N. M., The Babylonian Theory of the Planets, Princeton
Thompson [1972] - Thompson, J. E. S., A Commentary on the Dresden Codex, MAPS 93, Philadelphia
Thompson [1900] - Thompson, R. C., The Reports of the Magicians and Astrologers of Nineveh and Babylon, 2 vols., London
Thureau-Dangin [1913] - Thureau-Dangin, F., "Distances entre étoiles fixes d'après une tablette de l'époque des Séleucides. Première partie," RA 10, 215-225
Thureau-Dangin [1931a] - Thureau-Dangin, F., "Mesures de temps et mesures angulaires dans l'astronomie babylonienne," RA 28, 111-114
Thureau-Dangin [1931b] - Thureau-Dangin, F., "La tablette astronomique de Nippur," RA 28, 85-88
Toomer [1988] - Toomer, G., "Hipparchus and Babylonian Astronomy," A Scientific Humanist: Studies in Memory of Abraham Sachs, Philadelphia, 353-362
Tuckerman [1962] - Tuckerman, B., Planetary, Lunar, and Solar Positions 601 B.C. to A.D. 1 at Five-day and Ten-day Intervals, Philadelphia
Ungnad [1910] - Ungnad, A., "Zur Schaltungs-Praxis in der Ḫammurapi-Zeit," OLZ 13, 66-67
Ungnad [1940] - Ungnad, A., Die Venustafeln und das neunte Jahr Samsuilunas (1741 v. Chr.), Mitteilungen der Altorientalischen Gesellschaft 13/3, Leipzig 1940, reprinted Osnabrück 1972
Vettius Valens, Anthologies: Vettii Valentis Antiocheni Anthologiarum libri novem, ed. D. Pingree, Leipzig 1986
Villard [1991] - Villard, P., "Un rapport astrologique du Louvre", Mélanges Paul Garelli, Paris, 129-136
Virolleaud [1905-12] – Virolleaud, Ch., L'astrologie chaldéenne, le livre intitulé «enuma (Anu) $^{ilu}$Bêl», Paris
van der Waerden [1940] - van der Waerden, B. L., "Die Voraussage von Finsternissen bei den Babyloniern," BSAW Leipzig, Math.-phys. Kl. 92, 107-114

van der Waerden [1941] - van der Waerden, B. L., "Zur babylonischen Planetenrechnung," Eudemus 1, 23-48

van der Waerden [1943a] - van der Waerden, B. L., "Berechnung der ersten und letzten Sichtbarkeit von Mond und Planeten und die Venustafeln des Ammiṣaduqa," BSAW Leipzig, Math.-phys. Kl. 94, 23-56

van der Waerden [1943b] - van der Waerden, B. L., "Plaudereien zur babylonischen Astronomie I. Die Venusbeobachtungen unter Ammiṣaduqa," Die Himmelswelt 1943, Heft 10/12, 1-7

van der Waerden [1944] - van der Waerden, B. L., "Plaudereien zur babylonischen Astronomie II. Der Kalender von Nippur und die Zwölfmal drei Sterne," Die Himmelswelt, 1944, Heft 7-9, 1-4

van der Waerden [1945/1948] - van der Waerden, B. L., "On Babylonian Astronomy I. The Venus Tablets of Ammiṣaduqa," JEOL 10, 414-424

van der Waerden [1949] - van der Waerden, B. L., "Babylonian Astronomy II. The Thirty-Six Stars," JNES 8, 6-26

van der Waerden [1950] - van der Waerden, B. L., "Dauer der Nacht und Zeit des Mondunterganges in den Tafeln des Nabû-zuqup-GI.NA," ZA 49, 291-312

van der Waerden [1952/1953] - van der Waerden, B. L., "History of the Zodiac," AfO 16, 216-230

van der Waerden [1957] - van der Waerden, B. L., "Babylonische Planetenrechnung," Vierteljahrsschrift der Naturforschenden Gesellschaft in Zürich 102, Abh. 2, 39-60

van der Waerden [1963] - van der Waerden, B. L., "Das Alter der babylonischen Mondrechnung," AfO 20, 97-102

van der Waerden [1968] - van der Waerden, B. L., "The Date of Invention of Babylonian Planetary Theory," AHES 5, 70-78

van der Waerden [1972] - van der Waerden, B. L., "Aegyptische Planetenrechnung," Centaurus 16, 65-91

van der Waerden [1974] - van der Waerden, B. L., Science Awakening II. The Birth of Astronomy, Leiden - New York

Walker [1982] - Walker, C. B. F., "Episodes in the History of Babylonian Astronomy," Bull. Can. Soc. Mesop. St. 5, 10-26

Walker [1984] - Walker, C. B. F., "Notes on the Venus Tablet of Ammiṣaduqa," JCS 36, 64-66

Walker [1989] - Walker, C. B. F., "Eclipse Seen at Ancient Ugarit," Nature 338, 204-205

Walker [1995] - Walker, C. B. F., "The Dalbanna Text: a Mesopotamian Star-List," WO 26, 27-42

Walker-Hunger [1977] - Walker, C. B. F. and Hunger, H., "Zwölfmaldrei," MDOG 109, 27-34

Weidner [1911] - Weidner, E., Beiträge zur babylonischen Astronomie, Leipzig (= BA 8/4)

Weidner [1912a] - Weidner, E., "Zum Alter der babylonischen Astronomie," Babyloniaca 6, 129-133

Weidner [1912b] - Weidner, E., "Studien zum Kalender der Hethiter und Babylonier," Babyloniaca 6, 164-181

Weidner [1912c] - Weidner, E., "Mondlauf, Kalender und Zahlenwissenschaft," Babyloniaca 6, 8-28

Weidner [1913a] - Weidner, E. F., "Zu der neuen Sternliste in CT XXXIII," OLZ 16, 149-152

Weidner [1913b] - Weidner, E. F., "Beiträge zur Erklärung der astronomischen Keilschrifttexte. 4. Die ὑψώματα der griechischen Astrologie sind babylonischen Ursprungs," OLZ 16, 209-212

Weidner [1913c] - Weidner, E. F., "Kannten die Babylonier die Phasen des Mars?," OLZ 16, 303-304

Weidner [1914a] - Weidner, E., "Die Entdeckung der Präzession, eine Geistestat babylonischer Astronomen," Babyloniaca 7, 1-19
Weidner [1914b] - Weidner, E., Alter und Bedeutung der babylonischen Astrallehre, Leipzig
Weidner [1915] - Weidner, E., Handbuch der babylonischen Astronomie, Band I, Teil I, Leipzig 1915, reprinted Leipzig 1976
Weidner [1919] - Weidner, E. F., "Babylonische Hypsomatabilder," OLZ 22, 10-16
Weidner [1924] - Weidner, E., "Ein babylonisches Kompendium der Himmelskunde," AJSL 40, 186-208
Weidner [1927] - Weidner, E., "Eine Beschreibung des Sternenhimmels aus Assur," AfO 4, 73-85
Weidner [1931/1932] - Weidner, E. F., "Der Tierkreis und die Wege am Himmel," AfO 7, 170-178
Weidner [1941/1944] - Weidner, E., "Die astrologische Serie Enûma Anu Enlil," AfO 14, 172-195, 308-318
Weidner [1952/1953] - Weidner, E., "Die Bibliothek Tiglatpilesers I.," AfO 16, 197-215
Weidner [1954/1956] - Weidner, E., "Die astrologische Serie Enûma Anu Enlil," AfO 17, 71-89
Weidner [1959/1960] - Weidner, E., "Ein astrologischer Sammeltext aus der Sargonidenzeit," AfO 19, 105-113
Weidner [1963] - Weidner, E., "Astrologische Geographie im Alten Orient," AfO 20, 117-121
Weidner [1967] - Weidner, E., Gestirn-Darstellungen auf babylonischen Tontafeln, Wien
Weidner [1968/1969] - Weidner, E., "Die astrologische Serie Enûma Anu Enlil," AfO 22, 65-75
von Weiher [1983] - von Weiher, E., Spätbabylonische Texte aus Uruk, Teil II, Berlin
von Weiher [1988] - von Weiher, E., Spätbabylonische Texte aus Uruk, Teil III, Berlin
von Weiher [1993] - von Weiher, E., Uruk. Spätbabylonische Texte aus dem Planquadrat U 18, Teil IV, Mainz am Rhein
von Weiher [1998] - von Weiher, E., Uruk. Spätbabylonische Texte aus dem Planquadrat U 18, Teil V, Mainz am Rhein
Weir [1972] - Weir, J. D., The Venus Tablets of Ammizaduqa, Istanbul
Weissbach [1901] - Weissbach, F. H., "Über einige neuere Arbeiten zur babylonisch-persischen Chronologie," ZDMG 55, 195-220
Weissbach [1909] - Weissbach, F. H., "Zum babylonischen Kalender," Hilprecht Anniversary Volume, Leipzig, 281-290
Weissbach [1925] - Weissbach, F. H., "Zur assyrisch-babylonischen Chronologie," ZA 36, 55-65
Wiseman [1953] - Wiseman, D. J., The Alalakh Tablets, London
Wiseman-Black [1996] - Wiseman, D. J. and Black, J. A., Literary Texts from the Temple of Nabû, London

# INDICES

## 1. SUBJECT INDEX

## 2. TABLET NUMBERS


80-7-19,273 44
82-5-22,501 43
83-1-18,608 50
85-4-30,15 53

A 3405 177, 219, 269
A 3406+ 211
A 3408 240
A 3409 219, 264
A 3410 217
A 3412+ 225
A 3415 256
A 3417+ 211
A 3418 246
A 3424+ 179, 219
A 3425 264
A 3426 253
A 3427 198
A 3456 178-180, 201, 219

AO 6475 216
AO 6476 215, 264
AO 6477 215, 219, 246
AO 6478 84, 90
AO 6480 215
AO 6481 215
AO 6492 240
AO 7540 57
AO 17649 156

BE 13918 44

BM 17175+ 50
BM 29371 80
BM 30739 143
BM 30830 142
BM 31051 162
BM 31476 142
BM 31539 143
BM 31581 142
BM 31583 142
BM 31804 142
BM 31847 142
BM 32143 142
BM 32154 171
BM 32209+ 176, 177
BM 32218 243, 251
BM 32222 169
BM 32231+ 178
BM 32234 181
BM 32238 181
BM 32247 161
BM 32286 171
BM 32299+ 203
BM 32327+ 201
BM 32349+ 143
BM 32351 231
BM 32363 182
BM 32572 142
BM 32590 177
BM 32597 143
BM 32599 212, 256
BM 32651 213, 226
BM 32845 182
BM 33066 174, 197
BM 33478 141
BM 33552 208
BM 33562A 182
BM 33593 226
BM 33618 219
BM 33631 231
BM 33633 164
BM 33643 182
BM 33671 142
BM 33746 166
BM 33758 214
BM 33784+ 165
BM 33797 162
BM 33801 244
BM 33808 142
BM 33812 182
BM 33837 140
BM 33869 213, 250
BM 33873 162
BM 33982 182
BM 33991 143
BM 33992 142
BM 34000 142
BM 34032 148, 159
BM 34033 148, 159, 212
BM 34034 197
BM 34037 213
BM 34041 212
BM 34042 165
BM 34049 143
BM 34050 142
BM 34051 164
BM 34053 169

| | |
|---|---|
| BM 34066+ | 212 |
| BM 34069+ | 212 |
| BM 34075 | 175, 197 |
| BM 34076 | 161 |
| BM 34078+ | 159, 160 |
| BM 34079+ | 225, 231 |
| BM 34080 | 160 |
| BM 34081+ | 244 |
| BM 34085 | 186 |
| BM 34088 | 212 |
| BM 34112 | 143 |
| BM 34121 | 164 |
| BM 34128+ | 214, 256 |
| BM 34134 | 226 |
| BM 34148 | 231 |
| BM 34221+ | 244 |
| BM 34227+ | 32 |
| BM 34232 | 164 |
| BM 34236 | 181 |
| BM 34237 | 217 |
| BM 34245+ | 231 |
| BM 34345 | 165 |
| BM 34368+ | 180 |
| BM 34497 | 227 |
| BM 34560 | 205 |
| BM 34562 | 143 |
| BM 34563 | 142 |
| BM 34564 | 142 |
| BM 34570 | 213 |
| BM 34571 | 213 |
| BM 34574 | 213 |
| BM 34575 | 213 |
| BM 34579+ | 170 |
| BM 34580 | 212, 214, 215 |
| BM 34581+ | 233 |
| BM 34582 | 213 |
| BM 34585 | 213, 219 |
| BM 34587 | 213 |
| BM 34589 | 214 |
| BM 34593 | 256 |
| BM 34594 | 213 |
| BM 34597 | 182, 183 |
| BM 34600 | 213 |
| BM 34604+ | 213 |
| BM 34606 | 228 |
| BM 34608 | 213 |
| BM 34616+ | 156 |
| BM 34617 | 213 |
| BM 34619 | 213 |
| BM 34621 | 213 |
| BM 34623 | 240 |
| BM 34625+ | 172 |
| BM 34631 | 172 |
| BM 34659 | 166 |
| BM 34667+ | 164 |
| BM 34676 | 246 |
| BM 34684 | 181 |
| BM 34702 | 142 |
| BM 34705+ | 192 |
| BM 34713 | 53, 198 |
| BM 34722 | 163 |
| BM 34737 | 226 |
| BM 34750+ | 176 |
| BM 34757 | 252 |
| BM 34758 | 160 |
| BM 34765 | 246 |
| BM 34771 | 213 |
| BM 34787 | 181 |
| BM 34790 | 99 |
| BM 34803+ | 228 |
| BM 34934 | 225 |
| BM 34940 | 182 |
| BM 34949 | 165 |
| BM 34963+ | 181 |
| BM 34991+ | 165 |
| BM 35039 | 165 |
| BM 35048 | 212 |
| BM 35076 | 235 |
| BM 35115 | 181 |
| BM 35118 | 214, 257 |
| BM 35125 | 235 |
| BM 35149 | 165 |
| BM 35195 | 141 |
| BM 35241 | 249 |
| BM 35253 | 226 |
| BM 35277 | 219 |
| BM 35318 | 213 |
| BM 35339+ | 172 |
| BM 35399 | 220, 235 |
| BM 35402 | 186, 203 |
| BM 35495+ | 256 |
| BM 35501+ | 177 |
| BM 35514 | 142 |
| BM 35525 | 142 |
| BM 35541 | 156 |
| BM 35564+ | 226 |
| BM 35651 | 142 |
| BM 35688 | 213 |
| BM 35730 | 213 |
| BM 35789 | 181 |
| BM 35853 | 256 |
| BM 35943 | 249 |
| BM 35979 | 142 |
| BM 36004 | 226 |
| BM 36006 | 170 |
| BM 36300+ | 243, 247 |
| BM 36301 | 189, 207 |
| BM 36311+ | 225 |

BM 36321 259
BM 36395 32
BM 36400 195
BM 36438+ 226
BM 36580 194
BM 36591 142
BM 36599+ 185, 189, 190, 192, 194, 222, 224
BM 36636 231
BM 36651+ 195, 259
BM 36699+ 224
BM 36700 224
BM 36705+ 188
BM 36712 211, 229
BM 36722 210, 223
BM 36723+ 242, 260, 264
BM 36731 151, 202
BM 36733 142
BM 36737+ 185, 189, 190, 192-194, 222, 224
BM 36744+ 189
BM 36754 182
BM 36758+ 32
BM 36762+ 243, 249
BM 36793 224
BM 36801 251
BM 36807 142
BM 36810+ 151, 201, 243, 255
BM 36811 201, 243, 255
BM 36814 243, 247
BM 36822+ 189, 222, 229
BM 36823 203
BM 36824+ 224
BM 36838 202
BM 36846+ 224, 226
BM 36857 142
BM 36908 224, 226
BM 36910+ 182, 183
BM 36922+ 260
BM 36957 243, 255
BM 36994 224
BM 37010 32
BM 37021 225
BM 37024 243
BM 37043+ 181
BM 37044 183
BM 37089 237
BM 37094 181
BM 37121+ 32
BM 37127 44
BM 37149 189
BM 37151+ 209
BM 37174 249
BM 37196 243, 255
BM 37203 224
BM 37236 243, 255
BM 37266 205
BM 37284 142
BM 37467 139
BM 37474 249
BM 37652 181
BM 38357 156
BM 38369+ 84, 89
BM 38462 181
BM 40057 142
BM 40067 142
BM 40069 142
BM 40082 210, 223
BM 40083 166
BM 40084 166
BM 40094 223, 224
BM 40101+ 166
BM 40113 205
BM 40119 142
BM 40122 141
BM 40574 142
BM 40754 233
BM 41004 198, 204
BM 41037 142
BM 41111 142
BM 41123 142
BM 41129 181
BM 41161 142
BM 41498 32
BM 41536 181
BM 41565 181
BM 41599 160
BM 41688 32
BM 41800 181
BM 41884+ 142
BM 41958 172
BM 41985 181
BM 41990 226
BM 42053 181
BM 42073 181
BM 42145 181
BM 42147 181
BM 42799+ 257
BM 45627 142
BM 45628 181
BM 45632 142
BM 45640 181
BM 45662 223
BM 45667 181
BM 45688 213, 227
BM 45694 213, 215, 238
BM 45696 160
BM 45698 163

BM 45707 215
BM 45728 203
BM 45768 142
BM 45774 142
BM 45777 214
BM 45782 142
BM 45818+ 240
BM 45821 44
BM 45845 181
BM 45851 213
BM 45930 224
BM 45953 163
BM 45980 179, 214, 219
BM 45982 166
BM 46015 224, 227
BM 46021 164
BM 46042 233
BM 46046 163
BM 46083 149
BM 46112 226
BM 46176 214
BM 46231 172
BM 47650 142
BM 47735 141
BM 47860 83
BM 47912 224
BM 48090 181
BM 48104 161
BM 54619 115
BM 55502 52
BM 55511 142
BM 55517 172
BM 55530 231
BM 55539 142
BM 55541 156
BM 55546 258
BM 71537 182
BM 76738+ 173
BM 77226 142
BM 77238 227
BM 77242 198
BM 78080 243, 248
BM 78161 90
BM 82923 53
BM 86378 57
BM 92682 169
BM 92688+ 140
BM 123340 112
BM 140677 156

BRM 4 6 6

CBS 17 170
CBS 1493 227, 233
CBS 11901 175, 186, 200

DT 143 166
DT 183 244

HS 229 54
HS 1897 52

IM 44152 180

K 90 44, 49
K 170+ 83
K 229 41
K 778 42
K 2070+ 100
K 2077+ 115
K 2079 100
K 2164+ 44, 83
K 2321+ 32
K 2670 83
K 2907 41
K 3105 32
K 3601 41
K 5963+ 32
K 6153 139
K 6427 44
K 6490 100
K 7072 32
K 7090 32
K 9794 84
K 12186 32
K 12344+ 32
K 14943+ 50

KTU 1.78 10

KUB 4 63 9
KUB 4 64 9
KUB 8 35 31 fn.

MLC 1860 161, 201
MLC 1866 63
MLC 1880 240
MLC 1885 161, 201
MLC 2195 166, 201
MLC 2205 226

MM 86.11.354 166
MM 86.11.405+ 233

MNB 1856 146, 198, 207
MNB 1879 142

Musée de Rouen HG-x 172

| | |
|---|---|
| Nv. 10 | 55 |
| PBS 2/2 123 | 12 |
| Rm 777 | 240 |
| Rm 839 | 231 |
| Rm 845 | 140 |
| Rm 2, 303 | 139 |
| Rm 2, 361 | 139 |
| Rm 2, 531 | 32 |
| Rm 4, 337 | 58 |
| Sm 162 | 50, 53-55 |
| Sm 174 | 32 |
| Sm 1171 | 100 |
| TCL 6 11 | 29, 197 |
| TCL 6 13 | 30 |
| TCL 6 14 | 30 |
| TCL 6 16 | 14 |
| TCL 6 19 | 29 |
| TCL 6 20 | 29 |
| U 96 | 240 |
| U 97+ | 255 |
| U 101 | 264 |
| U 102 | 264 |
| U 106 | 255 |
| U 107+ | 201 |
| U 122+ | 240 |
| U 136 | 246 |
| U 148+ | 243 |
| U 151+ | 219, 264 |
| U 152+ | 243 |
| U 169 | 219, 264 |
| U 174+ | 243 |
| U 179 | 242, 256 |
| U 180(3)+ | 161 |
| U 180(10) | 250 |
| U 180(27) | 264 |
| U 181 | 198 |
| U 193a | 201 |
| U 194 | 161, 201 |
| VAT 209 | 215 |
| VAT 290+ | 165 |
| VAT 1762 | 226 |
| VAT 1770 | 238 |
| VAT 4924 | 141, 156 |
| VAT 4956 | 141 |
| VAT 7809 | 215, 216 |
| VAT 7819 | 216 |
| VAT 7821 | 216 |
| VAT 7844 | 216 |
| VAT 7852 | 214 |
| VAT 9412+ | 57 |
| VAT 9416 | 52 |
| VAT 9428 | 65 |
| VAT 9527 | 57 |
| VAT 10218 | 40 |
| VAT 16436 | 84, 88 |
| VAT 16437 | 84 |
| W 1924.802 | 32 |
| W 22281a | 99 |
| W 22755/3 | 264 |
| W 22801+ | 202 |
| W 22925 | 202 |
| Warka X 40 | 219, 264 |

# HANDBUCH DER ORIENTALISTIK

## Abt. I: DER NAHE UND MITTLERE OSTEN

ISSN 0169-9423

**Band 1. Ägyptologie**

1. *Ägyptische Schrift und Sprache.* Mit Beiträgen von H. Brunner, H. Kees, S. Morenz, E. Otto, S. Schott. Mit Zusätzen von H. Brunner. Nachdruck der Erstausgabe (1959). 1973. ISBN 90 04 03777 2
2. *Literatur.* Mit Beiträgen von H. Altenmüller, H. Brunner, G. Fecht, H. Grapow, H. Kees, S. Morenz, E. Otto, S. Schott, J. Spiegel, W. Westendorf. 2. verbesserte und erweiterte Auflage. 1970. ISBN 90 04 00849 7
3. Helck, W. *Geschichte des alten Ägypten.* Nachdruck mit Berichtigungen und Ergänzungen. 1981. ISBN 90 04 06497 4

**Band 2. Keilschriftforschung und alte Geschichte Vorderasiens**

1-2/2. *Altkleinasiatische Sprachen [und Elamitisch].* Mit Beiträgen von J. Friedrich, E. Reiner, A. Kammenhuber, G. Neumann, A. Heubeck. 1969. ISBN 90 04 00852 7

3. Schmökel, H. *Geschichte des alten Vorderasien.* Reprint. 1979. ISBN 90 04 00853 5

4/2. *Orientalische Geschichte von Kyros bis Mohammed.* Mit Beiträgen von A. Dietrich, G. Widengren, F. M. Heichelheim. 1966. ISBN 90 04 00854 3

**Band 3. Semitistik**

*Semitistik.* Mit Beiträgen von A. Baumstark, C. Brockelmann, E. L. Dietrich, J. Fück, M. Höfner, E. Littmann, A. Rücker, B. Spuler. Nachdruck der Erstausgabe (1953-1954). 1964. ISBN 90 04 00855 1

**Band 4. Iranistik**

1. *Linguistik.* Mit Beiträgen von K. Hoffmann, W. B. Henning, H. W. Bailey, G. Morgenstierne, W. Lentz. Nachdruck der Erstausgabe (1958). 1967. ISBN 90 04 03017 4

2/1. *Literatur.* Mit Beiträgen von I. Gershevitch, M. Boyce, O. Hansen, B. Spuler, M. J. Dresden. 1968. ISBN 90 04 00857 8

2/2. *History of Persian Literature from the Beginning of the Islamic Period to the Present Day.* With Contributions by G. Morrison, J. Baldick and Sh. Kadkanī. 1981. ISBN 90 04 06481 8

3. Krause, W. *Tocharisch.* Nachdruck der Erstausgabe (1955) mit Zusätzen und Berichtigungen. 1971. ISBN 90 04 03194 4

**Band 5. Altaistik**

1. *Turkologie.* Mit Beiträgen von A. von Gabain, O. Pritsak, J. Benzing, K. H. Menges, A. Temir, Z. V. Togan, F. Taeschner, O. Spies, A. Caferoglu, A. Battal-Tamays. Reprint with additions of the 1st (1963) ed. 1982. ISBN 90 04 06555 5
2. *Mongolistik.* Mit Beiträgen von N. Poppe, U. Posch, G. Doerfer, P. Aalto, D. Schröder, O. Pritsak, W. Heissig. 1964. ISBN 90 04 00859 4
3. *Tungusologie.* Mit Beiträgen von W. Fuchs, I. A. Lopatin, K. H. Menges, D. Sinor. 1968. ISBN 90 04 00860 8

**Band 6. Geschichte der islamischen Länder**

5/1. *Regierung und Verwaltung des Vorderen Orients in islamischer Zeit.* Mit Beiträgen von H. R. Idris und K. Röhrborn. 1979. ISBN 90 04 05915 6

5/2. *Regierung und Verwaltung des Vorderen Orients in islamischer Zeit.* 2. Mit Beiträgen von D. Sourdel und J. Bosch Vilá. 1988. ISBN 90 04 08550 5

6/1. *Wirtschaftsgeschichte des Vorderen Orients in islamischer Zeit.* Mit Beiträgen von B. Lewis, M. Rodinson, G. Baer, H. Müller, A. S. Ehrenkreutz, E. Ashtor, B. Spuler, A. K. S. Lambton, R. C. Cooper, B. Rosenberger, R. Arié, L. Bolens, T. Fahd. 1977. ISBN 90 04 04802 2

**Band 7**

*Armenisch* und *Kaukasische Sprachen.* Mit Beiträgen von G. Deeters, G. R. Solta, V. Inglisian. 1963. ISBN 90 04 00862 4

**Band 8. Religion**

1/1. *Religionsgeschichte des alten Orients.* Mit Beiträgen von E. Otto, O. Eissfeldt, H. Otten, J. Hempel. 1964. ISBN 90 04 00863 2

1/2/2/1. Boyce, M. *A History of Zoroastrianism. The Early Period.* Rev. ed. 1989. ISBN 90 04 08847 4

1/2/2/2. Boyce, M. *A History of Zoroastrianism. Under the Achaemenians.* 1982. ISBN 90 04 06506 7

1/2/2/3. Boyce, M. and Grenet, F. *A History of Zoroastrianism. Zoroastrianism under Macedonian and Roman Rule.* With a Contribution by R. Beck. 1991. ISBN 90 04 09271 4

2. *Religionsgeschichte des Orients in der Zeit der Weltreligionen.* Mit Beiträgen von A. Adam, A. J. Arberry, E. L. Dietrich, J. W. Fück, A. von Gabain, J. Leipoldt, B. Spuler, R. Strothman, G. Widengren. 1961. ISBN 90 04 00864 0

**Ergänzungsband 1**

1. Hinz, W. *Islamische Maße und Gewichte umgerechnet ins metrische System.* Nachdruck der Erstausgabe (1955) mit Zusätzen und Berichtigungen. 1970. ISBN 90 04 00865 9

**Ergänzungsband 2**

1. Grohmann, A. *Arabische Chronologie* und *Arabische Papyruskunde.* Mit Beiträgen von J. Mayr und W. C. Till. 1966. ISBN 90 04 00866 7

2. Khoury, R. G. *Chrestomathie de papyrologie arabe.* Documents relatifs à la vie privée, sociale et administrative dans les premiers siècles islamiques. 1992. ISBN 90 04 09551 9

**Ergänzungsband 3**

*Orientalisches Recht.* Mit Beiträgen von E. Seidl, V. Korošc, E. Pritsch, O. Spies, E. Tyan, J. Baz, Ch. Chehata, Ch. Samaran, J. Roussier, J. Lapanne-Joinville, S. Ş. Ansay. 1964. ISBN 90 04 00867 5

**Ergänzungsband 5**

1/1. Borger, R. *Einleitung in die assyrischen Königsinschriften.* 1. Das zweite Jahrtausend vor Chr. Mit Verbesserungen und Zusätzen. Nachdruck der Erstausgabe (1961). 1964. ISBN 90 04 00869 1

1/2. Schramm, W. *Einleitung in die assyrischen Königsinschriften.* 2. 934-722 v. Chr. 1973. ISBN 90 04 03783 7

**Ergänzungsband 6**

1. Ullmann, M. *Die Medizin im Islam.* 1970. ISBN 90 04 00870 5

2. Ullmann, M. *Die Natur- und Geheimwissenschaften im Islam.* 1972. ISBN 90 04 03423 4

**Ergänzungsband 7**

Gomaa, I. *A Historical Chart of the Muslim World.* 1972. ISBN 90 04 03333 5

**Ergänzungsband 8**

Kornrumpf, H.-J. *Osmanische Bibliographie mit besonderer Berücksichtigung der Türkei in Europa.* Unter Mitarbeit von J. Kornrumpf. 1973. ISBN 90 04 03549 4

**Ergänzungsband 9**

Firro, K. M. *A History of the Druzes.* 1992. ISBN 90 04 09437 7

**Band 10**

Strijp, R. *Cultural Anthropology of the Middle East. A Bibliography.* Vol. 1: 1965-1987. 1992. ISBN 90 04 09604 3

**Band 11**

Endress, G. & Gutas, D. (eds.). *A Greek and Arabic Lexicon.* (*GALex*). Materials for a Dictionary of the Mediæval Translations from Greek into Arabic.

Fascicle 1. Introduction—Sources—ʾ – ʾ-kh-r. Compiled by G. Endress & D. Gutas, with the assistance of K. Alshut, R. Arnzen, Chr. Hein, St. Pohl, M. Schmeink. 1992. ISBN 90 04 09494 6

Fascicle 2. ʾ-kh-r – ʾ-ṣ-l. Compiled by G. Endress & D. Gutas, with the assistance of K. Alshut, R. Arnzen, Chr. Hein, St. Pohl, M. Schmeink. 1993. ISBN 90 04 09893 3

Fascicle 3. ʾ-ṣ-l – ʾ-l-y. Compiled by G. Endress, D. Gutas & R. Arnzen, with the assistance of Chr. Hein, St. Pohl. 1995. ISBN 90 04 10216 7

Fascicle 4. Ilā – inna. Compiled by R. Arnzen, G. Endress & D. Gutas, with the assistance of Chr. Hein & J. Thielmann. 1997. ISBN 90 04 10489 5.

**Band 12**

Jayyusi, S. K. (ed.). *The Legacy of Muslim Spain.* Chief consultant to the editor, M. Marín. 2nd ed. 1994. ISBN 90 04 09599 3

**Band 13**

Hunwick, J. O. and O'Fahey, R. S. (eds.). *Arabic Literature of Africa.* Editorial Consultant: Albrecht Hofheinz.

Volume I. *The Writings of Eastern Sudanic Africa to c. 1900.* Compiled by R. S. O'Fahey, with the assistance of M. I. Abu Salim, A. Hofheinz, Y. M. Ibrahim, B. Radtke and K. S. Vikør. 1994. ISBN 90 04 09450 4

Volume II. *The Writings of Central Sudanic Africa.* Compiled by John O. Hunwick, with the assistance of Razaq Abubakre, Hamidu Bobboyi, Roman Loimeier, Stefan Reichmuth and Muhammad Sani Umar. 1995. ISBN 90 04 10494 1

**Band 14**

Decker, W. und Herb, M. *Bildatlas zum Sport im alten Ägypten. Corpus der bildlichen Quellen zu Leibesübungen, Spiel, Jagd, Tanz und verwandten Themen.* Bd.1: Text. Bd. 2: Ab-bildungen. 1994. ISBN 90 04 09974 3 *(Set)*

**Band 15**

Haas, V. *Geschichte der hethitischen Religion.* 1994. ISBN 90 04 09799 6

**Band 16**

Neusner, J. (ed.). *Judaism in Late Antiquity.* Part One: The Literary and Archaeological Sources. 1994. ISBN 90 04 10129 2

**Band 17**

Neusner, J. (ed.). *Judaism in Late Antiquity.* Part Two: Historical Syntheses. 1994. ISBN 90 04 09799 6

**Band 18**

Orel, V. E. and Stolbova, O. V. (eds.). *Hamito-Semitic Etymological Dictionary.* Materials for a Reconstruction. 1994. ISBN 90 04 10051 2

**Band 19**

al-Zwaini, L. and Peters, R. *A Bibliography of Islamic Law, 1980-1993.* 1994. ISBN 90 04 10009 1

**Band 20**

Krings, V. (éd.). *La civilisation phénicienne et punique.* Manuel de recherche. 1995. ISBN 90 04 10068 7

**Band 21**

Hoftijzer, J. and Jongeling, K. *Dictionary of the North-West Semitic Inscriptions.* With appendices by R.C. Steiner, A. Mosak Moshavi and B. Porten. 1995. 2 Parts. ISBN *Set (2 Parts)* 90 04 09821 6 Part One: ʾ - L. ISBN 90 04 09817 8 Part Two: M - T. ISBN 90 04 9820 8.

**Band 22**

Lagarde, M. *Index du Grand Commentaire de Faḫr al-Dīn al-Rāzī.* 1996. ISBN 90 04 10362 7

**Band 23**

Kinberg, N. *A Lexicon of al-Farrāʾ's Terminology in his Qurʾān Commentary.* With Full Definitions, English Summaries and Extensive Citations. 1996. ISBN 90 04 10421 6

**Band 24**

Fähnrich, H. und Sardshweladse, S. *Etymologisches Wörterbuch der Kartwel-Sprachen.* 1995. ISBN 90 04 10444 5

**Band 25**

Rainey, A.F. *Canaanite in the Amarna Tablets.* A Linguistic Analysis of the Mixed Dialect used by Scribes from Canaan. 1996. ISBN *Set (4 Volumes)* 90 04 10503 4
Volume I. Orthography, Phonology. Morphosyntactic Analysis of the Pronouns, Nouns, Numerals. ISBN 90 04 10521 2 Volume II. Morphosyntactic Analysis of the Verbal System. ISBN 90 04 10522 0 Volume III. Morphosyntactic Analysis of the Particles and Adverbs. ISBN 90 04 10523 9 Volume IV. References and Index of Texts Cited. ISBN 90 04 10524 7

**Band 26**

Halm, H. *The Empire of the Mahdi.* The Rise of the Fatimids. Translated from the German by M. Bonner. 1996. ISBN 90 04 10056 3

**Band 27**

Strijp, R. *Cultural Anthropology of the Middle East.* A Bibliography. Vol. 2: 1988-1992. 1997. ISBN 90 04 010745 2

**Band 28**

Sivan, D. *A Grammar of the Ugaritic Language.* 1997. ISBN 90 04 10614 6

**Band 29**

Corriente, F. *A Dictionary of Andalusi Arabic.* 1997. ISBN 90 04 09846 1

**Band 30**

Sharon, M. *Corpus Inscriptionum Arabicarum Palaestinae (CIAP).* Vol. 1: A. 1997. ISBN 90 04 010745 2 Vol.1: B. 1999. ISBN 90 04 110836

**Band 31**

Török, L. *The Kingdom of Kush.* Handbook of the Napatan-Meroitic Civilization. 1997. ISBN 90 04 010448 8

**Band 32**

Muraoka, T. and Porten, B. *A Grammar of Egyptian Aramaic.* 1998. ISBN 90 04 10499 2

**Band 33**
Gessel, B.H.L. van. *Onomasticon of the Hittite Pantheon.* 1998.
ISBN *Set (2 parts)* 90 04 10809 2
**Band 34**
Klengel, H. *Geschichte des hethitischen Reiches* 1998. ISBN 90 04 10201 9
**Band 35**
Hachlili, R. *Ancient Jewish Art and Archaeology in the Diaspora* 1998. ISBN 90 04 10878 5
**Band 36**
Westendorf, W. *Handbuch der altägyptischen Medizin.* 1999.
ISBN *Set (2 Bände)* 90 04 10319 8
**Band 37**
Civil, M. *Mesopotamian Lexicography.* 1999. ISBN 90 04 11007 0
**Band 38**
Siegelová, J. and Souček, V. *Systematische Bibliographie der Hethitologie.* 1999.
ISBN *Set (3 Bände)* 90 04 11205 7
**Band 39**
Watson, W.G.E. and Wyatt, N. *Handbook of Ugaritic Studies.* 1999.
ISBN 90 04 10988 9
**Band 40**
Neusner, J. *Judaism in Late Antiquity, III,1.* 1999. ISBN 90 04 11186 7
**Band 41**
Neusner, J. *Judaism in Late Antiquity, III,2.* 1999. ISBN 90 04 11282 0
**Band 42**
Drijvers, H.J.W. and Healey, J.F. *The Old Syriac Inscriptions of Edessa and Osrhoene.* 1999.
ISBN 90 04 11284 7
**Band 43**
Daiber, H. *Bibliography of Philosophical Thought in Islam.* 2 Volumes.
ISBN *Set (2 Volumes)* 90 04 11347 9
Volume I. Alphabetical List of Publications 1999. ISBN 90 04 09648 5
Volume II. Index of Names, Terms and Topics. 1999. ISBN 90 04 11348 7
**Band 44**
Hunger, H. and Pingree, D. *Astral Sciences in Mesopotamia.* 1999. ISBN 90 04 10127 6